W0051347

Archives of Virology

Supplementum 4

C. De Bac, W. H. Gerlich,
G. Taliani (eds.)

Chronically Evolving Viral Hepatitis

Springer-Verlag Wien New York

Prof. Dr. C. De Bac
Dr. G. Taliani
Titolare II Cattedra Clinica
Malattie Tropicli e Infettive
Policlinico Umberto, Roma, Italy

Prof. Dr. W. H. Gerlich
Institut für Medizinische Virologie
Justus-Liebig-Universität
Giessen, Federal Republic of Germany

ISBN-13: 978-3-211-82350-7   e-ISBN-13: 978-3-7091-5633-9
DOI: 10.1007/978-3-7091-5633-9

This work is subject to copyright.
All rights are reserved, whether the whole or part of the material is concerned, specifically those of translation, reprinting, re-use of illustrations, broadcasting, reproduction by photocopying machine or similar means, and storage in data banks.
© 1992 by Springer-Verlag/Wien

Printed on acid-free paper

With 72 Figures

Foreword

Chronic viral hepatitis remains one of the major medical problems for which neither a cure nor eradication of the infectious agents is in sight. The hepatitis viruses type B, C and D lead very often to persistent infections. In fact, chronic infection is a much greater problem than the acute hepatitis. Chronic viral hepatitis often remains undiagnosed until the patients develop decompensated liver cirrhosis or hepatocellular carcinoma. Furthermore, unrecognized virus carriers are a source of infection by sexual and other close contacts as well as during many medical procedures.

Recent progress has been made in vaccination against hepatitis B virus (HBV), but large vaccination campaigns have yet to be realized in most of those regions where HBV is most prevalent. For eradication of HBV worldwide, vaccination schedules must be simplified and immunogenicity of the vaccine must be improved. Therapy of chronic hepatitis B patients is still very unsatisfactory. In part, this situation reflects the poor understanding of the pathogenesis of HBV.

A major breakthrough in medical virology was the recent discovery of the hepatitis C virus (HCV) genome. Application of molecular biology has allowed the rapid development of diagnostical reagents for HCV. The relationship between the results of the new assays for HCV infection and various forms of liver disease have, however, not yet been completely clarified.

Besides their medical importance, HBV, HCV and HDV are fascinating objects for molecular biology. The mechanism of viral replication, pathogenesis and oncogenesis at a molecular level belong to the most rapidly developing fields of modern life sciences. The interplay between viral and host gene expression is most dramatically exemplified by the induction of malignant growth. The relationship between HBV and liver cell carcinoma has long been recognized, but only recently has a direct role of the viral genes been determined. Very surprisingly, the RNA virus HCV may be an even more important agent for the development of liver cell carcinoma.

Three years ago we undertook an effort to bring together, in the beautiful town of Fiuggi south of Rome, clinicians, laboratory physicians, epidemiologists, pathologists and molecular biologists whose primary research interest is chronic viral hepatitis. The contributions from these quite divergent participants on one common theme were most stimulating.

The success of this meeting prompted us to organize a second meeting at end of 1990 in Siena (Italy). We could persuade most of the participants to prepare manuscripts on their new data which are compiled in this volume. The articles give evidence that the problems mentioned above are tackled successfully. We thank all authors for their contributions, Dr. Bruce Boschek for his tremendous editorial help and Prof. Klenk and Springer Verlag for publishing this book.

Carlo de Bac, Wolfram H. Gerlich, Gloria Taliani

Contents

Immune pathogenesis of viral hepatitis

Vallbracht, A., Fleischer, B.: Immune pathogenesis of hepatitis A 3

Prevot, S., Marechal, J., Pillot, J., et al.: Relapsing hepatitis A in Saimiri monkeys experimentally reinfected with a wild type hepatitis A virus (HAV) 5

Ferrari, C., Penna, A., Bertoletti, A., et al.: Immune pathogenesis of hepatitis B . 11

Barnaba, V., Franco, A., Paroli, M., et al.: T cell recognition of hepatitis B envelope proteins 19

Penna, A., Bertoletti, A., Cavalli, A., et al.: Fine specificity of the human T cell response to hepatitis B virus core antigen 23

Sylvan, S. P. E., Hellström, U. B., Fei, G., et al.: HBcAg induced T-cell independent anti-HBc production in chronic HBsAg carriers. 29

Hellström, U. B., Sylvan, S. P. E.: Divergent anti-HBc reactivities in HB-immune and chronic HBsAg carriers 36

Possehl, C., Repp, R., Heermann, K.-H., et al.: Absence of free core antigen in anti-HBc negative viremic hepatitis B carriers. 39

Volpes, R., van den Oord, J. J., Desmet, V. J.: Homing of T-lymphocytes in acute and chronic HBV positive inflammatory liver disease 42

Melegari, M., Scaglioni, P. P., Pasquinelli, C., et al.: Hepatitis B virus specific transcripts in peripheral blood mononuclear cells. 46

Repp, R., Mance, A., Keller, C., et al.: Detection of transcriptionally active hepatitis B virus DNA in peripheral mononuclear blood cells after infection during immunosuppressive chemotherapy using the polymerase chain reaction. 50

Oncogenic properties of hepatitis viruses

Avantaggiati, M. L., Balsano, C., Natoli, G., et al.: The hepatitis B virus X protein transactivation of c-*fos* and c-*myc* proto-oncogenes is mediated by multiple transcription factors 57

Kekulé, A. S., Lauer, U., Weiß, L., et al.: *Trans*-Activation by hepatitis B virus X protein is mediated via a tumour promotor pathway. 63

Natoli, G., Balsano, C., Avantaggiati, M. L., et al.: Truncated pre-S/S proteins transactivate multiple target sequences 65

Sangiovanni, A., Covini, G., Rumi, M. G., et al.: Hepatitis C virus and hepatocellular carcinoma 70

Baur, M., Hay, U., Novacek, G., et al.: Prevalence of antibodies to hepatitis C virus in patients with hepatocellular carcinoma in Austria 76

Hepatitis B virus variants

Bonino, F., Brunetto, M. R.: Variants of hepatitis B virus. 83

Cariani, E., Fiordalisi, G., Primi, D.: Hepatitis B virus genomic variations in chronic hepatitis 86

Tong, S. P., Diot, C., Gripon, P., et al.,: Replication of a molecularly cloned HBeAg negative hepatitis B virus variant in transfected HepG2 cells 90

Tong, S. P., Vitvitski, L., Li, J. S., et al.: Lack of pre-C region mutation in woodchuck hepatitis virus from seroconverted woodchucks. 95

Gerken, G., Paterlini, P., Kremsdorf, D., et al.: Clinical significance of polymerase chain reaction (PCR) assay in chronic HBV carriers 97

Serodiagnosis of hepatitis B

Petit, M. A., Capel, F., Zoulim, F., et al.: PreS antigen expression and anti-preS response in hepatitis B virus infections: relationship to serum HBV-DNA, intrahepatic HBcAg, liver damage and specific T-cell response 105

Garbuglia, A. R., Manzin, A., Budkowska, A., et al.: PCR analysis of HBV infected sera: relationship between preS antigens expression and viral replication 113

Norder, H., Hammas, B., John, L., et al.: Detection of HBV DNA by PCR in serum from an HBsAg negative blood donor implicated in cases of post-transfusion hepatitis B. 116

Korec, E., Gerlich, W. H.: HBc and HBe specificity of monoclonal antibodies against complete and truncated HBc proteins from E. coli 119

Diment, J. A., Tyrrell, J., Brown, J., et al.: Measurement of anti-HBc IgM levels using the Amerlite anti-HBc IgM assay. 122

Spiller, G. H., Stalham, A., Holian, J., et al.: Evaluation of a new enhanced luminescence immunoassay for confirming the presence of HBsAg in human serum or plasma 124

Hepatitis B surface proteins and vaccination

Gerlich, W. H., Heermann, K.-H., Lu Xuanyong: Functions of hepatitis B surface proteins 129

Machein, U., Nagel, R., Prange, R., et al.: Deletion and insertion mutants of HBsAg particles 133

Petre, J., Rutgers, T., Hauser, P.: Properties of a recombinant yeast-derived hepatitis B surface antigen containing S, preS2 and preS1 antigenic domains . 137

Iwarson, S.: Diverging policies for vaccination against hepatitis B. 142

Corradi, M. P., Tata, C., Marchegiano, P., et al.: Immunogenicity and safety of a recombinant hepatitis B vaccine produced in mammalian cells and containing the S and the preS2 sequences 147

Gesemann, M., Schröder, S., Scheiermann, N., et al.: Kinetics of anti-HBs after hepatitis B vaccination: a comparison of two recombinant and one plasma-derived vaccines. 154

Boxall, E. H.: Enhanced luminescent assays for hepatitis markers: Assessment of post vaccine responses 156

Characterization of hepatitis C virus

Heinz, F. X.: Comparative molecular biology of flaviviruses and hepatitis C virus   163

Cristiano, K., di Bisceglie, A. M., Hoofnagle, J. H., et al.: Hepatitis C viral RNA in serum of patients with chronic non-A, non-B hepatitis: Detection by the polymerase chain reaction using multiple primer sets.   172

Schreier, E., Fuchs, K., Höhne, M., et al.: Detection and characterization of hepatitis C virus sequence in the serum of a patient with chronic HCV infection.   179

Li, J. S., Tong, S. P., Vitvitski, L., et al.: Sequence analysis of PCR amplified hepatitis C virus cDNA from French non-A, non-B hepatitis patients.   184

Neri, P., Bonci, A., Campoccia, G., et al.: Antigenicity of synthetic peptides derived from C100 protein of hepatitis C virus   186

Infantolino, D., Chiaramonte, M., Zanetti, A. R., et al.: Localization of hepatitis C virus antigen(s) by immunohistochemistry on fixed-embedded liver tissue. . . .   191

Krawczynski, K.: Identification of HCV-associated antigen(s) in hepatocytes . . .   196

Hepatitis C virus and liver disease

Meyer zum Büschenfelde, K.-H., Gerken, G., Manns, M.: Hepatitis C virus (HCV) and autoimmune liver diseases   201

Bertolini, E., Marelli, F., Zermiani, P., et al.: Antibodies to hepatitis C virus in primary biliary cirrhosis.   205

Suárez, A., Riestra, S., Rodriquez, M., et al.: Prevalence of antibodies against hepatitis C virus in primary biliary cirrhosis and autoimmune chronic hepatitis   210

Brillanti, S., Masci, C., Siringo, S., et al.: HCV infection and chronic active hepatitis in alcoholics.   212

Piperno, A., D'Alba, R., Roffi, L., et al.: Hepatitis C virus infection in patients with idiopathic hemochromatosis (IH) and porphyria cutanea tarda (PCT).   215

Diagnosis of hepatitis C virus

Glazebrook, J. A., Rodgers, B. C., Corbishley, T., et al.: Diagnostic reagents for hepatitis C virus.   219

Gmelin, K., Kurzen, F., Kallinowski, B., et al.: Follow-up of patients with hepatitis non-A, non-B: incidence and persistence of anti-HCV depend on route of transmission.   222

Tanzi, E., Galli, C., Delaito, M., et al.: Confirmation of anti-HCV EIA reactivities by RIBA and neutralization assay among blood donors and patients with chronic liver disease and hepatocellular carcinoma   227

Taliani, G., Badolato, M. C., Lecce, R., et al.: Recombinant immunoblot assay for hepatitis C virus antibody in chronic hepatitis.   232

Li, J. S., Vitvitski, L., Tong, S. P., et al.: PCR detection of HCV RNA among French non-A, non-B hepatitis patients   234

Hepatitis C virus and blood donation

Reesink, H. W., van der Poel, C. L., Cuypers, H. T. M., et al.: HCV and blood transfusion.   241

Esteban, J. I., Lopez-Talavera, J. C., Genescà, J., et al.: Evaluation of anti-HCV positive blood donors identified during routine screening   244

Villa, E., Melegari, M., Ferretti, I., et al.: Presence of HCV RNA in serum of
asymptomatic blood donors involved in post-transfusion hepatitis (PTH) . . .   247

Agulles, O., Janot, C., et al.: Epidemiology of anti-HCV antibodies in France . .   249

Caspari, G., Beyer, H.-J., Gerlich, W. H., et al.: Assay of antibodies to hepatitis C
virus protein C100-3 in blood donors from Northern Germany   253

Chronic hepatitis in childhood

Maggiore, G., de Giacomo, C.: Chronic viral hepatitis in children   259

Mengoli, M., Balli, M. E., Tolomelli, S., et al.: Long-term outcome of chronic type B
hepatitis in childhood   263

Nigro, G., Taliani, G., Bartmann, U., et al.: Hepatitis in children with thalassemia
major.   265

Nigro, G., Mattia, S., Vitolo, R., et al.: Hepatitis in pre-school children: prevalent
role of cytomegalo-virus.   268

Giacchino, R., Timitilli, A., Cristina, E., et al.: Repeated course of interferon
treatment in chronic hepatitis B in childhood   273

Utili, R., Sagnelli, E., Giusti, G., et al.: Effect of prednisone priming followed by alfa-
interferon in treatment of children with chronic hepatitis B: an interim analysis of
a controlled trial   277

Giacchino, R., Nocera, A., Timitilli, A., et al.: Association between HLA class I
antigens and response to interferon therapy in children with chronic HBV
hepatitis   281

Therapy of chronic viral hepatitis

Schalm, S. W.: Treatment of chronic viral hepatitis anno 1990   287

Fattovich, G., Betterle, C., Brollo, L., et al.: Induction of autoantibodies during
alpha interferon treatment in chronic hepatitis B   291

Taliani, G., Furlan, C., Grimaldi, F., et al.: One course versus two courses of
recombinant alpha interferon in chronic C hepatitis   294

Diodati, G., Bonetti, P., Tagger, A., et al.: Interferon therapy of cryptogenic chronic
active liver disease and its relationship to anti-HCV   299

Gargiulo, M., Tarquini, P., di Ottavio, L., et al.: Interferon alpha 2-b therapy of
HCV and NonBNonC chronic hepatitis   304

Tocci, G., Antonelli, L., Boumis, E., et al.: Long-term effects of recombinant
leukocyte alpha interferon in the treatment of chronic delta hepatitis   306

Liver transplantation and hepatitis viruses

Williams, R., O'Grady, J. G., Davies, S. E., et al.: Liver transplantation and
hepatitis viruses.   311

Epidemiology of HCV and other blood-transmissible viruses: Selected abstracts

Rodréguez, M., Riestra, S., San Román, F., et al.: Prevalence of antibody to hepatitis
C virus in acute non-A, non-B hepatitis in patients from different epidemiological
categories   319

Giuberti, T., Marchelli, S., Degli Antoni, A., et al.: Prevalence of anti-HCV antibodies in patients with acute nonA-nonB viral hepatitis. 321

Rodriguez, M., Suárez, A., Cimadevilla, R., et al.: Antibody to hepatitis C virus in acute, self-limited, type B hepatitis. 323

Fatuzzo, F., Mughini, M. T., Cacopardo, B., et al.: HCV infection in HBsAg positive chronic liver disease 325

Rodriguez, M., Navascues, C. A., Martinez, A., et al.: Prevalence of antibody to hepatitis C virus in chronic HBsAg carriers 327

Botti, P., Pistelli, A., Gambassi, A., et al.: HBV and HCV infection in i.v. drug addicts; coinfection with HIV. 329

Cacopardo, B., Fatuzzo, F., Cosentino, S., et al.: HCV and HIV infection among intravenous drug abusers in Eastern Sicily 333

Guadagnino, V., Zimatore, G., Rocca, A., et al.: Anti-hepatitis C antibody prevalence among intravenous drug addicts in the Catanzaro area 335

Pizzaferri, P., Padrini, D., Viale, P., et al.: Hepatitis C virus (HCV) infection in the Piacenza dialysis center 337

Vandelli, L., Medici, G., Savazzi, A. M., et al.: Prevalence of hepatitis C virus (HCV) antibodies in hemodialysis patients 339

Mughini, M. T., Cacopardo, B., Fatuzzo, F., et al.: Preliminary investigation on intrafamilial spread of hepatitis C virus (HCV) 343

Riestra, S., Rodriguez, M., Suárez, A., et al.: Involved factors in the intrafamilial spread of hepatitis C virus 345

Ilardi, I., Errera, G., de Sanctis, G. M., et al.: Prevalence of anti-HCV in two Tanzanian villages 347

I   Immune pathogenesis of viral hepatitis

Arch Virol (1992) [Suppl] 4: 3–4
© Springer-Verlag 1992

Immune pathogenesis of hepatitis A

A. Vallbracht[1] and **B. Fleischer**[2]

[1]Abteilung Medizinische Virologie, Universität Tübingen
[2]I Medizinische Klinik, Universität Mainz, Federal Republic of Germany

Summary. In an effort to elucidate the mechanism of liver damage resulting from Hepatitis A virus (HAV) infection, we have studied infected skin fibroblasts and autologous lymphocytes from HAV patients. We report here that HLA-restricted virus-specific T cells play an essential role in HAV-related hepatocellular injury.

*

The pathogenetic mechanism leading to liver tissue injury in hepatitis caused by hepatitis A virus (HAV) is not well understood. Although HAV has been classified as a picornavirus belonging to the enterovirus group, it generally induces an inapparent and persistent rather than a cytolytic infection in cell cultures in vitro. Only a few cytolytic variants have been described. High concentrations of infectious HAV are produced by and released from these HAV-carrier cells without evidence of cell destruction. As demonstrated, no changes in the usual culture procedures are required to maintain this carrier system. Immunofluorescence studies revealed that, after the establishment of this carrier system, all the cells are infected with HAV. Subpassages in the presence of anti-HAV serum do not decrease the proportion of HAV-positive cells, although no infectious virus can be detected in the supernatant. These results, plus the fact that cultures in which 100% of the cells contain HAV antigen can still multiply at a rate close to that of uninfected cells, prove that cells can divide repeatedly and that the antigen-producing potential can be passed from a parent cell to daughter cell. It seems that infection of cells with HAV (including a hepatoma cell line) results in a persistent infection which leads to a balance between cell metabolism and virus replication.

These data on HAV persistence in vitro are not compatible with the clinical course of HAV infection. Despite reports demonstrating protracted cases of HAV infection, there is no convincing evidence that HAV gives rise to true persistent infection in vivo. Consequently, all in vitro data tend to indicate that the symptoms of HAV infection and the elimination of this virus in vivo are not due to cytocidal infection of hepatic cells, but rather to the elimination of infected hepatocytes by an immunologic mechanism.

To test whether cell-mediated cytotoxicity is an operative mechanism in human HAV infection, we developed an HLA-congruent system, using HAV-infected skin fibroblasts (targets) and autologous lymphocytes (effectors) taken from HAV patients during the acute phase of infection. We identified a lymphocyte population in the peripheral blood of HAV patients that specifically lyses HAV-infected autologous target cells. The highest HAV-specific cytolytic activity, however, was demonstrated in peripheral blood lymphocytes collected during the early convalescent phase of HAV infection, 2 to 3 weeks after the onset of icterus. In patients who had a protracted form of HAV infection in which elevated transaminases persisted for at least 5 months, the highest cytolytic activity was observed in peripheral lymphocytes obtained 8 to 12 weeks after the onset of symptoms.

Our autologous target cell system offers the possibility of demonstrating virus-specific cytolytic activity of liver-infiltrating cells in acute HAV-infection. Of 170 randomly established T cell clones from liver biopsies of two HAV patients, 54 CD8+ clones were studied to determine their cytolytic activity against HAV-infected autologous skin fibroblasts. Using a Cr-release assay, we demonstrated that 42% and 53%, respectively, of the liver-infiltrating CD8+ clones were HAV-specific and killed HAV-infected autologous target cells in an HLA-restricted manner.

HAV-specific cytotoxic CD8+ cells in contact with autologous HAV-infected target cells produce IFN gamma in an HLA-dependent manner. This capability of cytotoxic T lymphocytes to produce IFN gamma during virus infection may be important for the pathogenesis of hepatitis A. We demonstrated that IFN gamma may limit HAV infection by direct mediation of the antiviral effect in vitro. Moreover, the released IFN could prevent the spread of the virus to neighboring cells. HLA class I antigens are not or are only weakly expressed on the surface of normal human hepatocytes. Since IFN gamma induces the expression of HLA proteins it might be the decisive factor in HAV infection which induces changes in HLA class I display, resulting in enhancement of efficient T-cell-mediated immune attack.

Authors' address: Dr. A. Vallbracht, Abteilung Medizinische Virologie, Universität Tübingen, D-W-7400 Tübingen, Federal Republic of Germany.

Arch Virol (1992) [Suppl] 4: 5–10
© Springer-Verlag 1992

Relapsing hepatitis A in Saimiri monkeys experimentally reinfected with a wild type hepatitis A virus (HAV)

S. Prevot[2], J. Marechal[1], J. Pillot[1], and J. Prevot[1]

[1] Unité de Virologie Médicale, Unité d'Immunologie Microbienne, Institut Pasteur, Paris, France
[2] Service d'Anatomie et de Cytologie Pathologiques, Hôtel Dieu, Paris, France

Summary. Saimiri monkeys were inoculated three times with hepatitis A virus and observed in a follow-up study for sixteen months. The monkeys developed recurrent hepatitis involving liver damage and cycles of HAV antigen shedding in stools. The relapses were presumably due to immune response effects.

Introduction

Hepatitis A virus (HAV) infection in humans is usually a mild self-limiting disease. However, there are reports of protracted [2, 3] or relapsing forms [4, 12, 14] of hepatitis following acute HAV infection, although persistent infection has not been demonstrated in these cases. Chimpanzees [1, 9] marmosets [7] and other species of monkey [5, 8] are susceptible to HAV infection, and hepatic replication and shedding of the virus in the stool has been demonstrated. We have previously shown that Saimiri monkeys develop acute hepatitis when HAV was either orally or parenterally administered. This report describes the follow-up of HAV-infected Saimiri monkeys, when re-infected nine months and a year after their first infection. Relapses were investigated by testing for HAV-antigens (HAV-Ag) shed in stools, liver transaminase activities and serum abnormalities, and by examining liver biopsies.

Materials and methods

Wild-type HAV extracts for inoculation were prepared from HAV-RNA-positive and HAV-Ag-positive human faeces. The 0.1% (W/v) faecal extract was filtered through a 0.22 µm pore size filter and tested for infectious HAV particles on Vero cells using an immunofluorescence test (IF). The only enterovirus detected in the extracts was HAV. The second HAV strain used was a Vero cell-adapted strain.

The Saimiri monkeys (Saimiri Sciureus) were colony bred and supplied by the Pasteur Institute in French Guyana. Seven monkeys, seronegative for HAV, HBV and HCV antibodies were used in the study. All inoculations were performed intravenously. Group I (Saimiri 1348, 953 and 367) was inoculated with wild-type HAV at the start of the study (MO), and nine and twelve months later (M9 and M12). Group II (Saimiri nos. 1186, 613, 620 and 100) were inoculated with Vero cell-adapted HAV on MO and then with wild-type HAV on M9 and M13. One noninfected animal was kept as a control.

For 16 months on (from M0), stools were collected daily, and blood samples taken twice a week immediately following each HAV infection, and once a week thereafter.

The presence of HAV antibodies (anti-HAV) was assessed by several techniques. Sera were screened for the presence of HAV-IgM (HAVAB-M EIA—Abbott) and anti-HAV- by blocking enzyme linked immunosorbent assays [10]. This technique is more sensitive than the HAVAB-EIA (Abbott). Liver alanine aminotransferase (ALT) was assayed in serum samples. ELISA was used to test for HAV-Ag in stool samples, prepared as dilutions (at four concentrations) of a 1/20 faecal extract in PBS. Infective particles in the HAV-Ag-positive faeces were assayed on Vero cells using an IF test. Results are reported for the day of sampling after first (F), second (S) and third (T) inoculations.

Liver samples were taken with a pediatric liver biopsy needle on days F180 and F270, S35 and S65 and T28, T60 and T110. Small fragments of the liver samples were fixed in 10% neutral formalin or in aqueous Bouin's solution and were embedded in paraffin wax. Serial sections were examined and scored by a single observer in a blind procedure.

Results

Group I—Saimiri no. 1348 (see Fig. 1): HAV-IgM were first detected 14 days (day F14) after the first inoculation (day F0). Blocking antibodies were

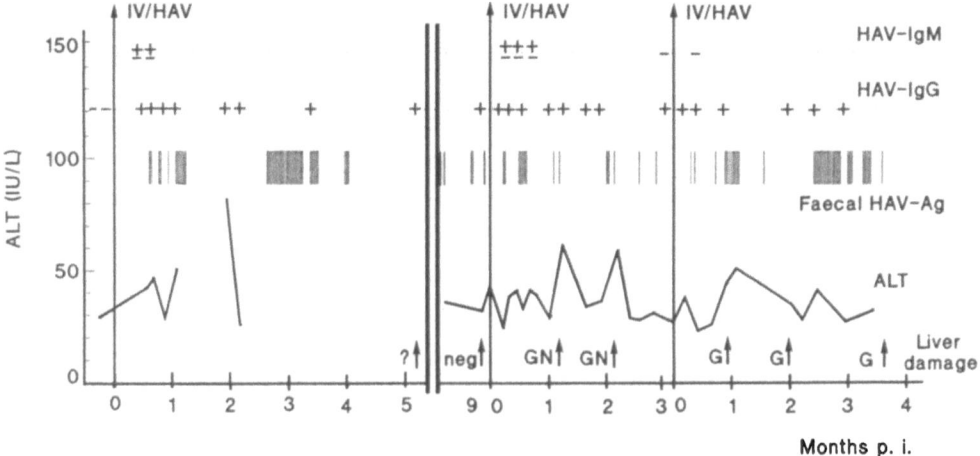

Fig. 1. Saimiri no. 1348, infected three times with wild-type HAV. Schedule of inoculation (*IV/HAV*); serological findings: HAV-IgM and HAV-blocking-antibodies (*HAV-IgG*); presence of faecal HAV-Ag; alanine aminotransferase (*ALT*) profile and liver damage observed in liver biopsy (arrows): foci of inflammatory cells, predominantly lymphocytes (*G*), small foci of hepatocytes necrosis (*N*)

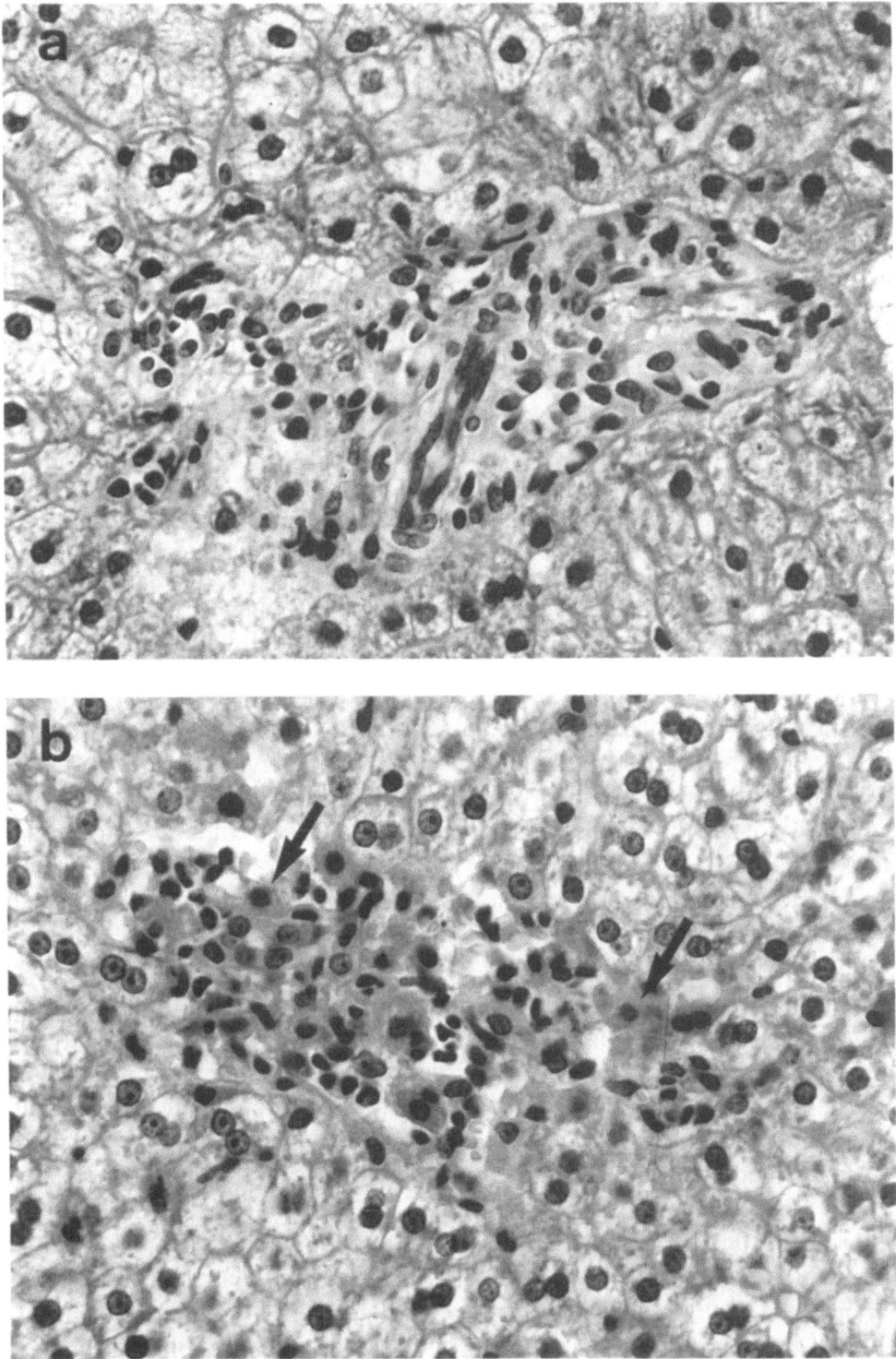

Fig. 2. Liver biopsy taken 60 days after the third wild type HAV infection. **a** Saimiri no. 1348: infiltrate of mononuclear cells in a portal tract and in the surrounding parenchyma. **b** Saimiri no. 367: infiltrate of mononuclear cells in the lobular zone with few necrotic hepatocytes (arrows). Hematoxylin eosin safran, original magnification 130 ×

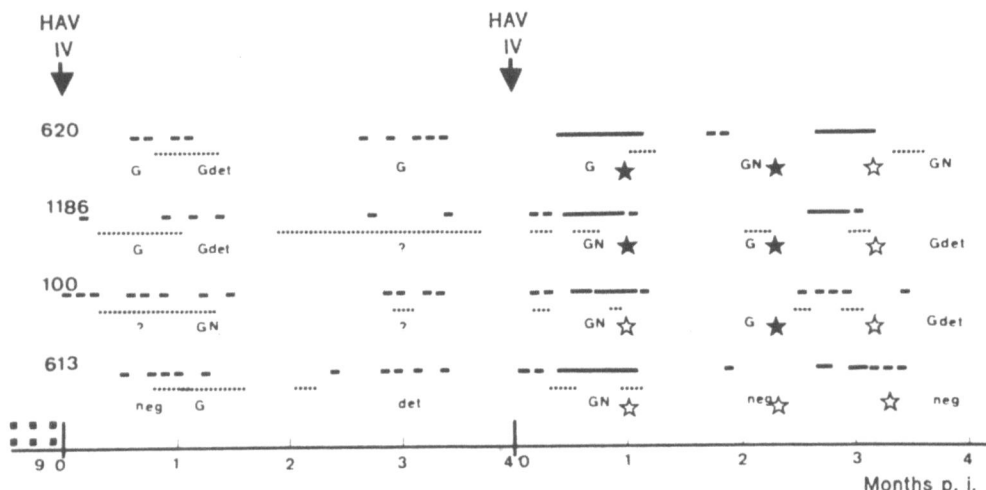

Fig. 3. Wild type HAV infections (IV/HAV) of the second group of monkeys preinfected with a Vero cell adapted HAV strain, nine months earlier: HAV-Ag shedding (——); abnormal ALT levels (·····); liver damage: presence of hepatocyte necrosis (N), clusters of inflammatory cells predominantly lymphocytes (G) or macrophages (det), and prothrombin time: normal (☆) and abnormal (★) levels

detected 3 days later (F17). HAV-Ag first appeared in faeces on day F20, and was thereafter present intermittently. Infective HAV particles were detected as late as four months after infection. The ALT activity peak was on day F45, which was after the first cycle of shedding, but fell to normal levels by month 6. Just prior to the second inoculation, HAV-blocking antibodies were present, HAV-Ag were sporadically detected (days F260 to F262, and day F267) and the liver biopsy was normal (although small and difficult to analyse). The corresponding liver biopsies from Saimiri nos. 953 and 367 contained foci of inflammatory cells. Saimiri 1348 suffered a recurrence of the disease after re-inoculation (day S0). HAV-IgM was detected on day S21 and several short periods of high-level HAV-Ag shedding were observed, around days S16, S34 and S60. Several small peaks of ALT activity were detected. Liver damage including hepatocyte necrosis was observed in liver biopsies taken on days S35 and S65. After the third inoculation, pronounced cycles of HAV-Ag shedding were detected, despite the continuing presence of anti-HAV. ALT activity was nearly normal. Liver biopsies taken on days T28, T60 and T110 displayed large clusters of mononuclear cells, most of which were lymphocytes, in the portal tract (Fig. 2a) or lobular zone of the liver parenchyma. The prothrombin time was longer, and hyperleucocytosis was still detected on day T120 (11,800 μl^{-1}, of which 60% lymphocytes).

Saimiri nos. 953 and 367: The results were similar to those of Saimiri no. 1348. Cycles of HAV-Ag shedding were observed, and increased ALT activities were detected. Liver biopsies contained foci of inflammatory cells and necrotic hepatocytes (Fig. 2b). The prothrombin time was longer.

Saimiri no. 953 died on day T13 from a sepsis syndrome. An autopsy was performed fifteen minutes after death, and the liver was removed. The liver was completely collapsed and displayed damage resulting from the viral infection: the sinuses were enlarged, and the tissue contained numerous lymphocytes and macrophages or Kuppfer cells with variable amounts of hemosiderin.

Group II (see Fig. 3): The results for the four group II monkeys were very similar to one another, and similar to those for group I. However, the response of group II to reinfection with wild-type HAV (S0) was faster than that of group I. HAV-Ag were detected early in the faeces (on day S5) followed by an ALT activity peak on day S20. The third inoculation was followed by a relapse with hepatocyte necrosis observed in three out of four biopsies taken on day T21.

Conclusion

Re-infection of Saimiri monkeys with wild-type HAV, up to 9 months after the first inoculation caused recurrence of the symptoms of the disease. Relapse was rapid in all cases. ALT activities were low, which is consistent with previous reports of relapse episodes in tamarins [6, 15]. However, this observation contrasts with the transaminase activity abnormalities observed in HAV-infected humans [11]. Re-infection of Saimiri with HAV caused a rapid response of sensitized lymphocytes, already present in liver paren-chyma. Mediated cell lysis of HAV-preinfected hepatocytes may then result, a phenomenon previously described in humans [13]. However, in Saimiri, this cell-mediated lysis, followed by the elimination of HAV through the digestive tract, may not be enough to induce significant ALT activity increases.

The protracted hepatitis induced in all the Saimiri used in this study may explain the prolonged cycles of HAV-Ag and HAV infective particle shedding, even when HAV-blocking antibodies were present.

Acknowledgements

The authors are very grateful to A. Cosson for liver biopsy punctures.

References

1. Cohen JI, Feinstone S, Purcell RH (1989) Hepatitis A virus infection in a chimpanzee: Duration of viremia and detection of virus in saliva and throat swabs. J Infect Dis 160: 887–890
2. Fagan E, Yousef G, Brahm J, Garelick H, Mann G, Wolstenholme A, Portmann B, Harrison T, Mowbray JF, Mowat A, Zuckerman A, Williams R (1990) Persistence of hepatitis A virus in fulminant hepatitis and after liver transplantation. J Med Virol 30: 131–136

3. Gordon SC, Reddy KR, Schiff L, Schiff ER (1984) Prolonged intrahepatic cholestasis secondary to acute hepatitis A. Ann Intern Med 101: 635–637
4. Jacobson IM, Nath BJ, Dienstag JL (1985) Relapsing viral hepatitis type A. J Med Virol 16: 163–169
5. Karayiannis P, Jowett T, Enticott M, Moore D, Pignatelli M, Brenes F, Scheuer PJ, Thomas HC (1986) Hepatitis A virus replication in Tamarins and host immune response in relation to pathogenesis of liver cell damage. J Med Virol 18: 261–276
6. Karayiannis P, Chitranukroh R, Fry M, Petrovic LM, Moore D, Scheuer PJ, Thomas HC (1990) Protracted alanine aminotransferase levels in tamarins infected with hepatitis A virus. J Med Virol 30: 151–158
7. Mathiesen LR, Moller AM, Purcell RH, London WT, Feinstone SM (1980) Hepatitis A virus in the liver and intestine of marmosets after oral inoculation. Infect Immun 28: 45–48
8. Miller Keenan C, Lemon SM, Le Duc JW, McNamee GA, Binn LN (1984) Pathology of hepatitis A infection in the owl monkey (Aotus trivirgatus). Am J Pathol 115: 1–8
9. Murphy BL, Maynard JE, Bradley DW, Ebert JW, Mathiesen LR, Purcell RH (1978) Immunofluorescence of hepatitis A virus antigen in chimpanzees. Infect Immun 21: 663–665
10. Prevot J, Kopecka H (1988) HAV detection by molecular hybridization and ELISA. In: Zuckerman A (ed) Viral hepatitis and liver disease. Alan R Liss, New York, pp 70–73
11. Raimondo G, Longo G, Caredda F, Saracco G, Rizetto M (1986) Prolonged, polyphasic infection with hepatitis A. J Infect Dis 153: 172–173
12. Sjogren MH, Tanno H, Fay O, Sileoni S, Cohen BD, Burke DS, Feighny RJ (1987) Hepatitis A virus in stool during clinical relapse. Ann Intern Med 106: 221–226
13. Vallbracht A, Maier K, Stierhof YD, Wiedmann KH, Flehmig B, Fleischer B (1989) Liver-derived cytotoxic T cells in hepatitis A virus infection. J Infect Dis 160: 209–217
14. Van den Anker JN, Sukhai RN, Dumas AM (1988) Relapsing hepatitis in a child associated with isolation of hepatitis A virus antigen from the liver. Eur J Pediatr 147: 333
15. Zamyatina NA, Andzhaparidze AG, Balayan MS, Sobol AV, Titova IP, Karetnyi YV, Poleschuk VF (1990) Development of infection in monkeys as a result of successive natural and experimental infection with hepatitis A virus. Vopr Virusol 35: 122

Authors' address: Dr. J. Prevot, Institut Pasteur, Virologie Médicale, F-75724 Paris Cedex 15, France

Arch Virol (1992) [Suppl] 4: 11–18
© Springer-Verlag 1992

Immune pathogenesis of hepatitis B

C. Ferrari[1,2], A. Penna[1], A. Bertoletti[1,2], A. Cavalli[1], A. Valli[1], G. Missale[1], M. Pilli[1],
S. Marchelli[1], T. Giuberti[1], and F. Fiaccadori[1]

[1] Cattedra Malattie Infettive, Università di Parma, Parma, Italy
[2] Department of Molecular and Experimental Medicine,
Scripps Clinic and Research Foundation, La Jolla, CA, USA

Summary. Available information about the immune pathogenesis of HBV infection in man is very limited. However, the present availability of recombinant sources of the different HBV antigens expressed in the appropriate forms to induce activation of either HLA class I or HLA class II-restricted T cells, provides the necessary tools to investigate directly the mechanisms of liver damage, the role of the different cellular components of the immune system in HBV clearance and the specific nature of the immune defects potentially responsible for the chronic evolution of HBV infection. In addition, improved knowledge of HBV biology suggests a dynamic interpretation of the HBV-immune system interactions, based on which viral mutations as well as direct interferences of HBV with specific immune functions are believed to play a relevant role with respect to the outcome of HBV infection.

*

The immune pathogenetic mechanisms involved in liver damage and viral clearance during hepatitis B virus (HBV) infection in man still remain largely unknown. Based on available data derived from studies in the HBV and other viral systems, it is generally assumed that an HLA class I restricted cytotoxic T cell response to one or more HBV-encoded antigens displayed at the hepatocyte membrane is a major effector mechanism of hepatocellular injury and clearance of infected cells. In addition, it is known that a HLA class II-restricted T helper cell-dependent B cell immune response, especially directed toward HBV envelope determinants, is needed for the clearance of circulating viral particles.

Potential mechanisms of liver damage

The possibility that the HBV nucleocapsid antigens represent important target structures for a T cell-mediated injury of infected hepatocytes is suggested by the finding that autologous liver cells from patients with chronic HBV infection can be lysed in vitro by peripheral blood T cells and that this phenomenon can be selectively blocked by anti-hepatitis B core and anti-hepatitis B e antigen monoclonal antibodies [17, 22]. The importance of hepatitis B core antigen (HBcAg) in the activation of an anti HBV-specific T cell response is also indicated by the observation that CD4+ and CD8+ T cells with HBcAg-specific helper and suppressor function, respectively, are present within the liver of patients with chronic HBV infection [10, 9].

The recent isolation of intrahepatic HLA class II as well as class I-restricted preS2-specific T cell clones in chronic active hepatitis B demonstrates that also the HBV envelope antigens are important sensitizing immunogens for intrahepatic T cells in chronic HBV infection [1]. The potential importance of these antigens as target molecules for liver cell damage is further suggested by the observation in a transgenic mouse model that HLA class I restricted cytotoxic T cells can kill liver cells producing non-toxic amounts of hepatitis B surface antigen (HBsAg) [18]. This study provides the first definitive evidence that an endogenously synthesized HBV antigen can be processed by liver cells and presented to HLA class I-restricted cytotoxic T lymphocytes.

Although an HLA class I-restricted cytotoxic T cell response is likely to play a central role in liver damage, recent data suggest that other immune mechanisms may also be implicated in the pathogenesis of the hepatocellular injury during HBV infection. The finding of a surface expression of hepatitis B e antigen (HBeAg) in HBeAg-producing cells, in a conformation which is recognizable by anti-HBe antibodies, suggests that this antigen can serve as a target for antibody-mediated elimination of HBV-infected cells [24].

The recent demonstration that HBV envelope-specific, cytotoxic CD4+ HLA class II-restricted T cells isolated from hepatitis B vaccine recipients can recognize not only exogenous [7] but also endogenously synthesized viral antigens shows that even these lymphocytes are potentially able to participate in the clearance of virus-infected liver cells [21]. The observation that hepatocytes express HLA class II molecules following HBV infection provides additional support to this possibility [29].

In addition to antigen-specific cellular interactions, also antigen non-specific mechanisms are implicated in the clearance of HBV-infected liver cells. This is suggested by the direct lytic effect of soluble lymphokines, such as gamma interferon and tumor necrosis factor alpha, on liver cells of transgenic mice expressing HBV large envelope proteins (Gilles PN et al., personal communication).

Role of the B cell and helper T cell functions in HBV clearance

Recovery from HBV infection is not only related to the clearance of infected cells, but also to the antibody neutralization of free viral particles needed to avoid the spread of the infection to uninfected cells.

Available data clearly demonstrate the virus neutralizing function of anti-envelope antibodies [3], whereas the importance of the antibody response to the nucleocapsid antigens and to the HBV non-structural proteins [8, 12, 26, 27, 31, 33] is still a debated issue. It is generally accepted that anti-HBc antibodies do not express virus neutralizing effects but protection of chimpanzees against HBV infection by passive immunization with human immunoglobulin containing anti-HBe antibodies has been observed, suggesting a role for these antibodies in virus neutralization [28].

For the development of an adequate antibody response an efficient helper T cell function is generally required, implying that a functional helper T cell population is needed for the final recovery from viral infection. We know that the anti-envelope and the anti-HBe antibody responses are strictly T cell-dependent [14]. In contrast, the anti-HBc antibody response can also be T cell independent [14].

A direct demonstration of the importance of the helper T cell response for the final outcome of the hepadnavirus infection has recently been provided in the woodchuck system, by studying the effect of cyclosporin A (CsA; an immunosuppressive agent that inhibits antigen-specific, IL2-dependent immune responses initiated by helper T cells) on woodchuck hepatitis virus (WHV) infection (Cote PJ et al., personal communication). Animals infected with WHV in the absence of CsA developed a typical self-limited infection with a transient detection of WHV-DNA in their plasma. In contrast, animals similarly infected with WHV but treated with CsA developed a chronic infection and remained DNA positive during the time of follow-up, suggesting that a suppression of the helper function during the immediate post exposure period in hepadnavirus infection may represent a potential risk factor for the evolution to chronicity.

Immune defects and progression to chronicity

If the development of an adequate cellular and humoral immune response to the different HBV antigens is required for HBV clearance, it follows that specific defects of such immune mechanisms could lead to a chronic infection. Available information concerning these hypothetical defects is very limited mostly due to the absence of suitable animal models for immune pathogenetic studies, to the lack of appropriate target systems for the study of the class I-restricted cytotoxicity in man and to the limited availability of lymphomononuclear cells derived from the liver of HBV infected patients.

In an attempt to gain new insights into this field we have recently compared the immune response to HBV antigens in patients with self-limited hepatitis B who successfully clear the virus with respect to patients with chronic HBV infection [11]. For this purpose, the peripheral blood T cell response to the S, preS2 and preS1 HBV envelope region encoded poly-peptides as well as to HBcAg and HBeAg has prospectively been analyzed in a large group of patients with different stages of HBV infection. The results of this study show that the HLA class II-restricted peripheral blood T cell response to HBV nucleocapsid antigens in patients with chronic HBV infection is dramatically lower than that displayed by patients with acute HBV infection. In addition, the appearance of a detectable level of class II restricted T cell sensitization to nucleocapsid antigens is temporally associated with the clearance of viral particles from the sera of subjects with self-limited acute HBV infection, suggesting a role for HBcAg-specific CD4+ T cells in HBV clearance. The demonstration that HBcAg-specific helper T cells can directly cooperate in vivo with envelope-specific B cells support-ing the production of virus neutralizing anti-envelope antibodies may provide an explanation of this temporal association [15].

The immunological basis of the defective T cell response to HBV nucleocapsid antigens as well as the cause of the defective production of neutralizing anti-envelope antibodies in patients with chronic HBV infection are completely unknown. However, recent progress in the understanding of the general mechanisms of T and B cell activation provides the appropriate tools to start addressing these issues.

Since HLA class II-restricted T cells usually recognize processed forms of exogenous antigens in association with HLA class II determinants on the surface of the antigen-presenting cells, the first potential immune defect in chronic patients could be at the level of the processing and presentation of specific viral antigens, leading either to defective generation of immunogenic peptides or to an inefficient association of the processed peptide fragments with the HLA molecules. In either case, the final result will be an impaired clonal expansion of antigen-responsive T cells and the development of a defective T cell repertoire.

The lack of an adequate HBV-specific T cell response could also be related to mechanisms of immune tolerance or clonal energy, acting directly on the effector antigen-specific T cell populations. The cellular basis for a mechanism of T cell tolerance to HBeAg has recently been characterized in a transgenic mouse model of neonatal tolerance [16]. Not only HBeAg-expressing transgenic mice but also their non-transgenic littermates were tolerant to both HBeAg and HBcAg at the T cell level. In non-transgenic littermates, tolerance was likely the result of transplacental exposure to HBeAg. Since HBeAg is a secreted soluble protein, it could gain access to the thymus through the circulation, thereby leading to clonal deletion of MHC class II-restricted HBeAg-helper T cells as a result of T cell recognition of HBeAg-derived peptides in association with HLA determinants on the thymic stroma. This observation suggests that a similar mechanism of T helper cell tolerance could be operative also in neonates born to HBeAg positive carrier mothers.

Hyperactivity of HBV-specific suppressor T cells has also been suggested as a potential cause of non-responsiveness to HBV antigens in patients with chronic HBV infection [25, 30, 32] but the real relevance of this phenomenon remains to be demonstrated in better defined systems. With respect to the cellular mechanisms of immune suppression, the recent isolation in chronic active hepatitis B of CD8+, HLA class I-restricted, preS2-specific T cell clones capable of efficiently lysing HBV envelope-specific B cells suggests a possible suppressive role on the envelope-specific antibody response played by these cells in vivo [2].

Finally, a specific B cell defect responsible for the inefficient production of neutralizing anti-envelope antibodies in chronic patients, who usually do not develop a detectable anti-HBs response, cannot be at present excluded.

Viral interference with the host immune response

Even though the individual immune response to the different viral proteins certainly represents the major determinant for the final outcome of HBV infection, the potential importance of emerging virus variants cannot be underestimated [4, 5, 6, 19]. Specific viral mutations could alter the sequence or the conformation of antigen regions relevant to the activation of immunodominant anti-virus T or B cell responses, thereby allowing the escape of the virus from the immune surveillance. Direct demonstration of this mechanism of immunological escape has recently been provided in different viral systems [23].

In addition, the possibility of a direct effect of HBV infection on the function of specific cellular components of the immune system must be considered, especially in light of the established capacity of other viruses [13] to interfere with the host immune system. In line with this possibility is the recent demonstration that HBV can alter the cellular response to interferon, inducing a low expression of HLA molecules on infected cells with potential consequences to antigen recognition by T cells [20].

In conclusion, based on our present understanding of HBV biology, the interaction between the immune system and HBV must be regarded as a dynamic process whose final balance probably depends on the capacity of the immune system to mount an anti-viral response sufficiently wide to face the continuous viral attempts to escape recognition by the immune cells through genetic changes.

16 C. Ferrari et al.

Acknowledgements

This work was supported in part by the Italian Ministry of the National Education, Project on Liver Cirrhosis 1989–1990 and by National Institute of Health, USA, Grant AI 26626.

References

1. Barnaba V, Franco A, Alberti A, Balsano C, Benvenuto R, Balsano F (1989) Recognition of hepatitis B virus envelope proteins by liver-infiltrating T lymphocytes in chronic HBV infection. J Immunol 143: 2650–2655
2. Barnaba V, Franco A, Alberti A, Benvenuto R, Balsano F (1990) Selective killing of hepatitis B envelope antigen-specific B cells by class I restricted, exogenous antigen specific T lymphocytes. Nature 345: 250–253
3. Beasley RP, Hwang LY, Stevens CE, Lin CC, Hsieh FJ, Wang KJ, Sun TS, Szmuness W (1983) Efficacy of hepatitis B immune globulin for prevention of perinatal transmission of the hepatitis B virus carrier state. Hepatology 3: 135–141
4. Brunetto MR, Stemler M, Bonino F, Schödel F, Oliveri F, Rizzetto M, Verme G, Will H (1990) A new hepatitis B virus strain in patients with severe anti-HBe positive chronic hepatitis B. J Hepatol 10: 258–261
5. Carman WF, Jacyna MR, Hadziyannis S, Karayiannis P, McGarvey MJ, Markis A, Thomas HC (1989) Mutation preventing formation of hepatitis B e antigen in patients with chronic hepatitis B infection. Lancet ii: 588–591
6. Carman WF, Zanetti AR, Karayiannis P, Waters J, Manzillo G, Tanzi E, Zuckerman AJ, Thomas HC (1990) Vaccine-induced mutant of hepatitis B virus. Lancet ii: 325–329
7. Celis E, Ou D, Otvos L (1988) Recognition of hepatitis B surface antigen by human T lymphocytes. J Immunol 140: 1808–1815
8. Chang LJ, Dienstag J, Ganem D, Varmus H (1989) Detection of antibodies against hepatitis B polymerase antigen in hepatitis B virus-infected patients. Hepatology 10: 332–335
9. Ferrari C, Mondelli MU, Penna A, Fiaccadori F, Chisari FV (1987) Functional characterization of cloned intrahepatic, hepatitis B virus nucleoprotein-specific helper T cell lines. J Immunol 139: 539–544
10. Ferrari C, Penna A, Giuberti T, Tong MJA, Ribera E, Fiaccadori F, Chisari FV (1987) Intrahepatic, nucleocapsid antigen-specific T cells in chronic active hepatitis B. J Immunol 139: 2050–2058
11. Ferrari C, Penna A, Bertoletti A, Valli A, Degli Antoni A, Giuberti T, Cavalli A, Petit MA, Fiaccadori F (1990) Cellular immune response to hepatitis B virus (HBV) encoded antigens in acute and chronic HBV infection. J Immunol 145: 3442–3449
12. Feitelson MA, Millman I, Duncan GD, Blumberg BS (1988) Presence of antibodies to the polymerase gene product(s) of hepatitis B and woodchuck hepatitis virus in natural and experimental infections. J Med Virol 24: 121–136
13. McChesney MB, Oldstone MBA (1987) Virus perturb lymphocyte functions: selected principles characterizing virus-induced immunosuppression. Ann Rev Immunol 5: 279–304
14. Milich DR, McLachlan A (1986) The nucleocapsid of hepatitis B virus is both a T-cell-dependent and a T-cell-independent antigen. Science 234: 1398–1401
15. Milich DR, McLachlan A, Thornton GB, Hughes JL (1987) Antibody production to the nucleocapsid and envelope of the hepatitis B virus primed by a single synthetic T cell site. Nature 329: 547–549
16. Milich DR, Jones JE, Hughes JL, Price J, Raney A, McLachlan A (1990) Is a function of

the secreted hepatitis B e antigen to induce immunological tolerance in utero? Proc Natl Acad Sci USA 87: 6657–6661

17. Mondelli M, Mieli-Vergani G, Alberti A, Vergani D, Portmann B, Eddleston ALWF, Williams R (1982) Specificity of T lymphocyte cytotoxicity to autologous hepatocytes in chronic hepatitis B virus infection: evidence that T cells are directed against HBV core antigen expressed on hepatocytes. J Immunol 129: 2773–2777

18. Moriyama T, Guilhot S, Klopchin K, Moss B, Pinkert CA, Palmiter RD, Brinster RL, Kanagawa O, Chisari FV (1990) Immunobiology and pathogenesis of hepatocellular injury in hepatitis B virus transgenic mice. Science 248: 361–364

19. Okamoto H, Yotsumoto S, Akahane Y, Yamanaka T, Miyazaki Y, Sugai Y, Tsuda F, Tanaka T, Miyakawa Y, Mayumi M (1990) Hepatitis B viruses with precore region defects prevail in persistently infected hosts along with seroconversion to the antibody against e antigen. J Virol 64: 1298–1303

20. Onji M, Lever AML, Saito I, Thomas HC (1989) Defective response to interferons in cells transfected with the hepatitis B virus genome. Hepatology 9: 92–96

21. Penna A, Fowler P, Bertoletti A, Guilhot S, Moss B, Margolskee RF, Cavalli A, Valli A, Fiaccadori F, Chisari FV, Ferrari C (1992) Hepatitis B virus (HBV)-specific cytotoxic T-cell (CTL) response in humans: characterization of HLA class II-restricted CTL that recognize endogenously synthesized HBV envelope antigens. J Virol 66: 1193–1198

22. Pignatelli M, Waters J, Lever A, Iwarson S, Gerety R, Thomas H (1987) Cytotoxic T cell response to the nucleocapsid proteins of HBV in chronic hepatitis. J Hepatol 4: 15–21

23. Pircher H, Moskophidis D, Rohrer U, Burki K, Hengartner H, Zinkernagel RM (1990) Viral escape by selection of cytotoxic T cell-resistant virus variants in vivo. Nature 346: 629–633

24. Schlicht HJ, Shaller H (1989) The secretory core protein of human hepatitis B virus is expressed on the cell surface. J Virol 63: 5399–5404

25. Shirai M, Hanada H, Kurokouchi K, Watanabe S, Nishioka M (1990) Peripheral blood CD4-mediated enhancement and CD8-mediated suppression in the presence of recombinant hepatitis B virus core antigen. J Infect Dis 161: 420–425

26. Stemler M, Hess J, Braun R, Will H, Schroder CH (1988) Scrological evidence for expression of the polymerase gene of human hepatitis B virus in vivo. J Gen Virol 69: 689–693

27. Stemler M, Weimer T, Tu SX, Wan DF, Levrero M, Jung C, Pape GR, Will H (1990) Mapping of B-cell epitopes of the human hepatitis B virus X protein. J Hepatol 64: 2802–2809

28. Stephan W, Prince A, Brotman B (1984) Modulation of hepatitis B infection by intravenous application of an immunoglobulin preparation that contains antibodies to hepatitis B e and core antigens but not to hepatitis B surface antigen. J Virol 51: 420–424

29. Van Den Oord J, de Vos R, Desmet VJ (1986) In situ distribution of major histocompatibility complex products and viral antigens in chronic hepatitis B virus infection: evidence that HBc-containing hepatocytes may express HLA-DR antigens. Hepatology 6: 981–989

30. Vento S, Hegarty JE, Alberti A, O'Brein CJ, Alexander GJM, Eddleston ALWF, Williams R (1985) T lymphocyte sensitization to HBcAg and T cell-mediated unresponsiveness to HBsAg in hepatitis B virus-related chronic liver disease. Hepatology 5: 192–197

31. Weimer T, Weimer K, Tu ZX, Jung C, Pape GR, Will H (1989) Immunogenicity of human hepatitis B virus P-gene derived proteins. J Immunol 143: 3750–3756

32. Yamauchi K, Nakanjshi T, Chiou S, Obata H (1988) Suppression of hepatitis B antibody synthesis by factor made by T cells from chronic hepatitis B carriers. Lancet i: 324–326

33. Yuki N, Hayashi N, Kasahara A, Katayama K, Ueda K, Fusamoto H, Kamada T (1990) Detection of antibodies against the polymerase gene product in hepatitis B virus infection. Hepatology 12: 193–198

Authors' address: Dr. C. Ferrari, Cattedra Malattie Infettive, Università di Parma, Via Gramsci 14, I-43100 Parma, Italy.

Arch Virol (1992) [Suppl] 4: 19–22
© Springer-Verlag 1992

T cell recognition of hepatitis B envelope proteins

V. Barnaba, A. Franco, M. Paroli, R. Benvenuto, I. Santilio, and **F. Balsano**

Fondazione "Andrea Cesalpino", I Clinica Medica, Università "La Sapienza", Roma,
Italy

Summary. We have studied the T-cell processing pathways of Hepatitis B antigens and the role of specific B lymphocytes. It could be shown that some form of processing by specific B cells is required for class I CTLs. This mechanism differs from class II endosomal processing. In addition, it could be shown that lysis of HBsAg-specific B cells may be partly responsible for chronic HBV carrier states.

*

The recent progress in understanding antigen processing and presentation to T lymphocytes has been of great help in the comprehension of the mechanisms involved in virus recognition by T cells in the course of viral diseases. It has been clearly established that CD4+ helper T lymphocytes recognize antigen fragments or peptides associated with class II MHC molecules on the surface of antigen-presenting cells (APC) and that these peptides generally derive from processing of exogenous soluble antigens entering APC. In contrast, CD8+ cytotoxic T cells recognize peptides in association with class I molecules on the surface of epithelial cells: these peptides usually derive from processing of endogenously synthesized viral antigens by infected host cells. Two separate pathways have been proposed for antigen processing and presentation to T cells, allowing exogenous antigens selective association to class II molecules and endogenous antigens to class I molecules [1, 2, 3]. Exogenous antigens enter APC by receptor-mediated endocytosis or by phase fluid pinocytosis, are processed in the acidic environment of endosomes, and bind class II molecules as a result of vesicular fusion at a trans-Golgi or post-Golgi location. Endogenously synthesized antigens are processed in the cytosol, probably in the endoplasmic reticulum or in the cis-Golgi region, where they interact with class I molecules.

Recently, exceptions to the rule of class II/class I discrimination have been demonstrated. Indeed, the delivering of endogenous antigens (as measles virus antigens, haemagglutinin or HBsAg) to both class I and class II processing pathways no longer seems to be an exception [4, 5, 6]. In contrast, the converse situation, that is the delivering of exogenous antigens to the class I pathway, has been reported for only a few antigens, including the exogenous form of HBsAg [7, 8, 9].

Previously, we demonstrated that soluble HBsAg can induce not only CD4+ T clones, but also CD8+ T clones, which recognized the antigen in a class II- or class I-restricted manner, respectively [9]. Treatment of APC, during pulsing with HBsAg, with the lysosomotropic agent chloroquine, which inhibits the endosomal processing pathway, drastically reduced the antigen presentation to class II-restricted T cells, but not to class I-restricted T cells, suggesting that the processing pathway for exogenous HBsAg presentation to CD8+ T cells is different from the class II endosomal pathway. A possible mechanism, by which exogenous HBsAg enters class I pathway, might involve the fusion of HBsAg with the membrane of APC and the delivery of the antigen into cytoplasmic class I-processing pathway. The fusion mechanism might play a role in delivering exogenous HBsAg into the cytosol, because this antigen is synthesized as a transmembrane protein inserted on the endoplasmic reticulum and contains lipids in addition to S and pre-S proteins [10, 11].

These data seem to have a biological significance, explaining the hepatocyte-lysis during HBV infection: lysis could occur not only for infected, but also for uninfected hepatocytes, because they could process exogenous antigen and present their products to specific class I-restricted cytotoxic T cells, which in turn can kill the uninfected hepatocytes.

At this point, we wondered whether specific B lymphocytes bearing Ig receptors specific for HBsAg were capable of presenting HBsAg to class I-restricted, as well as to class II-restricted T cells. To determine this possibility, we tested the ability of specific class II-restricted CD4+ T clones and of class I-restricted CD8+ T clones to lyse MHC-restricted B cells bearing Ig receptor specific for HBsAg, prepulsed with different concentrations of antigen. We know that specific B cells can act as very efficient APC: they bind very low concentrations of antigen with high affinity by their Ig receptors, process it through class II pathway, and present its fragments to class II-restricted T cells [12]. Thus, using HBsAg-specific B cells as APC, we would expect that they could present HBsAg only to class II-restricted, but not to class I-restricted T cells. This was, however, not the case. Lysis of HBsAg-specific B cells by both CD4+ and CD8+ T cells was obtained in an MHC-restricted fashion and with a concentration of antigen 5000× less than the concentration required to induce a comparable level of killing of non-specific B cells [13].

Chloroquine treatment of specific B cells, during pulsing with HBsAg, drastically reduced their own killing by class II-restricted, but not by class I-restricted T cells. In contrast, paraformaldehyde-fixation of specific B cells, before the antigen pulsing, abolished the antigen presentation to both types of CTLs. Control experiments showed that specific B cells, pulsed with antigen at least 1 h before fixation or pulsed with a synthetic peptide (mimicing the processing product binding MHC molecules) after fixation, were killed by both CTLs.

All these data, taken together, revealed that class I-CTLs required some form of antigen-processing by specific B cells, and that it is different from the class II endosomal processing pathway.

In conclusion, in those exceptional cases, where an exogenous antigen is allowed access to the class I pathway, the specific focusing of the antigen (as HBsAg) by surface Ig can permit the presentation of processed fragments not only to class II-restricted CTLs by the endosomal processing pathway, but also to class I-restricted CTLs by a cytoplasmic processing pathway. Both class II- and class I-CTLs could be responsible for selective killing of specific B cells suppressing the specific antibody response. The lysis of HBsAg-specific B cells might participate in the establishment of the chronic HBV carrier state.

References

1. Germain RN (1986) The ins and outs of antigen processing and presentation. Nature 322: 687–689
2. Morrison LA, Lukacher AE, Braciale VL, Fan DP, Braciale TJ (1986) Differences in antigen presentation to MHC class I- and class II-restricted influenza virus-specific cytolytic T lymphocyte clones. J Exp Med 163: 903–921
3. Bevan MJ (1987) Class discrimination in the world of immunology. Nature 325: 192–194
4. Jin Y, Sih JW, Berkower I (1988) Human T cell response to the surface antigen of hepatitis B virus (HBsAg). J Exp Med 168: 293–306
5. Sekaly RP, Jacobson S, Richert JR, Tonnelle C, McFarland HF, Long EO (1988) Antigen presentation to HLA class II-restricted measles virus-specific T-cell clones can occur in the absence of the invariant chain. Proc Natl Acad Sci USA 85: 1209–1212
6. Nuchtern JG, Biddison WE, Klausner RD (1990) Class II MHC molecules can use the endogenous pathway of antigen presentation. Nature 343: 74–76
7. Staerz UD, Karasuyama H, Garner AM (1987) Cytotoxic T lymphocytes against a soluble protein. Nature 329: 449–451
8. Moore MW, Carbone FR, Bevan MJ (1988) Introduction of soluble protein into the class I pathway of antigen processing and presentation. Cell 54: 777–785
9. Barnaba V, Franco A, Alberti A, Balsano C, Benvenuto R, Balsano F (1989) Recognition of hepatitis B virus envelope proteins by liver infiltrating T lymphocytes in chronic HBV infection. J Immunol 143: 2650–2655
10. Eble BE, Lingappa VR, Ganem D (1986) Hepatitis B surface antigen: an unusual secreted protein initially synthesized as a transmembrane polypeptide. Molec Cell Biol 6: 1454–1463

11. Standring DN, Ou JH, Rutter WJ (1986) Assembly of viral particles in Xenopus oocytes pre-surface-antigen regulate secretion of the hepatitis B viral surface envelope particle. Proc Natl Acad Sci USA 83: 9338–9343
12. Lanzavecchia A (1985) Antigen-specific interaction between T and B cells. Nature 314: 537–539
13. Barnaba V, Franco A, Alberti A, Benvenuto R, Balsano F (1990) Selective killing of hepatitis B envelope antigen-specific B cells by class I-restricted, exogenous antigen-specific T lymphocytes. Nature 345: 258–260

Authors' address: V. Barnaba, MD, I Clinica Medica, Università "La Sapienza", I-00161 Roma, Italy.

Arch Virol (1992) [Suppl] 4: 23–28
© Springer-Verlag 1992

Fine specificity of the human T cell response to hepatitis B virus core antigen*

A. Penna[1], A. Bertoletti[1,2], A. Cavalli[1], A. Valli[1], G. Missale[1], M. Pilli[1], S. Marchelli[1], T. Giuberti[1], P. Fowler[2], F. V. Chisari[2], F. Fiaccadori[1], and C. Ferrari[1,2]

[1] Cattedra Malattie Infettive, Università di Parma, Parma, Italy
[2] Department of Molecular and Experimental Medicine, Scripps Clinic and Research Foundation, LA Jolla, CA, USA

Summary. The fine specificity of the human T cell response to the hepatitis B virus core antigen (HBcAg) was investigated in 23 patients with acute hepatitis B virus (HBV) infection using a panel of short synthetic peptides covering the entire core region. An immunodominant T cell epitope which was recognized by all except one patient, was identified within the core sequence 50–69. Two further important T cell recognition sites were represented by the amino acid sequences 1–20 and 117–131, which were stimulatory for the T cells of 69% and 73% of the patients, respectively.

T cell recognition of the synthetic peptides was HLA class II restricted because the peptide-induced T cell proliferation was inhibited by anti-HLA class II but not by anti-HLA class I monoclonal antibodies.

These findings may be relevant to the development of future preventive and therapeutic strategies against HBV infection.

*

Several lines of experimental evidence suggest that the development of an adequate immune response to HBV nucleocapsid antigens can be important for HBV clearance [1–4, 6]. For this reason, the identification of immunodominant T cell epitopes within the core molecule could theoretically be useful for the design of more effective alternative vaccines against HBV

---

* Some of this material has been presented in a different format elsewhere (manuscript submitted).

infection and possibly to plan future strategies to manipulate the immune response to HBV in subjects who do not spontaneously clear the virus.

The aim of our study was to identify the amino acid sequences within the core molecule that are involved in the activation of nucleocapsid-specific T cells during HBV infection in man, by using a panel of short synthetic peptides covering the entire HBV core region.

Materials and methods

Twenty-three patients with acute self-limited hepatitis B were studied. The diagnosis was based on the finding of elevated SGPT values (more than 10 times the upper limit of the normal range) associated with the detection of IgM anti-HBc antibodies in their serum. Thirteen subjects without evidence of previous exposure to HBV were studied as normal controls.

Peripheral blood lympho-mononuclear cells (PBMC) were isolated by Ficoll-Hypaque gradient centrifugation, resuspended in RPMI 1640 with 10% human AB serum and stimulated with different concentrations of recombinant (r) HBcAg [7] or short synthetic peptides (10–20 amino acids) covering the complete sequence of the core and pre-core region encoded polypeptides (Fig. 1).

For the production of polyclonal T cell lines, PBMC were cultured either with the whole nucleocapsid molecule or with individual synthetic peptides for 7 days. Interleukin 2 (IL2) was then added and growing cells were restimulated after further 5–7 days of culture. T cell lines were maintained at a cell concentration of $3 \times 10^5 - 1 \times 10^6$ and were restimulated weekly with the appropriate antigen in the presence of irradiated (3000R) autologous PBMC as antigen presenting cells (APC) in RPMI 1640 with 10% fetal calf serum and IL2.

For the proliferation assays, T cell lines were extensively washed and then incubated (5×10^4 T cells/well) for 3 days with autologous irradiated PBMC as APC (1×10^5/well) and different concentrations of antigen or peptides in RPMI 1640 with 10% human AB serum. Proliferation assays were performed in triplicate and ^{3}H-thymidine was added 18 hrs before harvesting.

Results and discussion

Analysis of patients with different HLA haplotypes reveals that several sequences within the core molecule can induce significant levels of T cell response in HBcAg-sensitized individuals (Fig. 1), as previously reported in the mouse system [5]. In addition, more than one peptide fragment is usually recognized by T cells of individual patients (not shown).

However, the most relevant finding of our study is the identification of an immunodominant amino acid sequence (residues 50–69) which was recognized by all but one patient (95%) (Fig. 1). Even though the detection of a significant response to native HBcAg was generally associated with the presence of a significant T cell response to peptide 50–69, in a minority of patients studied serially during the course of the disease, T cell recognition of peptide 50–69 was only transient and undetectable in a few time points when T cells were still able to recognize the native HBcAg and other core peptides.

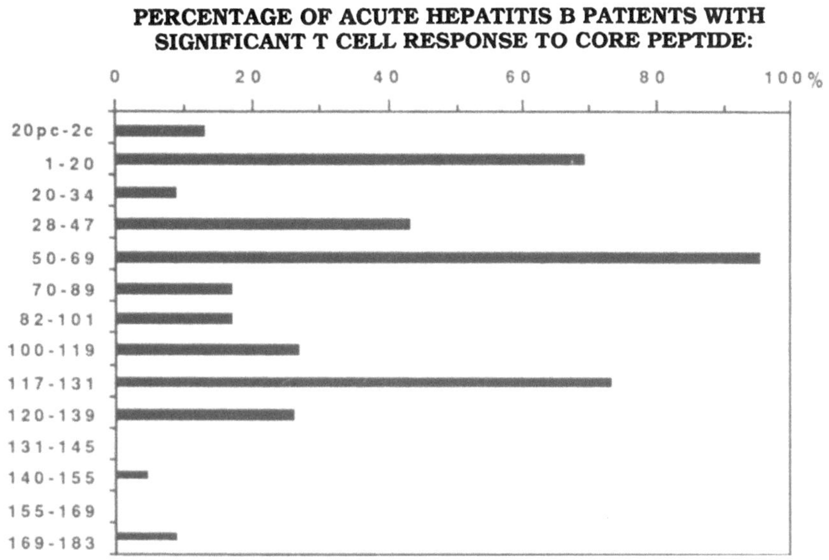

**PERCENTAGE OF ACUTE HEPATITIS B PATIENTS WITH
SIGNIFICANT T CELL RESPONSE TO CORE PEPTIDE:**

Fig. 1. PBMC proliferative response to HBcAg synthetic peptides in 23 patients with acute
hepatitis B. Stimulation indexes higher than 3 were considered as significant values of
proliferative response. The stimulation index is the ratio between ^{3}H-thymidine incorporated by T cells cultured in the presence of antigen and that incorporated by T cells cultured
in medium alone. No significant levels of PBMC proliferation to core peptides were
observed in 13 normal control subjects

This observation suggests that in a few patients the sequence 50–69 can be
recognized but does not represent the major T cell recognition site within the
core molecule.

Two additional peptides corresponding to amino acids 1–20 and
117–131 of the core molecule were also stimulatory for a large proportion of
patients (69% and 73% respectively) (Fig. 1).

When the peptide 50–69 was not the dominant T cell epitope, T cell
recognition of the core molecule was mainly focussed on the 1–20 and
117–131 sequences. Taken together these results suggest that T cell immunization with these immunodominant core peptides can potentially induce
significant T cell activation in a large proportion of the population irrespective of the genetic background.

Our results partially differ from those reported by Milich et al. [5] who
showed that the fine specificity of the murine T cell response to HBcAg is
much more dependent on the MHC haplotype of the responder strain. In
addition, the immunodominant T cell epitope located within residues 50–69
was not identified.

The relevance of specific amino acid sequences contained within peptides
1–20, 50–69 and 117–131 to the activation of a HBcAg-specific T cell
response is suggested by experiments with peptide or HBcAg primed

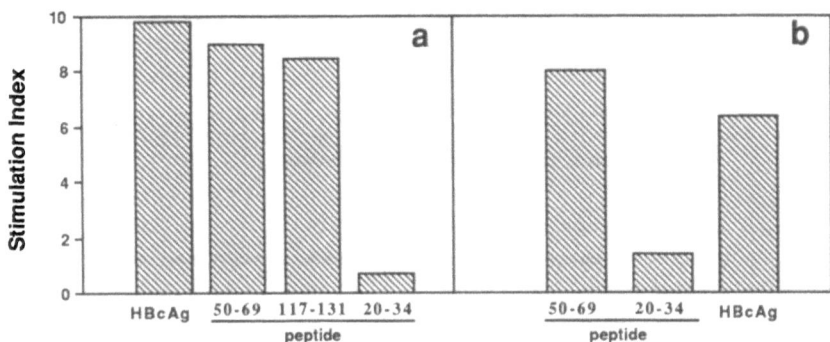

Fig. 2. **a** A polyclonal T cell line produced by PBMC stimulation with HBcAg can recognize the relevant peptides 50–69 and 117–131, which are stimulatory also for the parental PBMC. The proliferative response to an irrelevant core peptide (AA 20–34) is shown. No significant level of proliferative response was observed following stimulation with the other 11 core synthetic peptides (not shown). **b** A polyclonal T cell line produced by PBMC stimulation with peptide 50–69 can recognize the whole core molecule

polyclonal T cell lines. The observations that 1) antigen-specific polyclonal T cell lines produced by PBMC stimulation with HBcAg can be restimulated not only with the native nucleocapsid antigen but also with the relevant peptide analogs recognized by parental PBMC (Fig. 2a) and that 2) peptide-primed polyclonal T cell lines selected by PBMC stimulation with a single-core peptide can react with the whole core protein (Fig. 2b), provides evidence that AA sequences within the peptides 1–20, 50–69 and 117–131 are available to T cell recognition after processing of the native nucleocapsid molecule.

Amino acid sequences 1–20, 50–69 and 117–131 appear to preferentially activate CD4+ T cells which recognize peptide fragments in the context of HLA class II molecules as shown by blocking experiments with anti-HLA monoclonal antibodies (Fig. 3).

In conclusion, our study indicates the existence of an immunodominant T cell epitope (AA 50–69) within the core molecule which is recognized by more than 95% of patients with acute HBV infection and different HLA haplotypes. Two additional important T cell recognition sites were also identified at the aminoterminal end and within the carboxyterminal half of the core molecule. These T cell epitopes might be exploited to enhance the immunogenicity of the existing recombinant HBV envelope vaccines and to try to overcome non-responsiveness to hepatitis B surface antigen particles. Whether our results may also be useful for the design of alternative totally synthetic vaccines and for therapeutic strategies directed to manipulate the immune response to HBV in subjects with chronic HBV infection remains to be investigated.

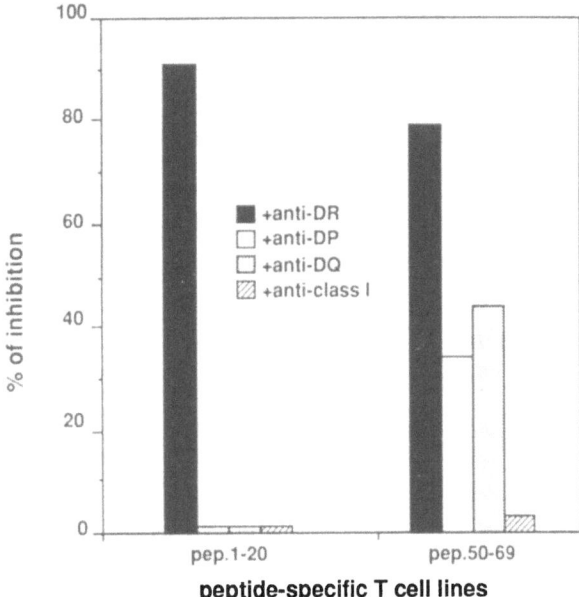

Fig. 3. HLA restriction of the T cell recognition of peptide 1–20 and 50–69. Autologous irradiated PBMC were incubated overnight with anti-HLA class I (W6/32) or anti-HLA class II (D1–12 specific for DR molecules; BT3/4 recognizing DQ molecules; B7/21 specific for DP antigens) monoclonal antibodies and the relevant synthetic peptide. The cells were washed and added to T cells from the peptide-primed T cell lines. Results are expressed as percent inhibition of the proliferative response induced by core peptides

Acknowledgements

The authors would like to thank Biogen S.A., Geneva, Switzerland, for providing rHBeAg; Sorin Biomedica Spa, Saluggia, Italy, for supplying rHBcAg; Marie Anne Petit, Inserm U 131, Clamart, France, for hHBcAg; Roberto Accolla, University of Verona, Italy, for the generous gift of anti-HLA monoclonal antibodies. This work was supported in part by the Italian Ministry of the National Education, Project on Liver Cirrhosis 1989–1990 and by National Institute of Health, USA, Grant AI 26626.

References

1. Ferrari C, Mondelli MU, Penna A, Fiaccadori F, Chisari FV (1987) Functional characterization of cloned intrahepatic, hepatitis B virus nucleoprotein-specific helper T cell lines. J Immunol 139: 539–544
2. Ferrari C, Penna A, Giuberti T, Tong MJA, Ribera E, Fiaccadori F, Chisari FV (1987) Intrahepatic, nucleocapsid antigen-specific T cells in chronic active hepatitis B. J Immunol 139: 2050–2058
3. Ferrari C, Penna A, Bertoletti A, Valli A, Degli Antoni A, Giuberti T, Cavalli A, Petit MA, Fiaccadori F (1990) Cellular immune response to hepatitis B virus (HBV) encoded antigens in acute and chronic HBV infection. J Immunol 145: 3442–3449
4. Milich DR, McLachlan A, Thornton GB, Hughes JL (1987) Antibody production to the nucleocapsid and envelope of the hepatitis B virus primed by a single synthetic T cell site. Nature 329: 547–549

5. Milich DR, McLachlan A, Moriarty A, Thornton GB (1987) Immune response to hepatitis B virus core antigen (HBcAg): localization of T cell recognition sites within HBcAg/HBeAg. J Immunol 139: 1223–1231
6. Mondelli M, Mieli-Vergani G, Alberti A, Vergani D, Portmann B, Eddleston ALWF, Williams R (1982) Specificity of T lymphocyte cytotoxicity to autologous hepatocytes in chronic hepatitis B virus infection: evidence that T cells are directed against HBV core antigen expressed on hepatocytes. J Immunol 129: 2773–2777
7. Pasek M, Goto T, Gilbert W, Zink B, Shaller H, Mackay P, Leadbetter G, Murray K (1979) Hepatitis B virus genes and their expression in E. Coli. Nature 282: 575–579

Authors' address: Dr. Amalia Penna, Cattedra Malattie Infettive, Università di Parma, Via Gramsci 14, I-43100 Parma, Italy.

Arch Virol (1992) [Suppl] 4: 29–35
© Springer-Verlag 1992

HBcAg induced T-cell independent anti-HBc production in chronic HBsAg carriers

S. P. E. Sylvan[1], U. B. Hellström[1], G. Fei[1]*, H. Norder[2], L. Magnius[2], and G. Lindh[1]

[1] The Elias Bengtsson Research Unit, Department of Infectious Diseases. Karolinska Inst, Roslagstull Hospital, Stockholm
[2] Department of Virology, The National Bacteriological Laboratory, Stockholm, Sweden

Summary. The capacity of the nucleocapsid protein of HBV to function as a T-cell independent antigen in man was studied. When T-cell depleted B-cell cultures were challenged with *E coli*-derived HBcAg, anti-HBc production was registered in culture supernatants from the majority of chronic HBsAg carriers in a quiescent stage of disease. In contrast, similarly prepared and stimulated cultures from donors with natural acquired immunity to hepatitis B or HB-susceptible controls were non-responsive. Addition of autologous T-cells effectively restored anti-HBc responsivencss in T-cell depleted B-cell cultures from HB-immune donors, demonstrating the T-cell dependency for anti-HBc induction in natural HBV-infection.

Introduction

It has been suggested that the nucleocapsid (hepatitis B core antigen, HBcAg) of the hepatitis B virus (HBV) is particularly suitable as an immunological carrier moiety for defined peptides in vaccine recipients who are immuno-compromised. This suggestion stems from the capacity of HBcAg to function as a T-cell independent antigen [3, 7, 14]. This T-cell independent capacity has only been demonstrated for nude mice [3, 6], a species in which HBV cannot replicate [14]. Hence, it seems critically important to determine whether T-cell independent B-cell activation also is elicited by HBcAg in humans and if this response is related to immunity to HB or is associated with persistent HBV-infection. In a first step to address this issue, we

---

* Present address: Jing An Central District Hospital, 287 Huangpi Bei Road, Shanghai 200003, China.

analyzed cells from donors with naturally acquired immunity to HB for T-cell independent HBcAg reactivity in vitro. We then compared these with a select group of patients with chronic HBsAg carriership. Our results suggest that T-cell independent anti-HBc reactivities are expressed in patients with chronic HBsAg carriership in a quiescent phase of the disease but not in HB-immune donors.

Patients and methods

Patients

Eight donors with naturally acquired immunity to HB as assessed by the presence of anti-HBs (1/10–1/10,000), anti-HBc (1/10–1/1000) and anti-HBe (1/10–1/100) were studied in comparison with 10 patients (nine anti-HBe$^+$ and one HBeAg$^+$) in a quiescent phase of HBV-disease (eg. normal transaminases) exhibiting anti-HBc (1/1000–1/1,000,000) and anti-HBe (1/100–1/10,000) titers. Three patients were positive for HBV-DNA in serum as assessed by dot-blot or PCR techniques [9]. All patients were seronegative for anti-delta. Seven healthy individuals without serological evidence of previous HBV-infection served as controls. All individuals were negative for HIV-infection and none received immuno-suppressive or antiviral therapy.

Preparation and culture of cells

Highly purified lymphocytes were isolated from heparinized blood by gelatin-sedimentation, iron treatment and Ficoll-Paque gradient centrifugation and separated into B-, T-, and monocytic(Mo)-cell enriched populations as previously described in detail [4]. The B-cell populations contained as a mean 80% (range 63–93%) B1-positive cells, <2% Mo and <1% rosette forming cells and the rest so called null cells, and was similar for the three donor groups. B-cell fractions from the three donor groups were functionally analyzed for residual T-cell contamination by stimulation with the T-cell mitogen ConA (1 ug/ml) for three days. The stimulation index (SI) never exceeded 2, whereas control T-cell cultures exhibited a mean SI of 48.

Quantities of 4×10^5 B-cells (1×10^6/ml) supplemented with 10% autologous Mo were cultured in round-bottomed tissue culture tubes (A/S NUNC, Roskilde, Denmark) in the absence or presence of autologous T-cells (T/B cell ratios 0.25–8.0) in Hepes buffered RPMI 1640 (Biocult Laboratories, Paisley, UK) supplemented with antibiotics, glutamine and 20% FBS (Gibco, Grand Island, NY, USA) in the absence or presence of E coli-derived core proteins (0.001–100 ng/ml) kindly provided by Dr I. Cayzer (Wellcome, Diagnostics, Beckenham, UK). After 4 days the cells were washed and resuspended in culture medium containing 5% FBS and cultured for additional 8 days.

Anti-HBc ELISA

Culture supernatants were analyzed for IgG anti-HBc content using microtitre plates (Dynatech, Plochingen, FRG) charged with HBcAg (100 ng/ml) in 0.05 M sodium carbonate buffer, pH 9.6. Post coating was done with 1%FBS in 10 mM Tris-HCl, 0.13 M NaCl and 1 mM EDTA at 37°C for 1 hr. Supernatants were diluted 1/5 and alkaline-phosphatase conjugated rabbit anti-human γ-chains (Sigma Chemical Co, Mo, USA) 1/1000 in PBS-Tween-1%FBS. The level of anti-HBc was calculated from a standard curve obtained by two-step dilution of immunosorbent purified anti-HBc.

Results

The anti-HBc response in T-cell depleted B-cell cultures from chronic carriers of HBsAg was highly antigen-dose dependent. The antigen dose for optimal antibody production varied between different individuals (Fig. 1 a–c). Similarly prepared and stimulated B-cell cultures from HB-immune donors were unresponsive Fig. 1 d–f). The optimal anti-HBc response for each individual in the three donor groups, obtained in the presence of antigen, are plotted with the corresponding control value obtained in the absence of antigen in Fig. 2. The results demonstrate that HB-immune donors and HB-susceptible controls do not exhibit a T-cell independent B-cell response to HBcAg in vitro. In contrast, a spontaneous secretion of anti-HBc antibodies was registered in 2/10 B-cell cultures from

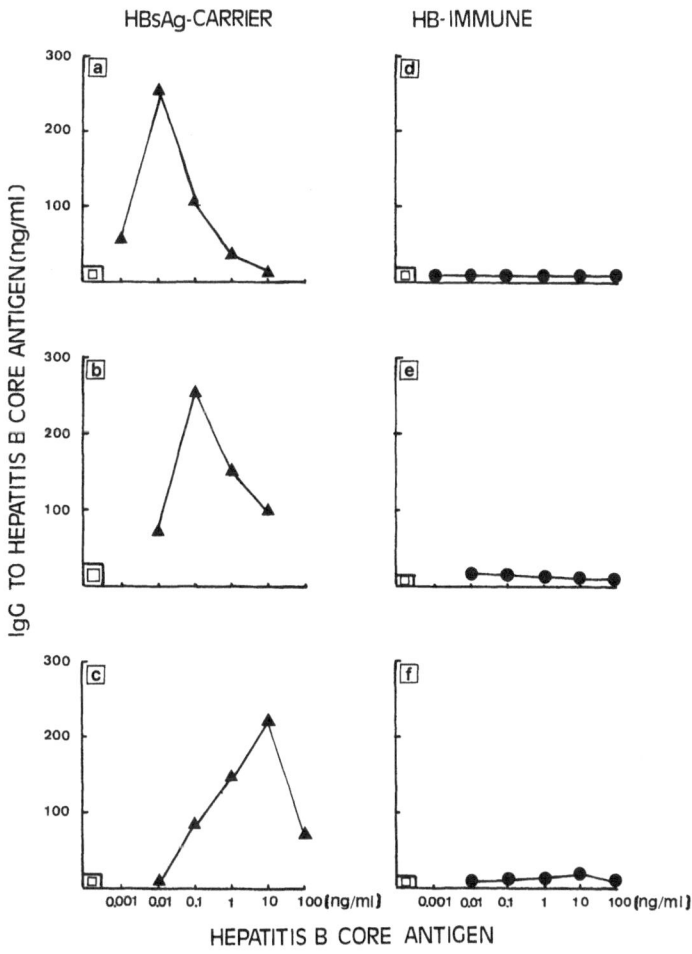

Fig. 1. Antigen dose dependency of HBcAg induced anti-HBc secretion. T-cell depleted B-cells from three chronic HBsAg carriers (a–c) and three HB-immune donors (d–f) were cultured in the absence (□) and presence (▲, ●) of different concentrations (0.001–100 ng/ml) of HBcAg for 12 days. Anti-HBc secretion is given as ng/ml/1 × 10^6 B-cells

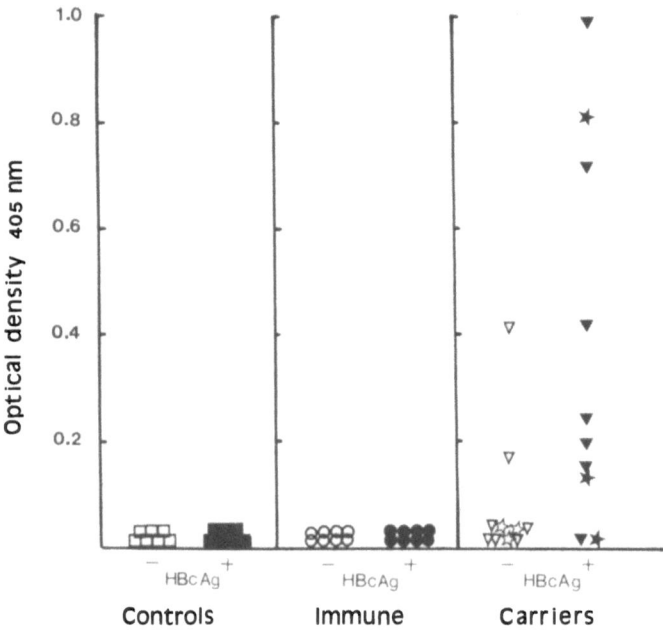

Fig. 2. Anti-HBc levels in supernatants of T-cell depleted B-cell cultures from seven control individuals (■), eight HB-immune donors (●) and ten chronic HBsAg carriers (▼) of whom three were HBV-DNA positive in serum (★) in the absence (open symbols) and presence (filled symbols) of optimal concentrations of HBcAg

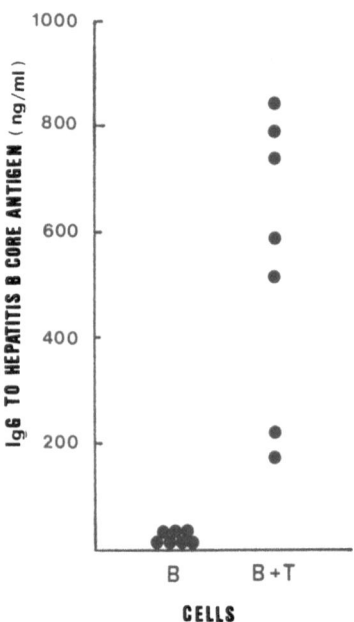

Fig. 3. T-cell depency of HBcAg induced anti-HBc IgG responses in HB-immune donors. T-cell depleted (B) and T-cell supplemented (T + B) B-cell fractions from seven HB-immune individuals were cultured at a density of 1×10^6 B-cells/ml in the presence of 1 ng HBcAg/ml

chronic HBsAg carriers. Addition of recombinant HBcAg significantly enhanced the anti-HBc production in 8/10 B-cell cultures.

No anti-HBc was detectable in unstimulated or stimulated B-cell cultures from two patients. When B-cells from the HB-immune donor population were stimulated with HBcAg in the presence of autologous T-cells at optimal numbers, anti-HBc antibodies were produced in all cell cultures (Fig. 3).

Discussion

In contrast to what has been reported in the murine system [3, 6], HBcAg did not function as T-cell independent antigen in HB-immune individuals. IgG class anti-HBc antibodies were not detectable in T-cell depleted B-cell cultures, when stimulated with low concentrations of HBcAg in vitro. During acute resolving hepatitis B, however, preliminary data demonstrated that transiently appearing IgM anti-HBc antibodies are produced in antigen-stimulated cultures (data not shown), suggesting that IgG producing memory B-cells are not generated in the absence of T-cell help during natural HBV infection. Supplementation with autologous T-cells restored the B-cell responsiveness, demonstrating the T-cell dependency of IgG class anti-HBc production in HB-immune donors.

A disparate regulation of the humoral immune response to HBcAg was detected in chronic carriers of HBsAg in a quiescent phase of the disease, as B-cells from the majority of these patients effectively produced IgG class anti-HBc antibodies in the absence of T-cell helper functions after challenge with recombinant HBcAg in vitro. Peripheral B-cells from two chronic HBsAg carriers of whom one was HBV-DNA[+] in serum, lacked the capacity to produce anti-HBc antibodies when repeatedly tested in vitro. Tolerogenic sites within the HBcAg have recently been described in mice [5]. Because immature B-cell populations are unduely sensitive to tolerance induction [12], it could be speculated that the lack of responsiveness to HBcAg stimulation is a consequence of transplantal transmission of HBV-infection in these two patients.

The capacity of HBcAg to function as a T-cell independent antigen in chronic HBsAg carriers but not in HB-immune donors in vitro, could result from antigen activation of a distinct B-cell subpopulation(s) which is beyond the resting state due to the previous encounter with HBcAg in vivo [8]. Alternatively, the residual natural killer cells in the B-cell preparations [8] from chronic HBsAg carriers and HB-immune donors deviate in their capacity to secrete lymphokines such as IL-2, IFN-γ and IL-4 essential for the T-cell independent B-cell activation.

It is well recognized that T-cell independent polysaccharide antigens are poor inducers of immunological memory and long-lasting immunity, particularly in the young host [10]. At best, they stimulate short-lived IgM and IgG responses of low affinity [10]. It has been demonstrated that the affinity of anti-HBc antibodies in sera from asymptomatic carriers of HBsAg and patients with chronic liver disease type B were ten times lower compared with that obtained in pooled human immune sera [15]. So far, we do not know whether the anti-HBc antibodies produced in T-cell depleted B-cell cultures from chronic HBsAg carriers deviate in affinity from those induced in T-cell supplemented B-cell cultures from HB-immune donors.

Restricted subclass diversification has been reported for T-cell independent carbohydrate antigens in man [13]. A similar phenomenon has been reported for HBcAg induced anti-HBc antibodies in nude mice [6]. Whether this also holds true for T-cell independent HBcAg activation in man needs yet to be determined.

T-cell independent antigens are generally polymeric molecules of relatively high molecular weight with a repeating structure and antigenic unit [10], and HBcAg fulfills these criteria [6]. Recent data indicate that a single dominant HBc determinant induces a numerous population of anti-HBc antibodies [11].

The T-cell independently induced anti-HBc antibodies in chronic HBsAg carriers in this report did not react with the synthetic peptides corresponding to the 40–49, 75–84 and 132–147 sequences of the C-region of HBV (data not shown). Reactivity towards the short linear subsequence 73–85 of the basic HBcAg/HBeAg structure has, however, been described for some patients with chronic liver disease type B [2]. Whether this discrepancy in anti-HBc reactivity reflects a divergent mode of anti-HBc induction as far as T-cell dependency is concerned is not known. It is also not known whether predominantly conformation-dependent anti-HBc antibodies, as described for acutely infected and convalescent patients [11], are induced by T-cell independent means in chronic HBsAg carriers in an indolent stage.

The importance of T-cell independent pathways for the development of protective immunity is controversial [1]. The data reported here indicate that T-cell independent reactivities elicited by HBcAg are expressed in chronic HBsAg carriers in a quiescent stage of the disease but not in HB-immune donors. Therefore, before taking advantage of the T-cell independent carrier effect of HBcAg on antibody production against non- or low-immunogenic peptide sequences in chimeric vaccines intended for human use, as suggested by several groups [3, 7, 14], further delineation of the contribution of the T-cell independent capacity of HBcAg to protective immunity or chronically evolving hepatitis B is needed.

Acknowledgements

We are indebted to Mrs Ulla-Britt Sallbert and Bertit Hammas for technical assistance, to Dr Madeleine von Sydow from the Hepatitis Section at the Central Microbiological Laboratory of the Stockholm County Council, Stockholm, Sweden for performing the hepatitis B serology. This work was supported by the Swedish Society of Medicine and The Swedish Society for Medical Research.

References

1. Bazin H, Turk JL, Zanetti M, Glotz D, McGhee JR, Eldridge JH, Beagley KW, Kiyono H, Ernst PB, Bienenstock J, Colle J-H, Truffa-Bachi P, Bach JF (1988) 21st forum in immunology: Is antiinfectious defence thymus-dependent? Ann Inst Pasteur/Immunol 139: 119–222

2. Colucci G, Beazer Y, Cantaluppi C, Tackney C (1988) Identification of a major hepatitis B core antigen (HBcAg) determinant by using synthetic peptides and monoclonal antibodies. J Immunol 141: 4376–4380

3. Francis MJ, Hastings GZ, Brown AL, Grace KG, Rowlands DJ, Brown F, Clarke BE (1990) Immunological properties of hepatitis B core antigen fusion proteins. Proc Natl Acad Sci USA 87: 2545–2549

4. Hellström U, Sylvan S, Lundbergh P (1985) Regulatory functions of T- and accessory-cells for hepatitis B surface antigen induced specific antibody production and proliferation of human peripheral blood lymphocytes in vitro. J Clin Lab Immunol 16: 173–181

5. Milich DR, Jones JE, McLachlan A, Houghten R, Thornton GB, Hughes JL (1989) Distinction between immunogenicity and tolerogenicity among HBcAg T cell determinants. J Immunol 143: 3148–3156

6. Milich DR, McLachlan A (1986) The nucleocapsid of hepatitis B virus is both a T-cell-independent and a T-cell-dependent antigen. Science 234: 1398–1401

7. Milich DR, McLachlan A, Hughes JL, Jones JE, Stahl S, Wingfield P, Thornton GB (1989) Characterization of the hepatitis B virus nucleocapsid as an immunologic carrier moiety. In: Lerner RA, Ginsberg H, Chanock RM, Brown F (eds) Vaccines 1989: Modern approaches to new vaccines including prevention of AIDS. Cold Spring Harbor Laboratory, pp 37–42

8. Mond JJ, Brunswick M (1987) A role for IFN-γ and NK cells in immune responses to T cell-regulated antigens type 1 and 2. Immunol Rev 99: 105–117

9. Norder H, Hammas B, Magnius L (1990) Typing of hepatitis B virus genomes by a simplified polymerase chain reaction. J Med Virol 31: 215–221

10. Robbins JB, Schneerson R (1990) Polysaccharide-protein conjugates: a new generation of vaccines. J Infect Dis 161: 821–832

11. Salfeld J, Pfaff E, Noah M, Schaller H (1989) Antigenic determinants and functional domains in core antigen and e antigen from hepatitis B virus. J Virol 63: 798–808

12. Scott DW (1984) Mechanisms in immune tolerance CRC. Crit Rev Immunol 5: 1–25

13. Siber GR, Schur PH, Aisenberb AC, Weitzman SA, Schiffman G. Correlation between serum IgG-2 concentrations and the antibody response to bacterial polysaccharide antigens. N Engl J Med 303: 178–182

14. Stahl SJ, Murray K (1989) Immunogenicity of peptide fusions to hepatitis B virus core antigen. Proc Natl Acad Sci USA 86: 6283–6287

15. Wen YM, Duan SC, Howard CR, Frew AF, Steward MW (1990) The affinity of anti-HBc antibodies in acute and chronic hepatitis B infection. Clin Exp Immunol 79: 83–86

Authors' address: Dr. S. P. E. Sylvan, Department of Infectious Diseases, Karolinska Institute, Roslagstull Hospital, Box 5651, S-11489 Stockholm, Sweden.

Arch Virol (1992) [Suppl] 4: 36–38
© Springer-Verlag 1992

Divergent anti-HBc reactivities in HB-immune and chronic HBsAg carriers

U. B. Hellström and **S. P. E. Sylvan**

The Elias Bengtsson Research Unit, Department of Infectious Diseases, Karolinska Institute, Roslagstull Hospital, Stockholm, Sweden

Summary. We have monitored titers of anti-HBc antibodies in sera from acutely HBV-infected and chronic HBsAg carriers. Our data show that there is a divergence in the specificity of the antibodies in these two populations. We also present preliminary results showing that serum from HB-immune carriers contain antibodies that are multispecific and display autoimmune characteristics, reacting with human serum albumin.

*

In the current search for a vaccine against HIV much interest has been focused upon the HBcAg specific peptide sequence p120–140, which in some strains of mice has been demonstrated to function as a carrier moiety for a defined B-cell epitope from the envelope material of HIV [3]. Because neither HBV nor HIV infection occur as natural infections in mice, more information regarding the T- and B- cell recognition sites localized within the HBcAg/HBeAg basic protein complex in man are needed. Recently Howard et al. demonstrated the presence of low-affinity antibodies to the synthetic peptide corresponding to the 132–147 sequence in the C-region of HBV in human anti-HBV immunoglobulin [1]. Therefore we consecutively determined the serological titers of anti-HBc[#] 132–147 antibodies in sera from acutely HBV-infected individuals through their convalescence phase and repeatedly tested chronic HBsAg carriers. Moreover, we studied the capacity of peripheral B-cells from HB-immune donors to secrete IgG antibodies with specificity for this peptide in comparison with cells from chronic HBsAg carriers after stimulation with low concentrations of *E coli*-derived core proteins in vitro.

We found that antibodies of the IgM as well as IgG class against HBc[#] 132–147 are elicited during acute hepatitis B. These antibodies were, however, transiently induced in patients with an uncomplicated course of the disease (eg. normal P-prothrombincomplex, normotest values) and eliminated at the time of HBe/anti-HBe seroconversion. In patients with transient liver failure the elimination of anti-peptide antibodies were much slower and antibody titers were detectable during a prolonged period of time. Similarly, persistent and fluctuating titers of anti-HBc$^{\#}$ 132–147 were repeatedly found in sera from HBeAg^{+} chronic carriers with active disease (manuscript in preparation). The notion that anti-HBc$^{\#}$ 132–147 antibody production was related to viral replication was substantiated by in vitro studies demonstrating T-cell dependently induced IgG anti-HBc$^{\#}$ 132–147 antibodies in culture supernatants from 11/13 chronic HBsAg carriers but only in 1/6 HB-immune donors. Moreover, the highest levels of IgG anti-HBc$^{\#}$ 132–147 in vitro were registered with cells from chronic patients seropositive for HBV-DNA (manuscript in preparation). Consequently, our data suggest that T-cell dependent helper functions regulate the anti-HBc$^{\#}$ 132–147 production in chronic carriers, whereas this antibody secretion is under the influence of a suppressor signal in HB-immune donors. This notion is supported by the finding in the mouse system that p120–140 is immunogenic in some but tolerogenic in other strains [2].

The specificity of the anti-HBc antibodies in HB-immune donors seemed to be disparate compared with those in chronic carriers of HBsAg, as the IgG antibodies reactive with E coli-derived HBcAg, in contrast to the IgG anti-HBc$^{\#}$ 132–147, remained in circulation and increased in titers during the convalescence phase. Furthermore IgG anti-HBc (E coli-derived), but not anti-HBc$^{\#}$ 132–147 antibodies were detectable in supernatants from HBcAg stimulated immune B/T-cell cultures in vitro. Whether these antibodies recognize conformational HBc epitopes or linear sequences of the basic HBcAg/HBeAg structure outside the sequence 132–147 is still unknown.

Taken together, these data indicate that the specificity of circulatory anti-HBc antibodies are divergent in chronically HBV-infected individuals compared with those, who had developed a protective immunity to HB. Preliminary data from our laboratory have demonstrated that the pool of anti-HBc antibodies in chronic HBsAg carriers contain anti-HBc antibodies, which are multispecific and reactive with the host "self"-component human serum albumin and thus autoimmune in their nature. The albumin association with the amino-terminal domain of the capsid protein of HBV has been previously described in the literature, a finding which must be taken into consideration in man, in the search for protective HBc-determinants.

References

1. Howard CR, Stirk HJ, Brown SE, Steward MW (1988) Towards the development of synthetic hepatitis B vaccines. In: Zuckerman AJ (ed) Viral hepatitis and liver disease. Alan R Liss, pp 1094–1101

2. Milich DR, Jones JE, McLachlan A, Houghten R, Thornton GB, Hughes JL (1989) Distinction between immunogenicity and tolerogenicity among HBcAg T cell determinants. J Immunol 143: 3148–3156

3. Milich DR, McLachlan A, Hughes JL, Jones JE, Stahl S, Wingfield P, Thornton GB (1989) Characterization of the hepatitis B virus nucleocapsid as an immunologic carrier moiety. In: Lerner RA, Ginsberg H, Chanock RM, Brown F (eds) Vaccines 89: Modern approaches to new vaccines including prevention of AIDS. Cold Spring Harbor Laboratory, pp 37–42

Authors' address: Dr. U. B. Hellström, Department of Infectious Diseases, Karolinska Inst., Roslagstull Hospital, Box 5651, S-11489 Stockholm, Sweden.

Arch Virol (1992) [Suppl] 4: 39–41
© Springer-Verlag 1992

Absence of free core antigen in anti-HBc negative viremic hepatitis B carriers

Ch. Possehl[1], R. Repp[2], K.-H. Heermann[1], E. Korec[3], A. Uy[1], and W. H. Gerlich[1]

[1] Department of Medical Microbiology, University of Göttingen
[2] Pediatric Clinic, University of Giessen, Federal Republic of Germany,
[3] Institute of Molecular Biology, Prague, CSFR

Summary. Using enzyme immune assay and immune electron microscopy, we have examined the sera of immune-suppressed anti-HBc negative HBV-infected patients for the presence of HBcAg. Our results suggest that free HBV core particles are absent or present only in minute amounts in the blood of chronic carriers and that at the most, only minimal amounts of core antigen are found on the surface of the virus particles.

*

Hepatitis B core antigen (HBcAg) of Hepatitis B virus (HBV) is considered to be the most immunogenic structure of HBV. Antibodies to HBcAg are only absent in the early phase of infection or in severely immunodeficient patients.

The question of how the immune system recognizes HBcAg has not yet been resolved. Recently, it was suggested that circulating virions would expose HBcAg at their surface [1]. Alternatively, "naked" core particles may be secreted or liberated from hepatocytes, as was found in vitro [2]. The observation of free HBcAg in the serum of HBV carriers is usually hampered by the excess of anti-HBc which covers up any available HBcAg. However, under the conditions of severe immune suppression anti-HBc may be absent. We observed an absence of anti-HBc in certain children suffering from malignancies [3].

During cytostatic treatment, 74 children were inadvertantly HBV infected, and became persistent asymptomatic HBV carriers. After 6 to 8 years, 47 of the 54 survivors were still strongly positive for HBsAg, HBeAg and HBV DNA. Out of 36 tested children, 22 were constantly positive for anti-HBc, 8 showed variable reactions, and 6 were constantly negative. As their immune reactions to other antigens seem to be normal, these patients have obviously developed a selective immune-tolerance against the antigens of HBV.

We investigated 13 of these patients, 8 positive for anti-HBc and 5 negative. The titers for HBsAg and viral DNA in these patients were compared to those of asymptomatic HBsAg and HBeAg carriers who were detected upon blood donation. The titers for HBsAg and viral DNA in the blood donor carriers scattered over a wide range, while these parameters seemed to be more homogeneous in the immunotolerant children with no significant difference between anti-HBc negative and positive sera. We concluded that in an immunotolerant organism, a typical and possibly stable state of viral replication is established with normal values of 70 to 200 µg HBsAg and 100 to 200 million virus particles per ml.

We consequently examined the sera of the anti-HBc negative children for free HBcAg by enzyme immune assay. Centrifugation through a sucrose gradient separated the HBV particles from serum proteins. The concentrated HBV particles were analyzed for HBcAg either directly, or, as a control, after removal of the viral envelope. The HBcAg assay was performed in microplates which were coated with two different HBc-specific monoclonal antibodies [4]; as peroxidase conjugate we used polyclonal anti-HBc from sheep. In this assay, all anti-HBc negative sera were highly positive for core antigen when the HBsAg was stripped off from the virions, containing more than 20 ng HBcAg per ml. This finding showed that the absence of anti-HBc was not due to mutation of the HBc epitopes.

We did not, however, find any HBcAg in the untreated virus samples, irrespective of whether anti-HBc was present or not. As a positive control we used recombinant core particles and achieved a sensitivity of 3 ng per ml corresponding to 0.3 ng per ml in the tenfold concentrated serum samples. According to this enzyme immune assay, free core particles do not circulate in blood of chronic carriers at levels higher than 0.3 ng per ml, and only minute if any amounts at all of core antigen are present on the surface of the virus.

Although HBcAg was not detectable by enzyme immune assay, we wanted to confirm this result using another method. To this end, we chose immune electron microscopy (see Table 1). We tried to generate aggregates of HBV particles by addition of various antibodies. These aggregates were enriched by centrifugation, resuspended and identified electron microscopically. First, we used monoclonal or polyclonal anti-HBc in a wide range of concentrations. We did not, however, detect aggregated virions or core particles. Only in one of four sera did we observe clusters of what appeared to be virus particles. As we found similar clusters also in the negative control, this was possibly an artefact of the preparation. To test the reactivity of the anti-HBc, we mixed it with recombinant core particles. Monoclonal and polyclonal anti-HBc lead to aggregates which we found neither in the

Table 1. Aggregation of HBV-particles by antibodies

Specificity of antibody	HBV from patients sera					rHBc-particles
	neg.[a]	neg.	neg.	neg.	pos[b]	
untreated	−	−	−	−	−	−
buffer	+	(+)	(+)	−	−	−
anti-HBc	+	−	(+)	−	n.d.	+ + +
anti-SHBs	n.d.	+ + +	n.d.	+ + +	+ + +	n.d.
anti-preS2	n.d.	+	n.d.	n.d.	+ +	n.d.
anti-preS1	n.d.	+	+	+	+	n.d.

[a] anti-HBc absent, or   [b] present in the patients' sera

untreated, nor in the mock-treated control.

Finally, we wanted to determine whether viral particles from the sera could be aggregated by monoclonal antibodies at all. Therefore, we also used antibodies against the small, middle and large HBs protein. Anti-SHBs and anti-preS2 were able to aggregate virions, filaments and 20 nm particles. Although it has been reported that the middle protein may be absent on virus particles [5], in our study the antibody against the preS2 glycopeptide also complexed the virions. However, the anti-preS1 antibody agglutinated only filaments and virus particles. This supports our previous finding that 20 nm particles have only a very small proportion of the large HBs protein [6].

References

1. Möller B, Hopf U, Stemerowicz R, Henze G, Gelderblom H (1989) HBcAg expressed on the surface of circulating Dane particles in patients with hepatitis B virus infection without evidence of anti-HBc formation. Hepatology 10: 179–185
2. Jean-Jean O, Levrero M, Will H, Perricaudet M, Rossignol J-M (1989) Expression mechanism of the hepatitis B virus (HBV) C gene and biosynthesis of HBe antigen. Virology 170: 99–106
3. Bertram U, Fischer HP, Repp R, Willems WR, Lampert F (1990) Hepatitis Endemie bei cytostatisch behandelten Kindern. Dtsch Med Wochenschr 115: 1253–1254
4. Korec E, Korcová J, König J, Hlozánek I (1989) Detection of antibodies against hepatitis B core antigen using the avidin-biotin system. J Virol Meth 24: 321–326
5. Petit MA, Capel F, Riottot MM, Dauguet C, Pillot J (1987) Antigenic mapping of the surface proteins of infectious hepatitis B virus particles. J Gen Virol 68: 2759–2767
6. Heermann K-H, Goldmann U, Schwartz W, Seyffarth T, Baumgarten H, Gerlich WH (1984) Large surface proteins of hepatitis B virus containing the pre-S sequence. J Virol 52: 396–402

Authors' present address: Dr. W. Gerlich, Institute of Medical Virology, University of Giessen, Frankfurter Strasse 107, D-W-6300 Giessen, Federal Republic of Germany.

Arch Virol (1992) [Suppl] 4: 42–45
© Springer-Verlag 1992

Homing of T-lymphocytes in acute and chronic HBV positive inflammatory liver disease

R. Volpes, J. J. van den Oord, and **V. J. Desmet**

Department of Pathology, Laboratory of Histochemistry and Cytochemistry, University Hospital Sint-Rafaël, Catholic University of Leuven, Leuven, Belgium

Summary. The expression of immune adhesion molecules governing cell–cell and cell–matrix interactions allows an optimal migration and accumulation of lymphocytes in acute and chronic inflammatory liver diseases.

Introduction

Cell communication is essential for the development, tissue organization and function of all multicellular organisms. Cells communicate with each other and with their environment via soluble mediators and direct contact. In the immune system, lymphocytes are involved in a complex network of soluble mediators, cell–cell and cell–matrix interactions. The circulating lymphocytes that arrive at the site of injury must directly adhere to the endothelium and basement membrane during extravasation from the blood stream. Once in the extravascular compartment, lymphocytes will migrate in response to gradients of chemoattractants and to adhesive gradients. By adherence to a variety of cells, lymphocytes will move towards antigen-presenting or infected target cells, adhere to them and deliver an appropriate immunological response.

Lymphocyte "homing" in the liver: interactions with endothelium

In the past few years, we have focused on homing and adhesion processes of lymphocytes in the liver. In order to leave the sinusoidal compartment and to enter the liver parenchyma, lymphocytes have to interact with endothelial cells (EC) [6]. This emigration process involves a lymphocyte adhesion step to EC, which is mediated by particular receptors expressed on lymphocytes

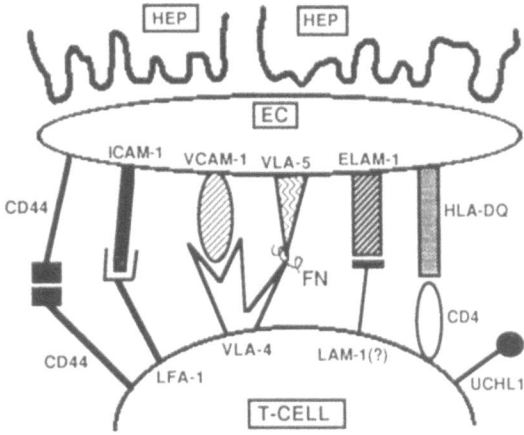

Fig. 1. Schematic illustration of the interactions between T-cell and endothelial cell (*EC*) in liver inflammation

("homing receptors") that bind to their corresponding natural ligands displayed by activated EC ("vascular selectins") [1, 2] (Fig. 1).

Using in situ immunohistochemistry and a panel of specific monoclonal antibodies, we have shown that both in acute and chronic HBV-related hepatitis, portal tracts as well as areas of periportal and intralobular inflammation consist virtually only of memory T-lymphocytes that strongly express LFA-1, VLA-4, CD44, and CD4 antigens. In addition, several vascular selectins are up-regulated (ICAM-1, CD44, HLA-DQ) or de-novo expressed (VCAM-1, ELAM-1) on sinusoidal endothelial cells in the same areas of inflammation. Moreover, the mechanism of accumulation of lymphocytes at sites of periportal "piecemeal necrosis" appears to be similar to that in intralobular "spotty necrosis", and to involve interactions with periportal sinusoidal endothelial cells. Expression of LAM-1 on portal (LAM-1+) but not on lobular (LAM-1−) T-cells suggests that homing of lymphocytes to portal tracts additionally involves the LAM-1 molecule, and that different homing mechanisms are operative in various compartments of the liver.

Lymphocyte "homing" in the liver: interactions with hepatocytes

Once they arrive in the lobular parenchyma, T-lymphocytes are involved in interactions with hepatocytes as potential target cells. This lymphocyte/hepatocyte interaction is mediated by several molecules expressed on both lymphocytes and hepatocytes (Fig. 2).

The HB core antigen (HBcAg) most likely represents the viral target antigen for cytotoxic T-cells in areas of inflammation, and studies on lymphocytes isolated from HBV-infected liver tissue have confirmed this view [3].

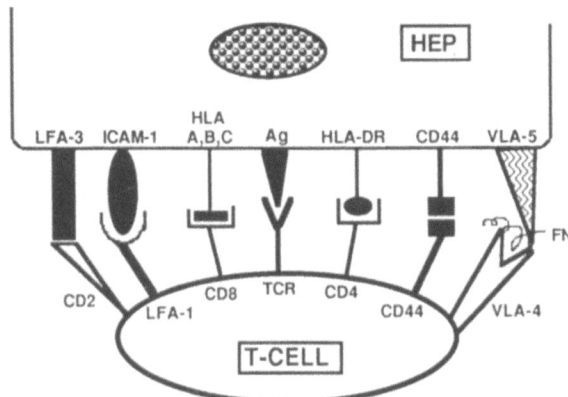

Fig. 2. Schematic illustration of the interactions between T-cell and hepatocyte (*HEP*) in liver inflammation

T-cell mediated cytotoxicity is restricted by products encoded by the Major Histocompatibility Complex (MHC) Class I (HLA A, B, C) and Class II (HLA-DR); both MHC-products are expressed by hepatocytes in areas of liver inflammation in acute and chronic HBV-infection [4]. Together with these HLA-antigens, the viral antigen is recognized by specific receptors expressed on the T-cell membrane (TCR).

An optimal T-cell mediated cytotoxic activity requires a strong adherence between T-cells and target cells. In addition to the antigen-dependent interactions, several adhesion molecules have been demonstrated to strengthen this adherence in an antigen-independent manner [5]. In contrast to normal liver, ICAM-1 and LFA-3 are expressed on hepatocyte membranes in HBV-infection in a diffuse (acute hepatitis) or focal distribution pattern (chronic hepatitis and cirrhosis). Moreover, a close topographical correlation exists between hepatocellular expression of ICAM-1 and LFA-3 on the one hand, and the presence of memory T-lymphocytes expressing the respective ligands LFA-1 and CD2 on the other. Hence, adhesion between hepatocytes and inflammatory cells is mediated by two pathways of cellular interaction, involving ICAM-1/LFA-1 and LFA-3/CD2.

In addition to ICAM-1 and LFA-3, other adhesion molecules are involved in the interaction between T-lymphocytes and hepatocytes. We could show that hepatocytes de-novo express the CD44 antigen in areas of inflammation, particularly in liver biopsies with acute hepatitis. Since inflammatory cells strongly express the CD44 antigen, it is likely that a CD44-mediated homotypic form of binding between T-cells and hepatocytes occurs in inflammatory liver diseases (Fig. 2). Moreover, the expression of the VLA-5 molecule is strongly up-regulated on the membrane of hepatocytes in both acute and chronic hepatitis. Since VLA-4, as well as VLA-5 are fibronectin receptors, recognizing two different domains of fibronectin (FN), a fibronectin-mediated binding between VLA-4+ lymphocytes and VLA-5+ hepatocytes is very likely.

Conclusions

The microenvironment in areas of intralobular inflammation (areas of "piecemeal" and "spotty" necrosis) allows an optimal egress of lymphocytes from the blood into the lobular parenchyma. Accumulation of lymphocytes in areas of liver inflammation involves interactions between homing receptors on lymphocytes, and corresponding ligands (vascular selectins) on sinusoidal endothelial cells. The expression of cytokine-inducible adhesion molecules (ICAM-1, ELAM-1, VCAM-1, HLA-DQ) by sinusoidal endothelial cells indicates that these cells actively modulate their phenotype in response to environmental factors, thus playing a key-role in the regulation of leukocyte traffic in inflammatory liver diseases.

Once lymphocytes arrive in the lobular parenchyma, their binding to hepatocytes is mediated by several pathways of adhesion molecule pairs, which strengthen this lymphocyte/hepatocyte interaction so that antigen-specific recognition can occur.

References

1. Berg EL, Goldstein LA, Jutila M, Nakache M, Picker LJ, Streeter PR, Butcher EC (1989) Homing receptors and vascular addressins: cell adhesion molecules that direct lymphocyte traffic. Immunol Rev 108. 5–18
2. Carlos TM, Harlan JM (1990) Membrane proteins involved in phagocyte adherence to endothelium. Immunol Rev 114: 5–28
3. Mondelli M, Chisari FV, Ferrari C (1990) The cellular immune response to nucleocapsid antigens in hepatitis B virus infection. Springer Semin Immunopathol 12: 25–31
4. van den Oord JJ, De Vos R, Desmet VJ (1986) In-situ distribution of Major Histocompatibility Complex products and viral antigens in chronic hepatitis B virus infection: evidence that HBc-containing hepatocytes may express HLA-DR antigens. Hepatology 6: 981–989
5. Volpes R, van den Oord JJ, Desmet VJ (1990) Immunohistochemical study of adhesion molecules in liver inflammation. Hepatology 12: 59–65
6. Yednock TA, Rosen SD (1989) Lymphocyte homing. Adv Immunol 44: 313–378

Authors' address: Dr. R. Volpes, Department of Pathology, Laboratory of Histochemistry and Cytochemistry, University Hospital Sint-Rafaël, Catholic University of Leuven, Minderbroedersstraat 12, B-3000 Leuven, Belgium.

Arch Virol (1992) [Suppl] 4: 46–49
© Springer-Verlag 1992

Hepatitis B virus specific transcripts in peripheral blood mononuclear cells

M. Melegari[1], **P. P. Scaglioni**[1], **C. Pasquinelli**[2], **F. Manenti**[1], and **Erica Villa**[1]

[1] Chair of Gastroenterology, University of Modena, Italy
[2] Department of Molecular and Experimental Medicine,
Research Institute of Scripps Clinic, La Jolla, California, USA

Summary. We report on the analysis of HBV transcription in peripheral blood mononuclear cells of chronically infected patients by polymerase chain reaction amplification. Our results suggest that in these cells gene expression occurs either as pregenomic or subgenomic transcripts.

Introduction

Hepatitis B Virus (HBV) infection of Peripheral Blood Mononuclear Cells (PBMC) has been shown to occur in all stages of liver disease. All information about the physical state of HBV in PBMC comes from Southern blot analysis and in situ hybridization [1, 3, 5, 6]. The virus is usually present as free monomeric or multimeric episomic forms at a very low copy number per cell. Moreover, the HBV infection of PBMC fluctuates in time in the same individual, independent of the presence of viremia [2, 7].

Usually, chronic carriers harbor a more stable infection than acute hepatitis patients [7], which can be detected either by Southern blot technique or, more sensitively, by Polymerase Chain Reaction (PCR). Whether HBV can undergo gene expression and replication in PBMC is still an open question. In the woodchuck animal model the presence of wood-chuck hepatitis virus (WHV) specific transcripts in PBMC and spleen cells has been clearly observed [4], suggesting that the WHV can replicate in these cells with a complete extrahepatic cycle.

In man, HBV intermediate replicative DNA forms have been seldom reported. HBsAg and preS proteins have been shown in the PBMC but one cannot rule out possible passive adhesion or receptor binding of surface proteins to the PBMC. The presence of the HBV specific transcripts in the cells seemed a better marker for a real replicative cycle. In order to discriminate whether HBV specific transcription and replication in PBMC occurs or whether PBMC simply represent a reservoir of the virus, we investigated a chronically infected group of patients by PCR amplification of HBV specific primed cDNA of total PBMC RNA.

Patients and methods

We investigated PBMCs from 15 patients (13 men, 2 women) with chronic active B hepatitis. All sera were tested for HBV, HCV, HDV and HIV markers with commercial radioimmuno-assays. Serum HBV DNA was detected by spot hybridization.

PBMC from 40 ml heparinized venous blood were separated on a Ficoll-Paque gradient, the cells were washed several times in phosphate buffered saline, the last washing medium kept for further processing.

DNA extraction was performed according to standard procedures. RNA extraction was performed using the guanidine isothiocyanate method.

To detect either pregenome or subgenomic transcripts, primers located in the core (C) or the X gene were employed. The positive control was liver RNA from an HBsAg + patient expressing only a 2.1 kb mRNA detectable in Northern blot by electrophoresing 10 μg of polyA + mRNA. To ascertain that no serum contamination took place, the PCR reaction was performed on the last washing medium of PBMC.

Total cellular RNA (3 μg), DNase treated, was vacuum dried with (−) strand primer and denatured at 80 °C for 3 min in a saline solution before cooling down to the annealing temperature for 1 hour. First-strand synthesis of cDNA occurred at 42 °C by adding 15 U of AMV RT and incubating for 1 hour. The samples were then set for the PCR procedure adding the (+) strand primer, the Taq buffer and the Taq DNA polymerase. The temperature cycling was 92 °C/1 min for denaturation, 55 °C/2 min for annealing and 72 °C/2 min for extension of the primed DNA. Portions of 10 μl of the PCR product were electrophoresed onto a 2% agarose gel and transferred to a nylon membrane. Hybridization was performed with a full length nick-translated HBV probe. Washings were performed under stringent conditions.

Results and discussion

The Southern blot analysis of PBMC DNAs, digested with EcoRI, detected positive signals at 3.2 kb, as for full length linearized HBV DNA, only in two viremic patients.

By PCR analysis, in all viremic patients the viral DNA was present in the PBMCs and transcriptionally active in most (Table 1) either as C RNA or subgenomic/coterminal RNA, while in the non-viremic patients the viral DNA was present in half of the group, but no C transcripts could be amplified. The patients who did not harbor viral DNA did not present any transcripts. The PCR conducted on the last washing medium in the same set of experiments was constantly negative.

We conclude that HBV transcription occurs in the PBMC of chronically infected patients with liver disease. Very few cells seemed to be infected since

Table 1. Scheme of the group studied, divided according to the presence of HBV DNA in the serum

	N.	PCR+ on DNA	PCR+ on RNA	X primers	C primers
Viremic patients	9	9	7	7	6
Non-viremic patients	6	3	3	3	0

PCR was performed with the same two sets of primers on DNA and RNA of PBMC

both viral DNA and RNA were detected via PCR. In most of the patients found to carry the viral genome in PBMC the viral genes were expressed. The presence of transcripts suggests that the replication of the HBV can take place also in non-hepatic tissue, as one may speculate from the presence of a C amplified transcript, although intermediate replicative forms of HBV have seldom been demonstrated in these cells.

In non-viremic patients HBV DNA was found in the PBMC in half of cases and only subgenomic transcripts were found. This suggests that HBV transcription in the PBMC may correlate with the hepatic replicative status of the virus.

Acknowkedgements

This study was partially supported by AIRC (Associazione Italiana per lo Studio sul Cancro)

References

1. Davison F, Alexander GJM, Anastssakos C, Fagan EA, Williams R (1987) Leucocyte hepatitis B virus DNA in acute and chronic hepatitis B virus infection. J Med Virol 22: 379–385
2. Davison F, Alexander GJM, Trowbridge R, Fagan EA, Williams R (1987) Detection of hepatitis B virus DNA in spermatozoa, urine, saliva and leucocytes of chronic HBsAg carriers. A lack of relationship with serum markers of replication. J Hepatol 14: 37–44
3. Hadchouel M, Pasquinelli C, Fournier JC, Hugon RN, Scotto J, Bernard O, Brechot C (1988) Detection of mononuclear cells expressing hepatitis B virus in peripheral blood from HBsAg positive and negative patients by in situ hybridization. J Med Virol 24: 27–32
4. Korba BE, Wells F, Tennant BC, Cote PJ, Gerin JL (1987) Lymphoid cells in the spleens of woodchuck hepatitis virus infected woodchucks are a site of active viral replication. J Virol 61: 1318–1324
5. Lie-Injo LE, Balasegaram M, Lopez CG, Herrera AR (1983) Hepatitis B virus DNA in liver and white blood cells of patients with hepatoma. DNA 4: 301–308
6. Pasquinelli C, Lauré F, Chatenoud L, Beaurin G, Gazengel C, Bismuth H, Degos F, Tiollais P, Bach JF, Brechot C (1986) Hepatitis B virus DNA in mononuclear blood cells. A frequent event in hepatitis B surface antigen positive and negative patients with acute and chronic liver disease. J Hepatol 3: 95–103

7. Pasquinelli C, Melegari M, Villa E, Scaglioni PP, Seidenari M, Mongiardo N, De Rienzo B, Manenti F (1990) Hepatitis B Virus infection of peripheral blood mononuclear cells is common in acute and chronic hepatitis. J Med Virol 31: 135–140

Authors' address: Dr. Margherita Melegari, Chair of Gastroenterology, Policlinico, via del Pozzo 71, I-41100 Modena, Italy.

Arch Virol (1992) [Suppl] 4: 50–53
© Springer-Verlag 1992

Detection of transcriptionally active hepatitis B virus DNA in peripheral mononuclear blood cells after infection during immunosuppressive chemotherapy using the polymerase chain reaction

R. Repp[1], A. Mance[1], C. Keller[1], S. Rhiel[1], W.H. Gerlich[2], and F. Lampert[1]

[1] Children's Hospital, University of Giessen, Federal Republic of Germany
[2] Department of Medical Microbiology, University of Göttingen,
Federal Republic of Germany

Summary. Using PCR we have studied mononuclear peripheral blood leucocytes (PMBLs) from HBV-infected immunosuppressed patients in order to detect the presence of HBV genomes. Our results indicate that non-transient PMBL infection is common in immunotolerant carriers. In addition, the presence of pregenomic mRNA sequences suggests that virus replication may take place in PMBLs, possibly implicating the latter as a source of virus after replication has ceased in the liver.

*

The Hepatitis B Virus (HBV) genome has already been detected in Peripheral Mononuclear Blood Leucocytes (PMBLs) and other non-hepatic cells in many stages of the clinical disease. The significance of these findings for viral pathogenesis, latency, and recurrence has, however, still not been established. In particular, it is unknown whether HBV genomes in PMBLs can be expressed in vivo as HBV particles and what their relative contribution to the pool of complete viral particles might be [1, 2, 3]. We, therefore, have studied HBV gene expression in PMBLs from a group of 74 children showing an uncommon outcome after HBV infection under uncommon circumstances.

These children, 3–18 years of age, had been infected with HBV during treatment for malignancies in 1983–1986 [4]. More than 90% have developed an immunotolerant carrier state without any clinical, serological, or histological signs of liver-cell destruction. They are serologically still positive for HBsAg, HBeAg and HBV-DNA and 20% have not even developed anti-HBc. The children who survived the malignant disease were immunologically normal except for the HBV infection.

Using the Polymerase Chain Reaction (PCR) [5, 6] we have looked for HBV DNA and HBV mRNA in PMBLs. PCR sets were carried out with 5 different pairs of primers enabling the detection of all parts of the HBV genome (Fig. 1). To assure specificity of the PCR reaction, adequate negative controls such as PMBL DNA from HBV seromarker negative patients were included in each PCR set. Furthermore, aliquots from the PCR products were blotted onto a Gene-Screen-Plus membrane and hybridized to specific $5'[^{32}P]$ labeled oligonucleotides under stringent conditions [6, 9]. All 5 different PCR fragments were completely sequenced in one case. To rule out possible serum contamination of PMBLs the washing fluid that had been used during the last step of the PMBL preparation was also tested. It was always negative. The high sensitivity of the PCR requires additional controls to rule out any HBV DNA contamination of mRNA preparations as the replication cycle of HBV generates RNA/DNA-hybrids [10]. We, therefore, performed two independent sets of reverse transcription reactions in which each of the two primers applied in PCRs of HBV DNA was used. In the first

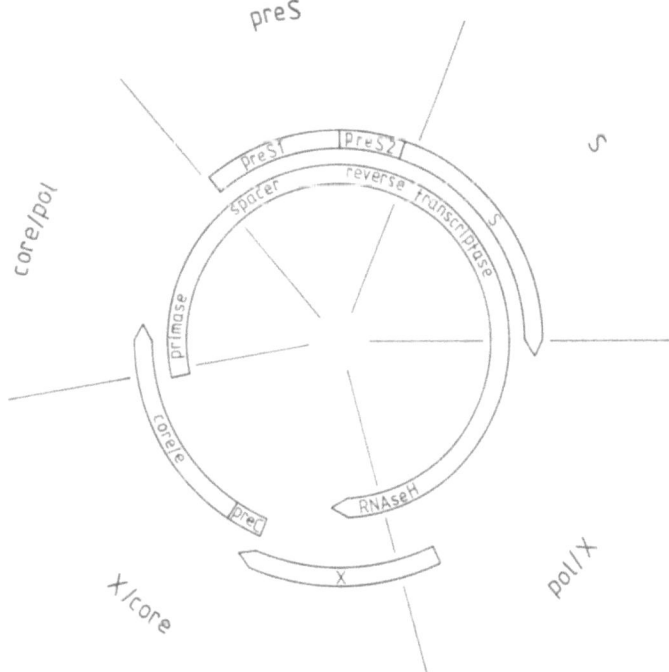

Fig. 1. Map of 5 different PCR amplifiable fragments within the HBV genome [Ref. 7, 8]. "*PreS*" 2819–193, "*S*" 169–845, "*pol/X*" 823–1450, "*X/core*" 1426–2379, "*core/pol*" 2355–2843. All PCR primers shared a 24 basepair homology to constant regions of the HBV genome and contained an additional 8 bases at their 5' end to provide recognition sites for restriction enzymes to facilitate cloning

set, the primer that was complementary to positive-sense HBV sequences was applied to prime cDNA synthesis. In the second reaction, the primer that was complementary to negative-sense HBV sequences was used for reverse transcription to detect potential DNA contamination.

PMBLs from all 10 HBsAg carriers out of this group of patients were positive for HBV DNA with all of the five primer pairs applied in PCR. To detect HBV specific transcriptional activity, PMBLs from five of these patients were tested for HBV mRNA. Two fragments of the HBV genome were amplified from cDNA by PCR, one specific for the HBV RNA pregenome and one covering the "S" gene region. Both HBV specific RNA sequences could be detected in all 5 HBs- and HBeAg positive carriers that had been tested.

PMBLs from the patients who had become serologically negative for viral antigens (less than 10% of our patients) were tested in only one case. Interestingly, this patient had an integrated HBV DNA fragment in a liver biopsy sample. Southern blot showed a monoclonal integration pattern with one HBV band in the HindIII digest of liver cell DNA [9, 11]. No replicative or episomal forms were visible. Furthermore, PCR amplified only fragments from the "pol/X" and the "X/core" region. This excludes HBV replication in liver cells, since the whole viral genome is required for this process. A possible source for the viral DNA in the patient's serum could be the mononuclear cells of the patient. DNA extracts of these cells after extensive washing reacted positively in PCR with all five primer pairs indicating complete HBV DNA. Moreover, HBV mRNA could also be detected in the patient's mononuclear cells by PCR.

These data suggest that HBV infection of PMBLs is common in an immunotolerant carrier state. HBV genomes are transcriptionally active in these cells and possibly support virus replication since HBV pregenomic mRNA sequences can be detected. In contrast to other reports [2], HBV infection of PMBLs seems not to be a transient event in these cases. PMBLs might still support HBV replication after cessation of virus replication in liver cells due to a subgenomic clonal integration of HBV DNA into the cellular chromosomes. A similar dissociation between PMBL and liver HBV infections has already been observed in HBsAg positive patients undergoing liver transplantation [12], but this has not yet been demonstrated under natural circumstances. The presence of anti-HBs does not seem to be sufficient to stop HBV replication in PMBLs after liver transplantation. It will be interesting to study the role of PMBL infection by HBV in the induction of selective immunotolerance.

References

1. Blum HE, Walter E, Teubner K, Offensperger WB, Offensperger S, Gerok BW (1989) Hepatitis B virus in non-hepatocytes. In: Bannasch P, Keppler D, Weber G (eds) Liver cell carcinoma. Kluwer, Dordrecht, pp 169–183

2. Pasquinelli C, Melegari M, Villa E, Scaglioni PP, Seidenari M, Mongiardo N, DeRienzo B, Manenti F (1990) Hepatitis B virus infection of peripheral blood mononuclear cells is common in acute and chronic hepatitis. J Med Virol 31: 135–140

3. Harrison TJ. Hepatitis B virus DNA in peripheral mononuclear blood leukocytes: A brief report. J Med Virol 1990; 31: 33–35

4. Willems WR, Bauer H, Westphal U, Lampert F (1984) Hepatitis-B-Endemie bei leukämiekranken Kindern. Dtsch Med Wochenschr 109: 1941

5. Saiki RK, Gelfand DH, Stoffel S, Scharf JS, Higuchi R, Horn GT, Mullis KB, Erlich HA (1988) Primer-directed enzymatic amplification of DNA with a thermostable DNA polymerase. Science 239: 487–491

6. Thiers V, Nakajima E, Kremsdorf D, Mack D, Schellekens H, Driss F, Goudeau A, Wands J, Sninsky J, Tiollais P, Brechot C (1988) Transmission of hepatitis B from hepatitis-B-seronegative subjects. Lancet ii: 1273–1276

7. Bichko V, Dreilina D, Pushko P, Pumpen P, Gren E (1985) Subtype ayw variant of hepatitis B virus. FEBS Lett 185: 208–212

8. Tiollais P, Pourcel C, Dejain A (1985) The hepatitis B virus. Nature 317: 489–495

9. Southern E (1975) Detection of specific sequences among DNA fragments separated by gel electrophoresis. J Mol Biol 98: 503–515

10. Miller RH, Marion PL, Robinson WS (1984) Hepatitis B viral DNA-RNA hybrid molecules in particles from infected liver are converted to viral DNA molecules during the endogenous DNA polymerase reaction. Virology 139: 64–72

11. Korba BE, Wells F, Tennant BC, Yoakum GH, Purcell RH, Gerin JL (1986) Hepadnavirus infection of peripheral blood lymphocytes in vivo: Woodchuck and chimpanzee models of viral hepatitis. J Virol 58: 1–8

12. Feray C, Zignego AL, Samuel D, Bismuth A, Reynes M, Tiollais P, Bismuth H, Brechot C (1990) Persistent hepatitis B virus infection of mononuclear blood cells without concomitant liver infection. Transplantation 49: 1155–1158

Authors' address: Dr. R. Repp, Pediatric Clinic, University of Giessen, D-W-6300 Giessen, Federal Republic of Germany.

II Oncogenic properties of hepatitis viruses

Arch Virol (1992) [Suppl] 4: 57–61
© Springer-Verlag 1992

The hepatitis B virus X protein transactivation of *c-fos* and *c-myc* proto-oncogenes is mediated by multiple transcription factors

M. L. Avantaggiati[1], C. Balsano[1], G. Natoli[1], E. De Marzio[1], H. Will[2], E. Elfassi[3], and
M. Levrero[1]

[1] I Clinica Medica and Fondazione Andrea Cesalpino Policlinico Umberto I, Rome,
Italy
[2] Max-Planck-Institut für Biochemie, Martinsried bei München,
Federal Republic of Germany
[3] Unitè d'Immunologie Microbienne, Institute Pasteur, Paris, France

Summary. We have constructed two expression vectors in order to study the action of the HBV 17 Kd X protein on the c-fos and c-myc promoters. The results show that the promoters contain multiple elements that respond to X protein, suggesting involvement of multiple transcription factors. The exact mechanism of the interaction remains elusive, but our data allow speculation about the factors that may be influenced.

*

The hepatitis B virus (HBV) X open reading frame encodes a 17 KD transcriptional transactivator (pX) which stimulates transcription of genes under the control of autologous [1] as well as other viral (SV40, RSV, HTLV-1, HIV1 and HIV2) [2, 3, 4, 5, 6, 7] and non-viral regulatory sequences (β-interferon gene, MHC II class genes) [2, 8]. During productive infection the X protein supplied from and acting upon the HBV genome augments the levels of HBV mRNAs required for viral replication [1]. Although this increase in the viral transcription is the major effect of pX in the infected hepatocytes, the broad spectrum of pX activity on transcription suggests the possibility that its expression during viral replication, or after HBV genome integration, may influence other important cellular genes.

To test the action of the X protein on the c-fos and c-myc promoter and to identify the X responsive sequences, we constructed two eukaryotic expression vectors, pMLP-X [9] and pSV-X [7] and cotransfected them in

HeLa cells, together with one of several plasmids containing the chloramphenicol acetyl transferase (CAT) gene under the control of truncated *c-fos* and *c-myc* regulatory sequences. As shown in Fig. 1B, pX stimulates by several fold the transcription from the entire c-fos promoter.

Several sequences important for transcription regulation have been mapped in the c-fos promoter: the Serum Response Element (SRE), located between −317 and −298, mediates the ability of the promoter to respond to serum stimulation, epidermal growth factor, phorbol 12-myristate 13-acetate and insulin. Downstream to SRE, between −303 and −283, there is

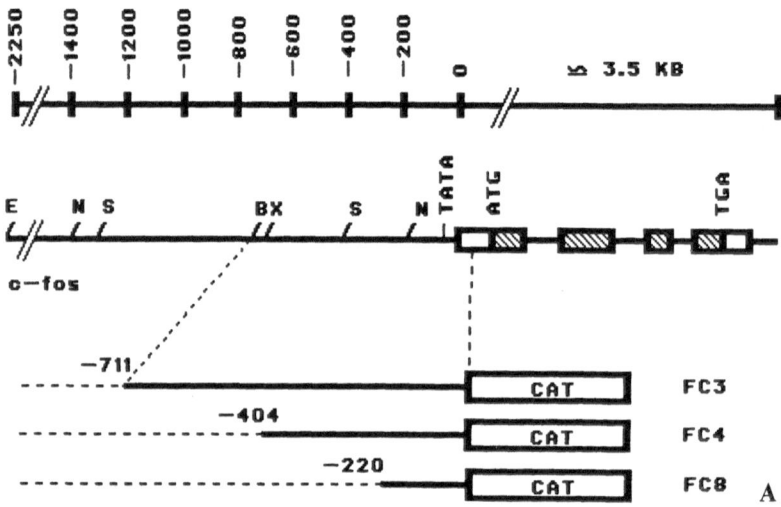

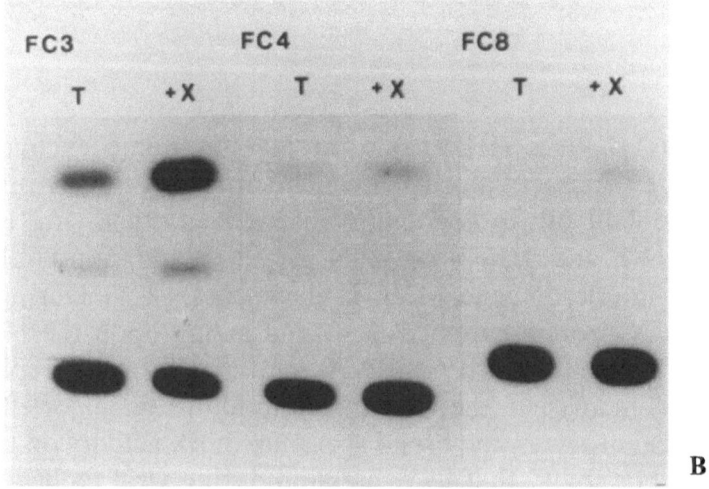

Fig. 1. A *c-fos* promoter Chloramphenicol Acetyl Transferase (*CAT*) constructs. Numbers indicate nucleotide position with respect to the transcription start site (+1) of the c-fos gene. **B** A representative CAT assay. HeLa cells were cotransfected with 2 μg of the CAT plasmids and 2 μg of either pUC13 (*T*) or pSV-X (+*X*)

a sequence with homology to AP-1 as well as an ATF site (AP-1/ATF). Another ATF site is located between -63 and -57, in the cAMP responsive element (CRE/ATF); it should be noted that this element is atypical and is also a target for AP-1. Two direct repeats (DR) between -97 and -76 seem to be important for the c-fos promoter. These DRs bind nuclear proteins and are also influenced by the retinoblastoma gene product [10]. Finally, there is a canonical TATA-box at position -30.

We found that the regions located between position -402 and -240, which contain both the SRE, target for the c-fos autoregulation, and a modified AP-1 site—activated by the jun-fos heterodimer—contribute to the pX mediated activation of the promoter, since both pFC4 (promoter truncation to -402) (Fig. 1a) and pFTK (containing the region of the c-fos promoter from -323 to -276 linked to a minimal TK promoter) were activated by pX (Fig. 1b). To better define the pX responsive sequences in pFTK, we used two derivatives in which either the SRE or the AP1 site was mutated. Our results show that both sites are activated by pX (data not shown). The mutant lacking the sequences located upstream of position -220, i.e. retaining only the DRs, the ATF/AP-1 site and the TATA box (pFC8, promoter truncation from -220, Fig. 1A) shows a drop in the basal level of expression but it is still activable, suggesting that a second pX responsive element is located downstream of position -220.

We next applied a similar approach in order to identify the regulatory sequences which respond to pX in the c-myc gene. The murine and the human c-myc gene are each composed of three exons, the first of which is non-coding. Transcription, which starts from two different initiation sites, P1 and P2, has been extensively studied by 5' deletions and in vivo competition experiments: these studies suggest that multiple negative and positive regulatory elements are responsible for the amount of c-myc mRNAs in different cells and tissues. We performed cotransfection experiments using c-myc regulatory sequence deletion mutants (Fig. 2A). The results obtained are compatible with the hypothesis that, as already demonstrated for the adenovirus E1A, the major target for the stimulation by pX could be the E2F site in the second promoter (Fig. 2B).

The presence of multiple elements in the *c-fos* and *c-myc* promoter, including the SRE and the AP1 site, that are responsive to pX, suggests that multiple transcription factors might be influenced by pX. However, the nature of the interaction between the pX and such factors remains unclear. pX could either form stable complexes with individual transcription factors or covalently modify one or more of such factors. Alternatively, pX could act on additional, so far unknown, proteins that mediate pX activation of nuclear transcription factors. The capability of a hybrid protein of pX fused to the DNA binding domain of C/EBP to activate the pX responsive element in the HBV enhancer suggests that pX could act as a transcriptional activating domain by directly interacting with the transcriptional machinery. On the

60 M. L. Avantaggiati et al.

other hand it has been recently reported that pX displays an intrinsic serine/threonine kinase activity and might activate transcription by catalyzing the phosphorylation—and subsequent activation—of cellular factors involved in transcription regulation.

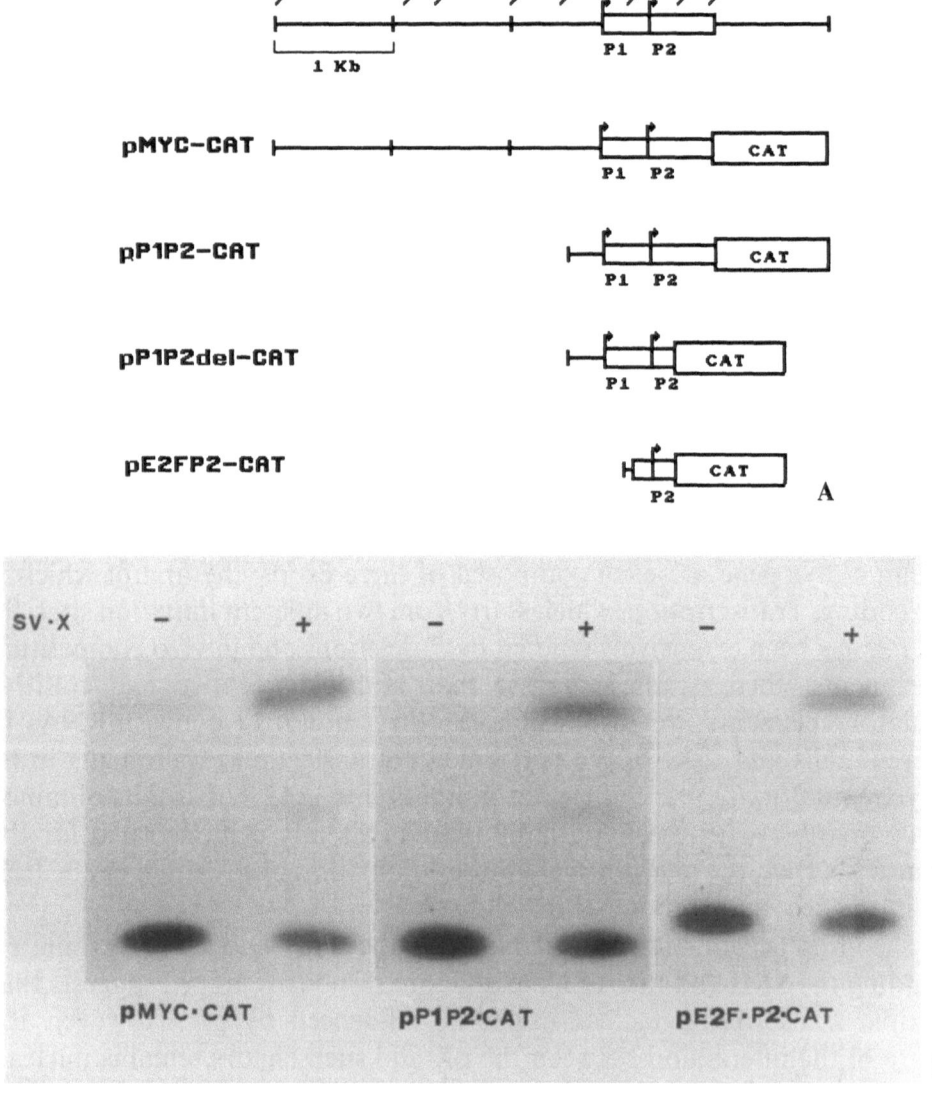

Fig. 2. A c-myc CAT construct and its deletion mutants. The top line shows the c-myc gene regulatory sequence organization. P1 (promoter 1), P2 (promoter 2) and the first exon (X) are indicated. CAT Chloramphenicol Acetyl Transferase; E EcoR1 restriction site in the c-myc promoter; C ClaI restriction site; K Kpn I restriction site; N Nco I restriction site. **B** A representative CAT assay showing the activation of the c-myc promoter deletion mutants by pX

References

1. Colgrove R, Gwynn S, Ganem D (1989) Transcriptional activation of homologous and heterologous genes by the hepatitis B virus X gene product in cells permissive for viral replication. J Virol 63: 4019–4026
2. Twu JS, Schloemer RH (1987) Transcriptional transactivating function of hepatitis B virus. J Virol 61: 3448–3453
3. Spandau DF, Lee CH (1988) Transactivation of viral enhancer by the hepatitis B virus X protein. J Virol 62: 427–434
4. Seto E, Yen T, Peterlin BM, Ou J (1988) Transactivation of the human immuno-deficiency virus long terminal repeats by the hepatitis B virus X protein. Proc Natl Acad Sci USA 85: 8286–8290
5. Zahm P, Hofschneider PH, Koshy R (1988) The hepatitis B virus X-ORF encodes a transactivator: a potential factor in viral hepatocarcinogenesis. Oncogene 3: 169–177
6. Wollersheim M, Debelka U, Hofschneider PH (1988) A transactivating function encoded in the hepatitis B virus X gene is conserved in the integrated state. Oncogene 3: 545–554
7. Levrero M, Balsano C, Natoli G, Avantaggiati ML, Elfassi E (1990) Hepatitis B virus X protein transactivates the long terminal repeats of the human immunodeficiency virus types 1 and 2. J Virol 64: 3082–3086
8. Hu KQ, Vierling JM, Siddiqui A (1990) Transactivation of HLA-DR gene by the hepatitis B virus X gene product. Proc Natl Acad Sci USA 87: 7140–7144
9. Levrero M, Jean-Jean O, Balsano C, Will H, Perricaudet M (1990) Hepatitis B virus X gene expression in human cells and anti-X antibodies detection in chronic HBV infection. Virology 174: 299–304
10. Robbins PD, Horowitz JM, Mulligan RC (1990) Negative regulation of human c-fos expression by the retinoblastoma gene product. Nature 346: 668–671

Authors' address: Dr. M. Levrero, Laboratory of Genetic Expression, Fondazione A. Cesalpino, I Clinica Medica, Policlinico Umberto I, I-00161 Rome, Italy.

Arch Virol (1992) [Suppl] 4: 63–64
© Springer-Verlag 1992

Trans-activation by hepatitis B virus X protein is mediated via a tumour promoter pathway

A. S. Kekulé, U. Lauer, L. Weiß, P. H. Hofschneider, and **R. Koshy**

Max-Planck-Institut für Biochemie, Martinsried, Federal Republic of Germany

Summary. We have studied the *c-myc* gene as a possible target of HBV X protein in liver carcinogenesis. Our results indicate that *trans*-activation by X protein occurs via PKC/AP1 signal transduction, suggesting a possible two-step mechanism in HBV related liver carcinogenesis.

*

The observation that the HBV X protein (HBx) exerts a transcriptional *trans*-activator function has led to the idea that *trans*-activation of cellular genes by integrated HBV DNA might contribute to liver carcinogenesis. But, as HBx does not bind to DNA, the question of whether and how cellular genes can be *trans*-activated is still open.

Investigating the *c-myc* gene as a possible oncogenic target of HBx *trans*-activation, we found after cotransfection of Chang liver (CCL13) cells an about 15 fold *trans*-activation of CAT reporter genes under the control of the human dual *c-myc* promoter. A minimal promoter construct containing the *myc* P2 initiation site and the AP1/fos binding site located upstream of P2 was still stimulatable, whereas deletion of the AP1/fos binding site rendered the promoter unresponsive to HBx. Introduction of a synthetic AP1/fos binding site into an otherwise non-stimulatable promoter caused it now to be *trans*-activatable by HBx. In F9 teratocarcinoma cells, which do not express detectable amounts of AP1 or fos, HBx displayed no *trans*-activation of reporter genes; cotransfection with vectors expressing AP1 and fos fully reconstituted the *trans* activity, showing that *trans*-activation by HBx is dependent on the presence of the AP1/fos transcription factor. Interestingly, also tumour promoters (e.g. the phorbolester TPA) influence cellular genes via TPA-responsive elements (TREs) such as AP1, AP2 or NF-kB binding sites. HBx has recently been shown to *trans*-activate a broad range of viral and cellular genes; a comparison of their 5′-regulatory regions revealed that common to these genes is the presence of one or more TREs. As the stimulation of TRE-dependent genes by tumour promoters is effected by an activation of the cellular protein kinase C (PKC), we investigated if *trans*-activation by HBx is also mediated via the PKC signal transduction pathway. In cotransfection experiments, three different PKC inhibitors each eliminated the *trans*-activating effect of HBx. Also, after depletion of PKC activity in CCL13 cells by long-term treatment with TPA, no more HBx *trans*-activation was detectable. Finally, direct measurement of PKC enzyme activity showed that PKC is translocated to the membrane after transfection with HBx expression vectors. We conclude that one way of mediating *trans*-activation by HBx occurs via the PKC/AP1 signal transduction pathway, suggesting that HBV could contribute to liver carcinogenesis as a tumour promoter in the sense of the two-step carcinogenesis mechanism.

Authors' address: Dr. A. S. Kekulé, Max-Planck-Institut für Biochemie, D-8033 Martinsried, Federal Republic of Germany.

Arch Virol (1992) [Suppl] 4: 65–69
© Springer-Verlag 1992

Truncated pre-S/S proteins transactivate multiple target sequences

G. Natoli[1], C. Balsano[1], M. L. Avantaggiati[1], E. De Marzio[1], M. Artini[1],
D. Collepardo[1], E. Elfassi[2], and M. Levrero[1]

[1] I Clinica Medica and Fondazione Andrea Cesalpino Policlinico Umberto I, Rome,
Italy
[2] Unitè d'Immunologie Microbienne, Institute Pasteur, Paris, France

Summary. In order to investigate the transactivational function of HBV truncated preS/S proteins we have constructed two sets of plasmids and have tested their transactivational potential on the *c-myc* regulatory sequences and the TPA-responsive element. We found that preS/S proteins only become transactivationally active when truncated at the carboxy terminal end. Furthermore, using immunofluorescence microscopy we determined that the proteins are located exclusively in the cytoplasm, apparently ruling out DNA binding and activation of factors in the nucleus.

*

Chronic hepatitis B virus (HBV) infection is associated with a high risk of developing hepatocellular carcinoma (HCC) [1, 2]: in HBV-induced HCC, integrated viral DNA sequences are virtually always present but only occasionally are genes controlling the cell growth and differentiation disturbed by these inserts [3–5]. Thus, since no common sites of viral DNA integration have been observed, the induction of cell growth by HBV enhancer or promoter insertion cannot be regarded as a general mechanism of transformation in human HCC. On the other hand HBV, like many other DNA viruses bears a transactivational potential, exerted both by the X protein [6–8] and, as recently demonstrated [9, 10] by preS/S proteins which have lost their carboxytermini.

In order to characterize this new transactivator function, we engineered two sets of plasmids carrying preS1/S2/S and preS2/S 3' deletion mutants, respectively, under the control of the adenoviral major late promoter (Fig. 1). The plasmids pMLP-S1 and pMLP-S2 contain the entire preS/S and the preS2/S region, respectively. Their derivatives del51, del74 and del194 contain 526, 458 and 94 bp of the S gene, respectively, thus encoding for proteins lacking 51, 74 and 194 amino acids from the carboxyterminus of the S-ORF. The transactivational properties of the preS/S truncated peptides were investigated in 293, Alexander and HeLa cells: the cells were transfected with 4 μg of a preS/S expression vector and 1 or 2 μg of a reporter plasmid carrying the CAT gene under the control of specific regulatory sequences. After 48 hrs, the cells were collected and extracts for the Chloramphenicol Acetyl Transaminase (CAT) assay were prepared as described [8]. As targets to test the transactivation phenomenon the c-myc regulatory sequences and the TPA-responsive element (TRE) were used. In the plasmid p111 (pMyc-CAT) a 2.9 kb Hind III-Pst I fragment, containing the 5' flanking sequences, the promoter and a nearly complete first exon of the human c-myc gene, has been inserted into the CAT vector pHP 34-CAT. The TRE-TK CAT plasmid contains three copies of the human collagenase TRE, followed by a minimal TK promoter. The activity of the full-length and carboxyterminal truncated HBV envelope proteins is summarized in Fig. 1A and B, and in Fig. 2 is shown a representative CAT assay. CAT expression levels were not augmented either by the full-length preS/S proteins or by proteins deleted of only 51 carboxyterminal amino acids; the transactivating function instead appeared when 23 additional amino acids were removed, both in the preS1/S2/S and in the preS2/S proteins.

The observation that the preS/S proteins acquire a transactivational activity only when deleted in the carboxyterminal part is relevant for several reasons. Firstly, the integration of the preS/S sequences into the host genome during a chronic infection often leads to the loss of the 3' of the gene, according to the published sequences of the viral integrants. Moreover, the deletion of the two carboxyterminal hydrophobic transmembrane domains that are believed to be involved in the assembly and in the endoplasmic reticulum processing of the HBV surface proteins, is needed to generate the transactivating activity. This observation suggests that the generation of the transactivating function of the truncated preS/S proteins after HBV integration could be due to an altered subcellular distribution and/or secretion of the preS/S proteins. To test this hypothesis we therefore performed indirect immunofluorescence experiments using monoclonal antibodies directed against the preS1 or preS2 region (a gift from A. Budkowska, Paris), in order to define the subcellular localization of the truncated preS/S protein. Briefly 293 cells were transfected, using the calcium phosphate method, with 10 μg of each preS/S expression vector; 16–18 hrs after transfection, the medium was replaced and 24 hrs later the cells were washed in phosphate buffered saline (PBS) and fixed for 5 min in cold methanol, followed by 8 washes in cold acetone. After abundant rinsing in PBS the cells were incubated for 30 min with the appropriate antibody. The cells were then washed twice in PBS containing 0.05% NP-40 and incubated for 20 min with a 1:100 dilution of a fluoresceine isothiocyanate-conjugated goat anti rabbit antiserum. After

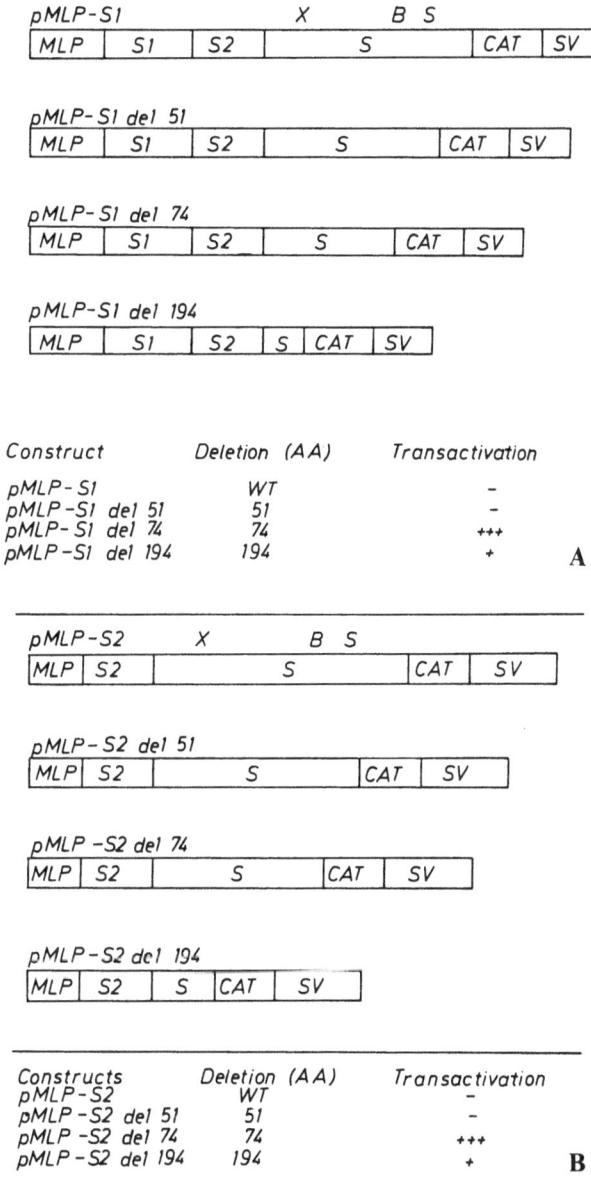

Fig. 1A and **1B.** HBV constructs used in transient transfection assays. *MLP* major late promoter [8]. *S1* preS1 surface protein gene. *S2* preS2 surface protein gene. *S* HBV major surface protein gene. *CAT* 3' 75 base pairs of the CAT gene. *SV* small intron and SV40 early region polyadenylation site. *X* Xba I restriction site in the HBV genome, position 249, *B* Bstx I (HBV genome, position 613), *S* Spe I (HBV genome, position 681)

further washes the cells were mounted in 90% glycerol and photographed using a Leitz microscope. All the carboxyterminally truncated preS/S proteins had a cytoplasmic distribution. Nuclear localization was not observed for any of the truncated preS/S proteins.

Although further experiments are needed to better define the association of the truncated preS/S proteins with the endoplasmic reticulum or other

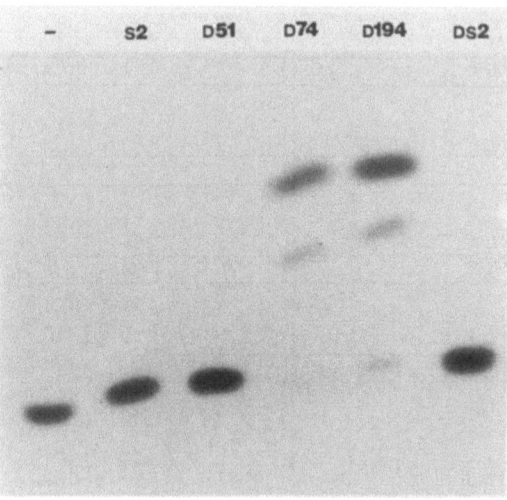

Fig. 2. Alexander cells were cotransfected with 2 μg of TRETKCAT plasmid (a gift from Paolo Sassone Corsi, CNRS, Strasburg), and the various preS2/S deletion mutants

subcellular structures, the cytoplasmic localization of these peptides seems to exclude both binding to the DNA and activation in the nuclear micro-environment of cellular transcription factors. We can presume that the way(s) by which the truncated preS/S proteins exert a transcriptional effect must involve the modification of the activity of signals starting from the periphery of the cell.

References

1. Beasley RP, Hwang Ly, Lin CC, Chien CS (1981) Hepatocellular carcinoma and hepatitis B virus: a prospective study of 22707 men in Taiwan. Lancet ii: 1129–1133
2. Ganem D, Varmus HE (1987) The molecular biology of hepatitis B viruses. Ann Rev Biochem 56: 651–693
3. Moroy T, Marchio A, Etiemble J, Trepo C, Tiollais P, Buendia MA (1986) Rearrangement and enhanced expression of c-myc in hepatocellular carcinoma of hepatitis virus infected woodchucks. Nature 324: 276–279
4. DeJean A, Bougueleret L, Grzeschik KH, Tiollais P (1986) Hepatitis B virus integration in a sequence homologous to v-erbA and steroid receptor genes in a hepatocellular carcinoma. Nature 322: 70–72
5. Wang J, Chenivesse X, Henglein B, Brechot C (1990) Hepatitis B virus integration in a cyclin A gene in a hepatocellular carcinoma. Nature 343: 555–557
6. Twu JS, Schloemer RH (1987) Transcriptional transactivating function of hepatitis B virus. J Virol 61: 3448–53
7. Colgrove R, Gwynn S, Ganem D (1989) Transcriptional activation of homologous and heterologous genes by the hepatitis B virus X gene product in cells permissive for viral replication. J Virol 63: 4019–4026
8. Levrero M, Balsano C, Natoli G, Avantaggiati ML, Elfassi E (1990) Hepatitis B virus X protein transactivates the long terminal repeats of the human immunodeficiency virus types 1 and 2. J Virol 64: 3082–86

9. Kekulè AS, Lauer U, Meyer M, Caselman WH, Hofschneider PH, Koshy R (1990) The preS2/S region of integrated hepatitis B virus DNA encodes a transcriptional transactivator. Nature 343: 457–461
10. Caselman WH, Meyer M, Kekulè AS, Lauer U, Hofschneider P, Koshy R (1990). A transactivator function is generated by integration of hepatitis B virus preS/S sequences in human hepatocellular carcinoma. Proc Natl Acad Sci USA 87: 2970–74

Authors' address: Dr. Levrero, Laboratory of Genetic Expression, Fondazione A. Cesalpino, I Clinica Medica, Policlinico Umberto I, I-00161 Rome, Italy.

Arch Virol (1992) [Suppl] 4: 70–75
© Springer-Verlag 1992

Hepatitis C virus and hepatocellular carcinoma

A. Sangiovanni, G. Covini, M. G. Rumi, R. Marcelli, and **M. Colombo**

Institute of Internal Medicine, University of Milan, Italy

Summary. Epidemiological, clinical and laboratory data point to a role of hepatitis C virus infection in hepatocellular carcinoma. The connection appears to be indirect and to be mediated by cirrhosis. Thus, geographical differences can be observed, based on the locally prevalent etiological factors for cirrhosis. In the end, prospective studies of hepatitis C virus infected persons will be needed to elucidate the role of this agent in liver cancer.

Introduction

Hepatocellular carcinoma (HCC) is a highly malignant tumor with an extremely poor prognosis and an estimated incidence of more than 250,000 cases per year. Worldwide, patients with cirrhosis are recognized as a population at risk for HCC [8]. Nearly all patients with HCC have cirrhosis and 30%–40% of cirrhotics are found at autopsy to have HCC. Why HCC frequently occurs in patients with cirrhosis is not clear. HCC could be either the inevitable consequence of a long-standing hepatic disease or an independent response to a hepatic insult common to HCC and cirrhosis. Epidemiological studies and animal experiments have indicated hepatitis B virus (HBV) as the most common etiologic factor for this tumor [1]. However, many patients have no HBV markers and many HBV carriers do not develop HCC. Chemicals, alcohol, tobacco, sex hormones and metabolic diseases have also been recognized as risk factor for developing HCC [2]. However, in a great proportion of HBV negative patients, none of these factors are present. Epidemiological, experimental and clinical studies suggest that hepatitis C virus (HCV), the major agent of blood-borne NANB hepatitis may play a role in some cases of HCC.

Epidemiological studies

The first piece of evidence connecting HCC with NANB hepatitis is the existence of HCC patients who have no detectable serologic markers of HBV

Table 1. Percent of patients without serum markers for HBV in hepatocellular carcinoma

Reference	Year	Country	Percent of HBV seronegative patients
Tabor	1977	Zambia	0
Beasley	1981	Taiwan	0
Chen	1986	Taiwan	0
Yeh	1985	China	2
Lingao	1981	Philippines	3
Larouzè	1976	Senegal	4
Tabor	1977	Uganda	11
Kubo	1977	Japan	19
Trichopoulos	1978	Greece	20
Tabor	1978	USA	26
Buscarini	1987	Italy	51
Bertrand	1987	France	57
Dunk	1988	United Kingdom	68

References are reported in the review by Tabor [18]

infection: HBsAg, anti-HBc or anti-HBs [18]. The percentages of HCC cases with no serologic markers for HBV have been reported to go from zero, in studies in Zambia and Taiwan, up to 68% in Great Britain (Table 1). In Japan, from 1970 to 1988, the incidence of HCC increased by 155% [13]. In the same period HBsAg positive patients with HCC decreased from 50% to 30%, while alcohol abuse remained stable at around 30%. Further investigations revealed that the increased proportion of HBsAg negative patients was not due to cryptic HBV infection. In fact, Sakamoto and coworkers [15], who tested for the presence of HBV-DNA in patients with HCC resected from 1970 to 1987, report that during this time there was a reduction not only in the prevalence of HBsAg positive patients (from 47% to 10%) but also in the prevalence of HBsAg negative patients with integrated HBV-DNA (from 12% to 4%).

Case reports

The second type of evidence connecting HCC with NANB hepatitis is the sequential development of cirrhosis and HCC that has been observed in patients with post-transfusion NANB hepatitis (reviewed by Tabor [18]). In these patients the time-lag between infection and development of HCC ranged from 7 to 18 years (Table 2). In 1983, Resnick first described the case of a 59 year old female who developed NANB hepatitis 5 weeks after blood transfusion and HCC after 7 years. In the case reported by Kiyosawa in 1984, 5 liver biopsies and the autopsy illustrated the histologic progression from

Table 2. Development of hepatocellular carcinoma in patients with NANB hepatitis

Modality of Infection	Reference	Time-lag (yr)
Transfusional	Resnik, 1983[a]	7
	Kiyosawa, 1984[a]	18
	Gilliam, 1984[a]	9
	Tremolada, 1990[b]	12
Sporadic	Jenkins, 1981[a]	9
	Ayoola, 1982[a]	2
	Cohen, 1987[a]	8

[a] References are reported in the review by Tabor [18]
[b] Reference No 19

acute NANB viral hepatitis to HCC via chronic persistent hepatitis, chronic active hepatitis and cirrhosis over a period of 18 years. In the same year Gilliam reported the case of a 63 year old man who developed chronic active post-transfusion hepatitis, cirrhosis and HCC over a period of 9 years. In 1990, Tremolada [19] described a patient with post-transfusion hepatitis who developed chronic active hepatitis, active cirrhosis and, after 12 year, HCC. Serum samples obtained 5, 29, 144, 166, and 240 months after transfusion reacted positively for antibody to HCV (anti-HCV). More recently, in a retrospective study of serum anti-HCV in 231 patients with post-transfusion hepatitis, Kiyosawa [11] demonstrated that the mean time-lags between transfusion and development of chronic hepatitis, cirrhosis and HCC were 10, 21, and 29 years, respectively. Additional evidence that there is a connection between NANB hepatitis and HCC was obtained in studies of patients with sporadic NANB hepatitis. Jenkins (1981), Ayoola (1982), and Chen (1987) reported cases of patients in whom HCC developed 9, 2, and 8 years after sporadic NANB infection, respectively (reviewed by Tabor [18]).

Experimental studies

Linke et al. [12] described the case of a chimpanzee, that developed chronic NANB hepatitis and HCC after experimental infection with human plasma, after a time-lag of 7 years. Chronic hepatitis was documented by intermittently elevated serum levels of transaminases and by repeated liver biopsies. A peculiar finding of this study was the absence of cirrhosis in the non-neoplastic liver tissue. The persistence of the NANB agent in the liver was documented by transmission of hepatitis to another chimpanzee by infectious homogenates from neoplastic and non-neoplastic liver tissue. Infection with HBV or hepatitis D virus was excluded by testing liver samples for HBV-DNA and HDV-RNA by molecular biology techniques (Southern blot, northern blot, dot blot).

Serological studies

The development of a commercial assay for serum antibodies to HCV made it possible to retrospectively investigate the role of HCV in the pathogenesis of HCC. Anti-HCV was demonstrated in a high proportion of patients with HCC worldwide (Table 3) [3–7, 10, 14, 16, 17, 20, 21]. In Europe there was a disproportionately higher prevalence of anti-HCV in patients with HCC superimposed on cirrhosis of different etiology than in comparable patients without HCC. In South Africa HCC is common in rural male blacks chronically infected with HBV and, in patients with HCC, anti-HCV is found more frequently in urban females older than 45 years. The pathogenetic importance of these studies according to current opinion is that in many of these patients anti-HCV is a marker for ongoing HCV infection. The finding of HCV-RNA in the tumor tissue of some patients with HCC further established a strong link between chronic HCV infection and neoplastic transformation [9]. It is unlikely that HCV plays a direct role in carcinogenesis, since no reverse transcriptase activity has been found in infected livers (M. Houghton, personal communication). Instead, HCV may be pathogenetic because it is an important cause of cirrhosis, which is the most relevant single factor for HCC [8].

Table 3. Prevalence of antibody to HCV in patients with hepatocellular carcinoma

Author	Country	No. of patients	Positive anti-HCV No	(%)	Associated factors to anti-HCV positivity
Bruix [3]	Spain	96	72	(75)	Alcoholic cirrhosis
Colombo [4]	Italy	132	86	(65)	anti HBc
Simonetti [17]	Italy	200	152	(76)	Cirrhosis
Sbolli [16]	Italy	78	46	(59)	HBsAg negative
Ducreux [6]	France	74	21	(28)	HBV markers
Vargas [20]	Spain	81	44	(54)	
Kew [10]	S. Africa	380	110	(29)	Female sex, age, urban resident
Dazza [5]	Mozambique	189	69	(37)	
Yu [21]	USA	51	15	(29)	
Hasan [7]	USA	87	35	(40)	
Saito [14]	Japan	253	138	(55)	HBsAg negative

Conclusions

Epidemiological, clinical and experimental studies support a pathogenetic linkage between HCV and HCC. This linkage is not a direct one, but is probably mediated by cirrhosis. In support of this evidence, in areas like Southern Europe, where HCV infection is an important cause of cirrhosis, the linkage between HCV and HCC is stronger than in areas of the world in which alcohol (France) and HBV (South Africa) are the main etiologic factors for cirrhosis. Only prospective studies of patients infected by HCV will be able to confirm the role of HCV in the pathogenesis of HCC.

References

1. Beasley RP (1988) Hepatitis B virus. The major etiology of hepatocellular carcinoma. Cancer 61: 1942–1956
2. Bosch FX, Munoz N (1988) Prospects for epidemiological studies on hepatocellular carcinoma as a model for assessing viral and chemical interactions. In: Bartsch H, Hemminki K, O'Neill IK (eds) Methods for detecting DNA-damaging agents in humans: applications in cancer epidemiology and prevention. IARC Scientific Publication 89, IARC, Lyon, pp 427–438
3. Bruix J, Barrera JM, Calvet X, Ercilla G, Costa J, Sanchez-Tapias JM, Ventura M, Vall M, Bruguera M, Bru C, Castillo R, Rodes J (1989) Prevalence of antibodies to hepatitis C virus in Spanish patients with hepatocellular carcinoma and hepatic cirrhosis. Lancet ii: 1004–1006
4. Colombo M, Kuo G, Choo QL, Donato MF, Del Ninno E, Tommasini MA, Dioguardi N, Houghton M (1989) Prevalence of antibodies to hepatitis C virus in Italian patients with hepatocellular carcinoma. Lancet ii: 1006–1009
5. Dazza MC, Menerses LV, Girard PM, Villaroel C, Brechot C, Larouzè B (1990) Hepatitis C antibody and hepatocellular carcinoma. Lancet i: 1216
6. Ducreux M, Buffet C, Dussaix EA, Pelletier G, Briantais MJ, Yvart Y, Jaques L, Etienne JP (1990) Antibody to hepatitis C virus in hepatocellular carcinoma. Lancet i: 301
7. Hasan F, Jeffers LJ, De Medina M, Reddy R, Parker T, Schiff ER, Houghton M, Choo Q, Kuo G (1990) Hepatitis C associated hepatocellular carcinoma. Hepatology 12: 589–591
8. Johnson PJ and Williams R (1987) Cirrhosis and aetiology of hepatocellular carcinoma. J Hepatol 4: 140–47
9. Kaneko S, Kuno K, Yanagi M, Unoura M, Hattori N, Murakami S, Kobayashi K (1991) Sequence analysis of hepatitis C virus genomes isolated from five patients with chronic non-A, non-B hepatitis. In: Hollinger FB, Lemon SM, Margolis H (eds) Viral hepatitis and liver disease. Williams & Wilkins, Baltimore, pp 364–367
10. Kew MC, Houghton M, Choo QL, Kuo G (1990) Hepatitis C virus antibodies in southern African Blacks with hepatocellular carcinoma. Lancet i: 873–874
11. Kiyosawa K, Sodeyama T, Tanaka E, Gibo Y, Yoshizawa K, Nakano Y, Furuta S, Akahane Y, Nishioka K, Purcell R, Alter H (1990) Interrelationship of blood transfusion, non-A, non-B hepatitis and hepatocellular carcinoma: analysis by detection of antibody to hepatitis C virus. Hepatology 12: 641–675
12. Linke HL, Miller MF, Peterson DA, Muchmore E, Lesniewski RR, Carrick RJ, Gagne GD, Popper H (1987) Documentation of hepatocellular carcinoma in a chimpanzee with non-A, non-B hepatitis. In: Robinson W, Koike K, Will H (eds) Hepadna viruses. Liss, New York, pp 357–370

13. Okuda K, Fujimoto I, Hanai A, Urano Y (1989) Changing incidence of hepatocellular carcinoma in Japan. Cancer Res 47: 4967–4972
14. Saito I, Miyamura T, Ohbayashi H, Harada H, Katayama T, Kikuchi S, Watanabe Y, Koi S, Onji M, Ohta Y, Choo QL, Houghton M, Kuo G (1990) Hepatitis C virus infection is associated with the development of hepatocellular carcinoma. Proc Natl Acad Sci USA 87: 6547–6549
15. Sakamoto M, Hirohashi S, Tsuda H, Ino Y, Shimosato Y, Yamasaki S, Makuuchi M, Hasegawa H, Terada M, Hosoda Y (1988) Increasing incidence of hepatocellular carcinoma possibly associated with non-A, non-B hepatitis in Japan, disclosed by hepatitis B virus DNA analysis of surgically resected cases. Cancer Res 48: 7294–7297
16. Sbolli G, Zanetti AR, Tanzi E, Cavanna L, Fornari F, Di Stasi M, Buscarini L (1990) Serum antibodies to hepatitis C virus in Italian patients with hepatocellular carcinoma. J Med Virol 30: 230–232
17. Simonetti RG, Cottone M, Craxì A, Pagliaro L, Rapicetta M, Chionne M, Costantino A (1989) Prevalence of antibodies to hepatitis C virus in hepatocellular carcinoma. Lancet ii: 1338
18. Tabor E (1989) Hepatocellular carcinoma: possible etiologies in patients without serologic evidence of hepatitis B virus infection. J Med Virol 27: 1–6
19. Tremolada F, Benvegnù L, Casarin C, Pontisso P, Tagger A, Alberti A (1990) Antibody to hepatitis C virus in hepatocellular carcinoma. Lancet 335: 300–301
20. Vargas V, Castells L, Esteban JI (1990) High frequency of antibodies to hepatitis C virus among patients with hepatocellular carcinoma. Ann Intern Med 3: 232–233
21. Yu MC, Tong MJ, Coursaget P, Ross RK, Govindarajan S, Henderson BE (1990) Prevalence of hepatitis B and C viral markers in black and white patients with hepatocellular carcinoma in the United States. J Nat Cancer Inst 82: 1038–1041

Authors' address: Dr. M. Colombo, Institute of Internal Medicine, Via Pace 9, I-20122 Milan, Italy.

Arch Virol (1992) [Suppl] 4: 76–80
© Springer-Verlag 1992

Prevalence of antibodies to hepatitis C virus in patients with hepatocellular carcinoma in Austria

[1] M. Baur, [2] U. Hay, [1] G. Novacek, [3] C. Dittrich, and [1] P. Ferenci

[1] I. Department of Gastroenterology and Hepatology, University of Vienna
[2] II. Department of Gastroenterology and Hepatology, University of Vienna
[3] Department of Chemotherapy, University of Vienna, Wien, Austria

Summary. In 12 of 54 (22%) patients with histologically verified hepatocellular carcinoma, antibodies to hepatitis C virus were found. In patients with hepatocellular carcinoma the frequency of anti-hepatitis C virus positivity was similar whether cirrhosis (6 of 22 patients (27%)) was present or not (2 of 15 (13%)). Out of 54, 23 patients (43%) were negative both for hepatitis B or C markers. Out of 53, 22 (42%) had positive hepatitis B markers, 8 of 22 were HBsAg positive. Patients with hepatocellular carcinoma and cirrhosis had a higher percentage of hepatitis B virus markers than patients with cirrhosis without hepatocellular carcinoma. Out of 70 patients with cirrhosis but without hepatocellular carcinoma, 24 (34%) had antibodies to hepatitis C virus. Our data of similar frequencies of antibodies to hepatitis C virus in patients with hepatocellular carcinoma or with liver cirrhosis but without hepatocellular carcinoma indicate that at least in Austrian patients, hepatitis C virus infections are not an important factor for development of hepatocellular carcinoma.

*

The etiology of hepatocellular carcinoma (HCC) is still unknown. Among the recognized factors associated with the development of HCC are chronic hepatitis B-virus (HBV) infection [1], cirrhosis [7] and alcohol abuse [13].

Since the development of an assay to detect antibodies against the hepatitis C virus (HCV) several authors have tested sera of patients with HCC to define the role of this virus in development of HCC. Recently, Kiyosawa [10] reported 21 cases of posttransfusion hepatitis with documented transition to chronic hepatitis C and finally to HCC. All patients were positive for anti-HCV at the time of diagnosis of HCC and in previous serial samples. In areas with a high incidence of chronic hepatitis B-virus (HBV) infections such as in South African blacks [9] the prevalence of anti-HCV in HCC was 29%. In contrast, in regions with intermediate high HBV incidence such as Italy [14] and Spain [2] a high percentage of antibodies to HCV was detected in patients with HCC (61 to 75%). In countries with even lower HBsAg carrier rate such as France, only 28% HCV positivity was found in HCC [6]. Austria has a low hepatitis B virus carrier rate and alcohol abuse accounts for most of the cases of cirrhosis. Alcohol intake itself seems to be a major risk factor for development of HCC, since chronic HBV infection occurs in only 7% to 19% of the patients with HCC [7, 16].

In this study the role of HCV infection in Austrian patients with HCC was investigated by testing their sera for anti-HCV.

Methods

Freshly obtained sera from 54 patients with histologically confirmed HCC (male n = 41, female n = 13; median age 59 a, range 15–81 a) and from 70 patients with cirrhosis without HCC (male n = 43, female n = 27; mean age 55 a) were tested for antibodies to HCV by ELISA (Ortho, Ortho Diagnostic Systems Inc., Raritan, NJ, USA).

Additionally, all sera were tested for HBsAg, anti-HBs, anti-HBc by ELISA (Organon, Organon Technika, Boxtel, Netherlands). Only one patient with HCC was not tested for HBV markers.

Of the patients with HCC, 22 had unequivocal liver cirrhosis proven at autopsy, laparotomy or by liver biopsy. The etiologies of liver cirrhosis were alcoholic: 10, posthepatitic: 7, hemochromatosis: 1, cryptogenic: 4. Alcoholic cirrhosis was defined if there was a documented history of alcohol abuse (more than 80 g ethanol/day for at least 5 years). In 15 patients no cirrhosis was present. In 16 patients mostly with history of chronic liver disease no information on the liver was available ("possible cirrhosis").

The etiology of cirrhosis in the patients without HCC was: alcoholic: 33, posthepatitic: 23, cryptogenic: 12, primary biliary: 2.

Results

In 12 (male n = 8, female n = 4) of 54 patients (22%) with HCC antibodies to HCV were detected and 22 (42%) patients had positive markers for HBV infection (see Table 1). Four patients were positive both for anti-HCV and HBV markers. The frequency of HCV markers did not differ whether liver cirrhosis was present or not (see Table 2). In the 70 patients with liver cirrhosis without HCC 24 (34%) had positive HCV antibodies and 15 (21%) had markers of HBV infection (see Table 1). The percentage of positive HBV markers in patients with HCC and cirrhosis was higher (45%) than in patients with liver cirrhosis without HCC (21%) (p = 0.05). This was mainly due to a higher percentage of positive HBsAg in patients with HCC (23%) than in patients with cirrhosis only (7%).

Table 1. Frequencies of positive anti-HCV and HBV markers in patients with HCC and with cirrhosis without HCC (first number represents n = cases, percentage in brackets)

	HCC n = 54	Cirrhosis n = 70
Anti-HCV +	12 (22)	24 (34)
HBV + (any marker)	22 (41)	15 (21)
HBsAg +	8	5
HBsAg − /anti-HBs + /anti-HBc +	8	1
HBsAg − /anti-HBs + or anti-HBc +	6	9

Table 2. Frequencies of positive anti-HCV and HBV markers in patients with HCC in relation to the presence or absence of cirrhosis (first number represents n = cases, percentage in brackets)

	Underlying liver disease		
	Cirrhosis n = 22	Possible cirrhosis n = 17[a]	No cirrhosis n = 15
Anti-HCV +	6 (27)	4 (24)	2 (13)
HBV + (any marker)	10 (45)	5 (29)[b]	7 (47)
HBsAg +	5	1	2
HBsAg − /anti-HBs + /anti-HBc +	3	1	4
HBsAg − /anti-HBs + or anti-HBc +	2	3	1

[a] Histology of nontumorous liver unknown
[b] One patient was not tested for HBV-markers

Discussion

In 22% of Austrian patients with HCC, antibodies to HCV were detected. This frequency of HCV antibodies was similar to that in a group of patients with cirrhosis without HCC. The observed frequency of anti-HCV in patients with HCC is similar that from South Africa [9] and France [6], but is considerably lower than in patients from Italy [3, 14, 15], Spain [2] or Senegal [11]. Several causes may account for these diverging results:

First, heterogenous HCV RNA sequences have been detected in various parts of the world. The French and the American HCV isolate only differed by 3.5% while the Japanese isolate showed a 21% discordance [12]. RNA sequences may vary in their oncogenic potency and may explain the

Table 3. Comparison of anti-HCV positivity in chronic liver disease and in HCC

Country	Author (Reference)		Anti-HCV+ in chronic liver disease (%)	Anti-HCV+ in HCC (%)
Japan	Kiyosawa	(10)	86	94
France	Ducreux	(6)	20	28
Italy	Chiaramonte	(3)	61	60
Mozambique	Dazza	(5)	52	37
Austria	Baur		34	22

inconsistent data on the frequency of HCV infection in patients with HCC throughout the world.

Second, the technique of HCV antibody testing is subject to false positive results. There may be interference with raised serum globulin concentrations [8]. Furthermore in long-term frozen sera, repeated thawing may be associated with false positive test results. Therefore in this study only fresh sera were used.

Third, selection of cases with HCC could influence the results on the incidence of anti-HCV positivity in HCC. If the patients studied have predominantly posthepatitic or cryptogenic liver cirrhosis anti-HCV positivity will be higher than in a group with a high percentage of alcoholic cirrhosis. Kiyosawa [10] tested only patients with posttransfusion nonA nonB hepatitis. Most authors tested consecutive patients with HCC, but the reasons for admissions to referral centers are unknown.

Fourth, it seems reasonable that in some countries a high incidence of anti-HCV positivity in patients with chronic liver diseases results also in a high frequency of HCV antibodies in patients with HCC (see Table 3). Only Bruix [2] found significantly higher anti-HCV positivity in HCC (75%) than in liver cirrhosis (56%). Thus, the differing frequency of HCV antibodies in HCC may simply reflect the varying importance of HCV infections for development of chronic liver disease in various areas of the world. Our data of similar frequencies of antibodies to HCV in patients with HCC or with liver cirrhosis without HCC indicate that, at least in Austrian patients HCV infections are not an important factor for development of HCC. In Austria the most obvious factor associated with HCC is the presence of cirrhosis, irrespective of its etiology.

References

1. Beasley RP, Hwang Lu-Yu, Lin GC, Chien CS (1981) Hepatocellular carcinoma and hepatitis B virus. A prospective study of 22 707 men in Taiwan. Lancet ii: 1129–1133

2. Bruix J, Calvet X, Costa J, Ventura M, Bruguera M, Castillo R, Barrera JM, Ercilla G, Sanchez-Tapias JM, Vall M, Bru C, Rodes J (1989) Prevalence of antibodies to hepatitis C virus in spanish patients with hepatocellular carcinoma and hepatic cirrhosis. Lancet ii: 1004–1006

3. Chiaramonte M, Farinati F, Fagiuoli S, Ongaro S, Aneloni V, De Maria N, Naccarato R (1990) Lancet 335: 301–302

4. Colombo M, Choo QL, Ninno ED, Dioguardi N, Kuo G, Donato MF, Tommasini MA, Houghton M (1989) Prevalence of antibodies to hepatitis C virus in Italian patients with hepatocellular carcinoma. Lancet ii: 1006–1008

5. Dazza MC, Meneses LV, Girard PM, Villaroel C, Brechot C, Larouzet B (1990) Hepatitis C virus antibody and hepatocellular carcinoma. Lancet 335: 1216

6. Ducreux M, Dussaix E, Briantais MJ, Jacques L, Buffet C, Pelletier G, Yvart J, Etienne LP (1990) Lancet 335: 301

7. Ferenci P, Dragosics B, Marosi L, Kiss F (1984) Relative incidence of primary liver cancer in cirrhosis in Austria. Etiological considerations. Liver 4: 7–14

8. Johnson PJ, Williams R (1990) Hepatitis C antibodies and hepatocellular carcinoma: new clues or a false trial? Editorial, J Natl Cancer Inst 82: 986–987

9. Kew MC, Houghton M, Choo QL, Kuo G (1990) Hepatitis C virus antibodies in southern African blacks with hepatocellular carcinoma. Lancet 335: 873–874

10. Kiyosawa K, Sodeyama T, Tanaka E, Gibo Y, Yoshizawa K, Nakamo Y, Furuta S, Akahane Y, Nishioka K, Purcell RH, Alter HJ (1990) Interrelationship of blood transfusion, non-A, non-B hepatitis and hepatocellular carcinoma: analysis by detection of antibody to hepatitis C virus. Hepatology 12: 671–675

11. Levrero M, Tagger A, Balsano C, Marzio ED, Avantaggiati ML, Natoli G, Diop D, Villa E, Diodati G, Alberti A (1991) Antibodies to hepatitis C virus in patients with hepatocellular carcinoma. J Hepatol 12: 60–63

12. Li JS, Tong SP, Vitvitski L, Trepo C (1992) Sequence analysis of PCR amplified hepatitis C virus cDNA from French non-A, non-B hepatitis patients. In: De Bac C, Gerlich WH, Taliani G (eds) Chronically evolving viral hepatitis. Springer, Wien, New York (Arch Virol, Suppl 4, pp 184–185)

13. Munoz N, Bosch X (1987) Epidemiology of hepatocellular carcinoma. In: Okuda K, Ishak KG (eds) Neoplasms of the liver. Springer, Berlin Heidelberg New York Tokyo, pp 3–19

14. Sbolli G, Zanetti AR, Tanzi E, Cavanna L, Civardi G, Fornari F, Di Stasi M, Buscarini L (1990) Serum antibodies to hepatitis C virus in Italian patients with hepatocellular carcinoma. J Med Virol 30: 230–232

15. Simonetti RG, Cottone M, Craxi A, Pagliaro L, Rapicetta M, Chionne P, Costatino A (1989) Prevalence of antibodies to hepatitis C virus in hepatocellular carcinoma. Lancet ii: 1338

16. Weiss W (1984) Das primäre Leberkarzinom. Facultas, Wien, p 45

Authors' address: Dr. Martina Baur, I. Department of Gastroenterology and Hepatology, University of Vienna, Lazarettgasse 14, A-1090 Wien, Austria.

III   Hepatitis B virus variants

Arch Virol (1992) [Suppl] 4: 83–85
© Springer-Verlag 1992

Variants of hepatitis B virus

F. Bonino and **M. R. Brunetto**

Division of Gastroenterology, Molinette Hospital, Torino, Italy

Summary. Variations in the course of disease caused by hepatitis B virus may often be attributed to genomic variants of the virus. It must be kept in mind, however, that other factors, i.e. immunocompetence of the host and new methods of detection such as PCR, may also result in apparently aberrant phenotypic expression. Examples of both situations are presented here and the need is stressed for combined virological, biochemical and clinical studies.

*

Many observations suggest the presence of hepatitis B virus (HBV) variants: detection of virus markers in unexpected circumstances, a peculiar pathogenicity or an altered response to therapy. However, factors other than genomic variations may explain some of the aberrant phenotypic expressions of HBV infection. Presence of HBsAg without anti-HBc can be due to lack of an anti-HBc immune response in immunocompromised hosts. Immunosuppression can be inherited as a genetic condition or induced by the infection of Human Immunodeficiency Virus (HIV). Detection of HBV-DNA in the absence of conventional markers of HBV is frequent using the new technique of polymerase chain reaction (PCR). PCR has brought the sensitivity of HBV-DNA detection from more than 500,000 genomes/ml down to a few genomes/ml. As a consequence we have to adapt to viruses the same diagnostic criteria used for bacteria. Both groups of agents can be amplified by PCR or culture before being analyzed, therefore detection of minute quantities of their nucleic acid does not bear a pathogenetic significance. Therefore, we need to define the number of HBVs (per ml of serum or per cell) which are clinically relevant. Finally, cross reactivities of monoclonal anti-HBs antibodies with host proteins can be mistaken as evidence for the existence of candidate HBV variants in individuals, with detectable HBsAg, but without other virus markers. Surgence of variant is warranted by conditions of selective pressure. Genetic heterogeneity may be an important element of pathogenesis providing a means to evade the host's immune response [1]. A mutant virus offering an altered target to the immunocytes may escape virus elimination more efficiently than wild virus [1]. Therefore, the host's immune response represents a major factor of selective pressure. Another condition favoring the emergence of new HBV variants is vaccination and one important piece of evidence of this has been recently presented [2]. Prevalence of a variant depends on the frequency with which it arises, on competitive growth with wildtype virus and infectivity. Identification of a variant requires cloning and sequencing of viral nucleic acid and a precise definition of the genomic variation which causes the phenotypic alteration. One example of a variant with important clinical implications is provided by HBV that lacks the capacity to release Hepatitis B e antigen (HBeAg minus) [3–9]. This variant prevails in patients with a peculiar, severe form of hepatitis B. In view of this, one may expect that cloning and sequencing of viral nucleic acid is necessary for diagnosis of Hepatitis B caused by HBeAg minus HBV. Recent data obtained using an oligonucleotide hybridization assay suggest that the virus population is frequently heterogenous (with both wild and HBeAg minus HBVs), but the prevalent virus determines the phenotype of hepatitis B (detection of HBeAg or anti-HBe). Presence of serum HBeAg identifies patients with classical chronic hepatitis B who show a yearly rate of 10–20% spontaneous seroconversion and a persistent response to IFN in the majority of cases (60–70% in immunocompetent individuals). In contrast, anti-HBe positive chronic hepatitis B has the propensity to a rapid transition to cirrhosis.

These patients have a widely fluctuating viremia, lower than that of HBeAg carriers (usually less than 20 pg of HBV genome equivalent per ml), intrahepatic HBcAg and low titres of IgM anti-HBc in serum. Therefore, a correct diagnosis of the two forms of chronic hepatitis B can only be posed on the basis of clinical, biochemical and virological features.

References

1. Mateu MG, Martinez MA, Rocha E, Andreu D, Parejo J, Giralt E, Sobrino F, Domingo E (1989) Implications of a quasispecies genome structure: effect of frequent naturally occurring amino acids substitutions on the antigenicity of foot-and-mouth disease virus. ad Sci USA 86: 5883–5887
2. Carman WF, Zanetti AR, Karayannis P, Waters J, Manzillo G, Tanzi E, Zuckerman AJ, Thomas HC (1990) Vaccine-induced escape mutant of hepatitis B virus. Lancet ii: 325–329
3. Brunetto MR, Stemler M, Schödel F, Will H, Ottobrelli A, Rizzetto M, Verme G, Bonino F (1989) Identification of HBV variants which cannot produce precore derived HBeAg and may be responsible for severe hepatitis. Ital J Gastroenterol 21: 151–154
4. Brunetto MR, Oliveri FR, Rocca G, Criscuolo D, Chiaberge E, Capalbo M, David E,

Verme G, Bonino F (1989) Natural course and response to interferon of chronic type B hepatitis, accompanied by antibody to hepatitis B e antigen. Hepatology 10: 198–203

5. Brunetto MR, Stemler M, Bonino F, Schödel F, Oliveri F, Rizzetto M, Verme G, Will H (1990) A new hepatitis B virus strain in patients with severe anti-HBe positive chronic hepatitis B. J Hepatol 10: 258–261

6. Carman WF, Jacyna MR, Hadziyannis S, Karayiannis P, McGarvey MJ, Makris A, Thomas HC (1989) Mutation preventing formation of Hepatitis B e antigen in patients with chronic Hepatitis B infection. Lancet ii: 588–591

7. Okamoto H, Yotsumoto S, Akahane Y, Yamanaka T, Miyazaki Y, Sugai Y, Tsuda F, Mayumi H (1990) Hepatitis B viruses with precore region defects prevail in persistently infected hosts along with seroconversion to the antibody against e antigen. J Virol 64: 1298–1303

8. Tong SP, Li JS, Vitvitski L, Trepo C (1989) The anti-HBe form of chronic active hepatitis B is associated with a stop codon in the pre-C region of the viral genome. In: Marion P, Schaller H (eds) Hepatitis B viruses. Cold Spring Harbor, NY, Cold Spring Harbor Lab, p 138

9. Raimondo G, Schneider R, Stemler M, Smedile A, Wilder G, Rodino G, Schödel F, Squadrito G, Will H (1991) Reactivation and latency of hepatitis B virus in chronically infected patients.

Authors' address: Dr. F. Bonino, Division of Gastroenterology, Molinette Hospital, In: Hollinger FB, Lemon SM, Margolis HM (eds) Viral hepatitis and liver disease. Williams & Wilkins, Baltimore, pp 210–211, Torino, Italy.

Arch Virol (1992) [Suppl] 4: 86–89
© Springer-Verlag 1992

Hepatitis B virus (HBV) genomic variations in chronic hepatitis

E. Cariani, G. Fiordalisi, and **D. Primi**

Consorzio per le Biotecnologie, Consiglio Nazionale delle Ricerche (CNR), Institute of
Chemistry, Medical School, Brescia, Italy

Summary. Using PCR we have examined the sequence of the pre-C/C region
of HBV from sera of anti-HBe positive, chronically HBV-infected patients.
The large majority of sera tested contained a mixture of heterogeneous pre-C
sequences with 1–3 non-randomly located point mutations. Some of the
resultant variant viruses are incapable of synthesizing immunogenic proteins
and may be involved in viral persistence in chronic carrier states.

*

Hepatitis B e antigen (HBeAg) is encoded by the pre-core/core (C) open
reading frame of HBV, starting from the first ATG located upstream from
the pre-C region. This protein is synthesized as a precursor which, after
proteolytic processing at the amino and carboxy terminus, is secreted into
the bloodstream [6, 4].

In chronic HBV infection, the development of anti-HBeAg reactivity is
generally associated with the clearance of serum HBV DNA. Some patients,
however, show persistence of serum HBV DNA, despite the seroconversion
to anti-HBe [1]. A translational stop codon at the 3' end of pre-C,
predictably preventing HBeAg synthesis, was reported to occur in HBV
isolated from HBeAg negative-HBV DNA positive patients [2, 3, 5]. A
nucleotide substitution at position 1899 was also observed in isolates from
several patients, but the functional implications of this mutation are still
undefined.

The mode of HBV replication, involving reverse transcription of an RNA
intermediate, may explain the accumulation of point mutations in the course
of chronic infection. However, the high frequency of mutations specifically
located in the last two codons of the pre-C region suggests that either this
region is a hot spot for mutations, or that mutations in this region represent

an advantage for viral persistence, and are thus selected during viral replication.

In order to analyze this point, we studied the nucleotide sequence of the pre-C/C region of HBV from the sera of anti-HBe, HBV DNA positive patients by the use of polymerase chain reaction (PCR). The primers used were located upstream of the pre-C start codon (5′-AGGAGTTGGGGGAGGAGATT-3′) and downstream from the C stop codon (5′-CTAACATTGAGATTCCCGAGA-3′). In order to analyze also the HBV genomes represented at low frequency, we chose the strategy of cloning the amplified DNA, since the alternative method, consisting of direct sequencing of the PCR products, only allows the detection of the prevalent molecular species.

The nucleotide sequence of the pre-C segment from 25 individual clones isolated from 11 patients demonstrated a high rate of point mutations (Table 1). Base substitutions were observed at 11 different sites, and different mutations were also detected in HBV clones isolated from the same subject. Indeed, only in two patients we did not find a mixture of variants, whereas most of the sera (80%) contained heterogeneous pre-C sequences with 1 to 3 different point mutations.

Several of the mutations were either silent or led to single amino acid substitutions (Table 1). These mutations were represented at low frequency, with the exception of the already reported G-A substitution at position 1899 that was observed in 15 isolates from 8 patients. Among the mutations likely to have functional transcriptional consequences, we observed a T-C substitution at position 1836 eliminating the stop codon of the X gene, and a G

Table 1. Point mutations in the pre-C region from anti HBe, HBV DNA positive chronic carriers

Mutation	Location	Consequences	No. of clones (patients)	(%)
T=C[a]	1815	pre-C initiation failure	2 (1)	8
T=G	1821	Leu=Arg	1 (1)	4
T=C	1822	silent	1 (1)	4
A=C	1827	His=Pro	1 (1)	4
T=C	1836	{Leu=Pro {X termination failure	1 (1)	4
A=G	1838	Ile=Val	1 (1)	4
T=A	1839	Ile=Asp	2 (1)	8
C=A	1843	silent	1 (1)	4
T=A	1895	Trp=Arg	1 (1)	4
G=A[a]	1896	STOP	18 (11)	72
G=A	1899	Gly=Asp	15 (8)	60
ΔG[a]	1899	frameshift	1 (1)	4

[a] Mutations preventing HBeAg synthesis

88 E. Cariani et al.

deletion at nucleotide 1899, generating a frameshift and a stop codon at codon 71. At least one clone isolated from each patient contained a mutation predictably preventing HBeAg synthesis. The G-A substitution generating a translational stop codon at position 1896 was detected in 18 clones. A second substitution predictably impairing HBeAg synthesis was observed in two clones isolated from the same serum. This mutation consisted of a T-C substitution at position 1815, eliminating the pre-C start codon.

The distribution of the mutations appeared to be non-random: in fact, they clustered in the first 10 and in the last 2 codons of the pre-C. This might indicate that either some sequences in the pre-C region are more prone to spontaneous mutations, or that mutations in the central region of the pre-C interfere with virus formation and, therefore, are not selected in the course of viral replication.

As a whole, these results are consistent with a high variability of pre-C sequence. However, the point mutations are not randomly located, and they differ in frequency; indeed, silent mutations are often associated with those preventing HBeAg synthesis, which are the most frequent. Since HBeAg is a major target for the immune system, the lack of HBeAg synthesis probably represents a selective advantage for viral persistence. Therefore, these results suggest that the mutants capable of avoiding the host immune system are clonally selected during chronic HBV infection.

Besides the ones involving the pre-C sequence, other mutations might play a role in the process of viral persistence in chronic carriers. We have recently observed the existence of a mixed HBV infection in a chronic carrier who seroconverted to anti-HBe. PCR with primers located in the C region allowed detection of a double amplification product in all the serum samples collected over a 1 year period, both before and after seroconversion to anti-HBe. Nucleotide sequence analysis showed the association of a wild-type HBV genome with a defective virus. The latter had a 37 bp deletion generating a stop codon in the C region. Moreover, a point mutation generated a stop codon in the pre-S1, and a 15 bp deletion involved the pre-S2 ATG.

These mutations predictably impair the synthesis of both core and pre-S proteins, leading to the hypothesis that this defective viral genome is incapable of autonomous replication and is probably enabled to propagate by the presence of the wild-type isolate, acting as a helper. In addition, this kind of defective viral genome, incapable of synthesizing immunogenic proteins such as core and pre-S, might be involved in viral persistence in HBV chronic carriers.

References

1. Bonino F, Hoyer B, Nelson J, Engle R, Verme G, Gerin JL (1981) HBV-DNA in the sera of HBsAg carriers: a marker of active hepatitis B virus replication in the liver. Hepatology 1: 386–391

2. Brunetto MR, Stemler M, Schödel F, Will H, Ottobrelli A, Rizzetto M, Verme G, Bonino F (1989) Identification of HBV variants which cannot produce precore derived HBeAg and may be responsible for severe hepatitis. Ital J Gastroenterol 21: 151–154
3. Carman WF, Jacyna MR, Hadziyannis S, Karayannis P, McGarvey MJ, Makris A, Thomas HC (1989) Mutation preventing formation of hepatitis B e antigen in patients with chronic hepatitis B infection. Lancet ii: 588–591
4. Jean-Jean O, Levrero M, Will H, Perricaudet M, Rossignol JM (1989) Expression mechanism of hepatitis B virus (HBV) C gene and biosynthesis of HBe antigen. Virology 170: 99–106
5. Okamoto H, Yotsumoto S, Akahane Y, Yamanaka T, Miyazaki Y, Sugai Y, Tsuda F, Tanaka T, Miyakawa Y, Mayumi M (1990) Hepatitis B viruses with precore region defects prevail in persistently infected hosts along with seroconversion to the antibody against e antigen. J Virol 64: 1298–1303
6. Schlicht HJ, Salfedt J, Schaller H (1987) The duck hepatitis B virus pre-C region encodes a signal sequence which is essential for synthesis and secretion of processed core proteins but not for virus formation. J Virol 61: 3701–3709

Authors' address: Dr. Elisabetta Cariani, Consorzio per le Biotecnologie, Consiglio Nazionale delle Ricerche (CNR), Institute of Chemistry, Medical School, P. le Spedali Civili 1, I-25123 Brescia, Italy.

Arch Virol (1992) [Suppl] 4: 90–94
© Springer-Verlag 1992

Replication of a molecularly cloned HBeAg negative hepatitis B virus variant in transfected HepG2 cells

S. Tong[1], C. Diot[2], P. Gripon[2], J. Li[1], L. Vitvitski[1], C. Trépo[1], and C. Gugen-Guillouzo[2]

[1] Unité Recherche sur les Hepatites Inserm 271, Lyon, France,
[2] Inserm Unité de Recherches Hépatologiques, U49, Hopital de Pontchaillou, Rennes, France

Summary. Transient expression in a hepatoma cell line of an HBV variant with a defective pre-C region suggested its ability to undergo full replication cycles except for the lack of HBeAg synthesis.

*

A hepatitis B virus (HBV) variant with mutated pre-C region has recently been detected in a subset of hepatitis B patients with anti-HBe antibody. The most prevalent mutation type observed is a TAG stop codon in the distal pre-C region with or without concomitant G to A transition in the succeeding GGC codon [1–5]. At present it is not known whether, in a manner similar to the related duck hepatitis B virus [6, 7], the naturally occurring HBV variants are replication competent but unable to synthesize HBeAg. To test this possibility, in the present study we have transfected a cloned HBV variant into Hep G2 cells. The pre-C variant HBV-α1 was cloned from serum of an anti-HBe patient into the SphI site of the pUC18 vector [4]. The complete 3182 nucleotide sequence of this HBV variant has been determined, revealing a TAG stop codon in the distal pre-C region and another G to A transition in the succeeding GGC codon [4]. For the transfection experiments a plasmid-integrated HBV tandem oligomer was constructed. The transfection procedure was performed essentially as described previously [8]. A wild type (WT) HBV [9] cloned into the EcoRI site of pBR322 as tandem trimer was also transfected to HepG2 cells as control.

We first sought evidence of HBV DNA replication in cells transfected with HBV-α1. Virus particles were pelleted from culture medium, and resuspended in TEN/CsCl. After centrifugation at 40,000 rpm for 65 hr, gradient was collected from the top into 13 fractions. Samples were treated with proteinase K, and DNA was extracted and electrophoresed through a 1% agarose gel. After Southern transfer, samples were hybridized with a ^{32}P labeled HBV probe. As shown in Fig. 1, hybridization signals peaked at fractions 7 and 10, with buoyant densities similar to those of Dane and core particles found in hepatitis B patients or produced by hepatocytes infected in vitro [10]. Fraction 7 had a 4 kb relaxed circular HBV DNA followed by a smear suggestive of varying length of the short strand; while fraction 10 had predominantly low molecular weight DNA indicating single stranded replicative DNA. These results suggest the formation of Dane and immature core particles following transfection of the pre-C region defective HBV variant. Moreover, we have also pelleted virus from culture medium, performed endogenous polymerase reaction with the incorporation of ^{32}P dCTP. Electrophoresis of the product in agarose gel followed by autoradiography revealed 4.0 and 3.2 kb bands corresponding to repaired full-length HBV DNA (data not shown).

Secondly, transcription of viral RNAs was analyzed from cells transfected with variant and with wild type HBV. RNA was extracted by guanidium

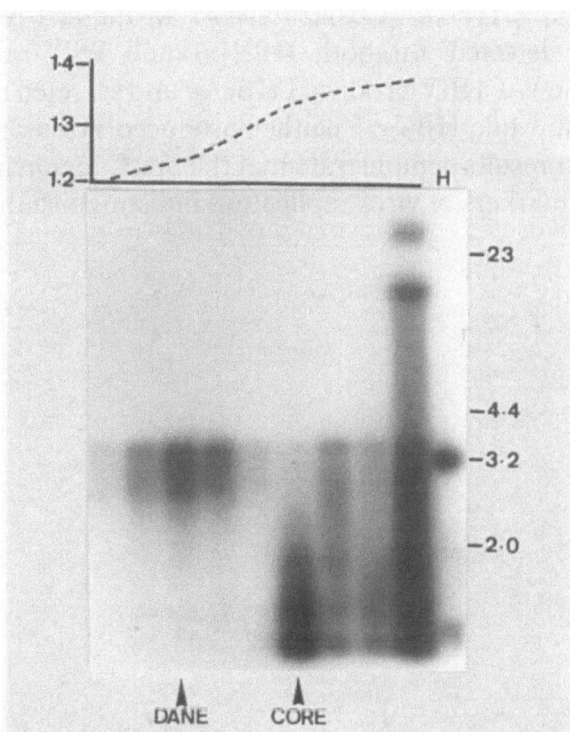

Fig. 1. Hybridization with HBV probe of CsCl gradient fractions of culture supernatant of HBV-α1 transfected HepG2 cells. Top: buoyant densities; Bottom: autoradiography of fractions 5–13 (left to right) and control linear HBV DNA (*H*). Fractions 7 and 10 show density values and hybridization patterns typical of Dane and core particles, respectively

cyanate [8], electrophoresed through a 1.5% agarose gel, transferred to a nitrocellulose filter, and hybridized with an HBV probe. Major viral transcripts of 3.5 and 2.4 kb were observed (Fig. 2), which represent the pregenomic RNA and messenger for HBsAg, respectively. The RNA patterns were indistinguishable between cells receiving HBV-α1 and WT HBV.

Finally, we studied HBV protein expression and secretion by transfected cells. High titres of HBsAg (P/N more than 100) and HBcAg (P/N more than 50) could be detected in the culture medium, which peaked at day 5 post-transfection. To ascertain whether an HBV variant with a defective pre-C region could express HBeAg, Western blotting was carried out. Culture medium was subjected to ultracentrifugation and divided into supernatant and pellet fractions. HBcAg or HBeAg in the supernatant fraction was immunoprecipitated by rabbit polyclonal anti-HBc conjugated to protein A-sepharose. After electrophoresis and transfer, the blot was reacted with anti-HBc followed by ^{125}I labeled protein A. The results are shown in Fig. 3. The pellet fraction of WT culture medium contained an HBcAg band and a weak HBeAg band (lane 3), while its supernatant fraction had a single band of HBeAg (lane 5). In contrast, only the HBcAg band was found in the pellet fraction of HBV-α1 medium (lane 2), and its supernatant fraction did not show any recognizable HBc/eAg band (lane 4). In the cell lysate only the HBcAg band was detected, for both HBV-α1 and WT transfected cells (lanes 6, 7). Thus, the WT HBV produced HBeAg and secreted most of it into the culture medium, while HBV-α1 neither produced nor secreted HBeAg.

In summary, our results demonstrate that the pre-C region mutated HBV variant showed all markers of virus replication indistinguishable from those

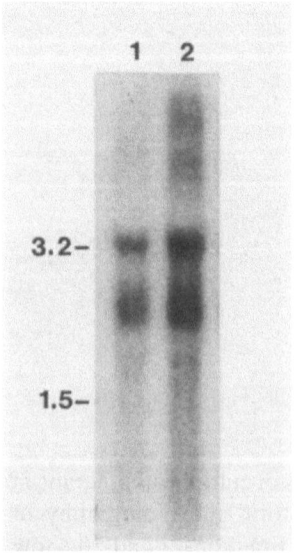

Fig. 2. Northern hybridization of cellular DNA of HepG2 cells transfected with HBV-α1 (lane *1*) or wild type HBV (lane *2*). Molecular size markers (in kb) are shown to the left of the gel

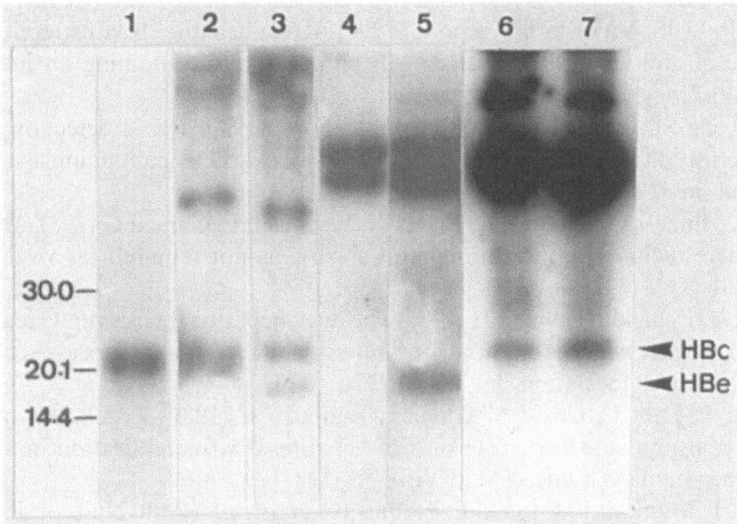

Fig. 3. Western blot analysis of the expression of HBcAg and HBeAg by HBV variant and WT HBV. *1* recombinant HBcAg; *2* pellet of HBV-α1 transfected cells; *3* pellet of WT HBV transfected cells; *4* supernatant of HBV-α1 transfected cells; *5* supernatant of WT HBV transfected cells; *6* lysate of HBV-α1 transfected cells; *7* lysate of WT HBV transfected cells. Molecular sizes (in kd) are shown to the left of the gel, and positions of HBcAg and HBeAg are indicated to the right of the gel

produced by the wild type HBV, except that it was unable to synthesize the conventional 17 kd HBeAg. These results are consistent with similar conclusions drawn from mutational studies of duck hepatitis B virus [6, 7], and suggest the infectivity of serum containing such HBV variants.

Acknowledgements

We thank C. Pichoud for ^{125}I labelling of protein A. S. Tong is a recipient of a Research Training Fellowship from the International Agency for Research on Cancer. C. Diot is a recipient of a fellowship from the Ligue Nationale Francaise contre le Cancer. J. S. Li is supported by the Merieux Fondation. This work was supported by the Institut National de la Santé et de la Recherche Médicale.

References

1. Brunetto MR, Stemler M, Schödel F, Will H, Ottoberlli A, Rizzetto M, Bonino F (1989) Identification of HBV variants which cannot produce precore derived HBeAg and may be responsible for severe hepatitis. Ital J Gastroenterol 21: 151–154
2. Carman W, Jacyna MR, Hadzıyannıs S, Karayıannıs P, McGarvey MJ, Makris A, Thomas HC (1989) Mutation preventing formation of hepatitis B e antigen in patients with chronic hepatitis B infection. Lancet ii: 588–591
3. Okamoto H, Yotsumoto S, Akahane Y, Yamanaka T, Miyazaki Y, Sugai Y, Tsuda F, Tanaka T, Miyakawa Y, Mayumi M (1990) Hepatitis B viruses with precore region defects prevail in persistently infected hosts along with seroconversion to the antibody against e antigen. J Virol 64: 1298–1303

 4. Tong SP, Li JS, Vitvitski L, Trépo C (1990) Active hepatitis B virus replication in the presence of anti-HBe is associated with viral variants containing an inactive pre-C region. Virology 176: 596–603

 5. Li JS, Tong SP, Vitvitski L, Zoulim F, Trépo C (1990) Rapid detection and further characterization of infection with hepatitis B virus variants containing a stop codon in the distal pre-C region. J Gen Virol 71: 1992–1997

 6. Chang C, Enders G, Sprengel R, Peters N, Varmus HE, Ganem G (1987) Expression of the precore region of an avian hepatitis B virus is not required for viral replication. J Virol 61: 3322–3325

 7. Schlicht HJ, Salfeld J, Schaller H (1987) The duck hepatitis B virus pre-C region encodes a signal sequence which is essential for synthesis and secretion of processed core proteins but not for virus formation. J Virol 61: 3701–3709

 8. Grippon P, Diot C, Corlu A, Guguen-Guillouzo C (1989) Regulation by dimethyl-sulfoxide, insulin, and corticosteroids of hepatitis B virus replication in a transfected human hepatoma cell line. J Med Virol 28: 193–199

 9. Galibert F, Mandart E, Fitoussi F, Tiollais P, Charnay P (1979) Nucleotide sequence of the hepatitis B virus genome (subtype ayw) cloned in *E coli*. Nature (London) 281: 646–650

10. Gripon P, Diot C, Thézé N, Fourel I, Loreal O, Brechot C, Guguen-Guillouzo C (1988) Hepatitis B virus infection of adult human hepatocytes cultured in the presence of dimethyl sulfoxide. J Virol 62: 4136–4143

Authors' address: Dr. C. Trépo, Unite Recherche sur les Hepatites Inserm 271, 151 Cours A-Thomas, F-69424 Lyon, Cédex 03, France.

Arch Virol (1992) [Suppl] 4: 95–96
© Springer-Verlag 1992

Lack of pre-C region mutation in woodchuck hepatitis virus from seroconverted woodchucks

S. P. Tong, L. Vitvitski, J. S. Li, C. Pichoud, and C. Trépo

INSERM U.271, Lyon, France

Summary. Woodchuck hepatitis virus, which shares a large degree of homology with human HBV, was examined for indications of mutational variants. No alteration in the pre-C region was found, but as in HBV, viral DNA could still be detected by PCR after seroconversion to anti-WHe.

*

A hepatitis B virus (HBV) variant with a nonsense mutation in the pre-C region has been detected in some chronic hepatitis B patients with anti-HBe. In the present study we have searched in seroconverted woodchucks for possible mutations in the pre-C region of woodchuck hepatitis virus (WHV), which shares a high degree of sequence homology with HBV, especially in the distal pre-C region where the nonsense mutation occurs in HBV [1]. Sera from 8 woodchucks (1 WHeAg positive, 4 anti-WHe positive, 3 negative for both WHeAg and anti-WHe) were studied. DNA was extracted, and the pre-C region sequences were amplified by PCR. A DNA band of expected size was amplified from 5/7 of WHeAg negative sera, albeit in amounts much lower than in the WHeAg positive serum. Direct sequencing of PCR products did not reveal any major alterations in the pre-C region of any of the samples, although there were 2 nucleotide differences among the samples suggesting infection of different animals by different viral strains. These preliminary results suggest that: 1. As in the human HBV infection [2] small amounts of viral DNA are still detectable by PCR in seroconverted woodchucks. 2. At least in most woodchucks there is no mutation in the pre-C region following seroconversion. This might be due to the simultaneous seroconversion in the WHV model to both anti-WHe and anti-WHs, which would explain why the mutation in the pre-C region alone does not result in the emergence of escape mutants.

References

1. Galibert F, Chen TN, Mandart E (1982) J Virol 41: 51–65
2. Kaneko S, Miller RH, Feinstone SM, Unoura M, Kobayashi K, Hattori N, Purcell RH (1989) Proc Natl Acad Sci USA 86: 312–316

Authors' address: Dr. S. P. Tong, INSERM U.271, 151 Cours A-Thomas, F-69003 Lyon, France.

Arch Virol (1992) [Suppl] 4: 97–101
© Springer-Verlag 1992

Clinical significance of the polymerase chain reaction (PCR) assay in chronic HBV carriers

G. Gerken[1,2], P. Paterlini[2], D. Kremsdorf[2], M. A. Petit[3], M. Manns[1], K.-H. Meyer zum Büschenfelde[1], and C. Brechot[2,4]

[1] I. Medizinische Klinik und Poliklinik der Johannes Gutenberg-Universität, Mainz, Federal Republic of Germany
[2] INSERM U 75 CHU Necker, Paris
[3] INSERM U 131, Clamart
[4] Laboratoire Hybridtest, Institut Pasteur, Paris, France

Summary. PCR was evaluated as a clinical tool for use in accurate identification of the specific etiologic agent in chronic HBV carriers. The method was found to be valuable in diagnosis and for monitoring therapy, as well as for elucidation of genotypic variants of HBV in chronic HBV cases. By this means an HBV defective variant with alterations in the preS1/preS2 sequence was detected and is consequently described here.

Introduction

Chronic HBV infection leads to distinct clinical outcomes varying from an asymptomatic HBsAg carrier state without HBV replication and normal liver function to symptomatic and progressive liver disease with high viremia and poor liver function [5]. Epidemiologic evidence supports a close relationship between chronic persistent HBV infection, liver cirrhosis, and the development of primary hepatocellular carcinoma [1]. Recently, in vitro amplification of HBV-DNA by means of the polymerase chain reaction (PCR) has become available [7]. The aim of our study was to analyze the clinical value of PCR to detect more accurately serum HBV-DNA in different groups of chronic HBV carriers. Furthermore, we report the discovery of a defective form of HBV in an HBsAg positive patient with liver cancer.

Patients and methods

Blood samples were obtained from patients with HBeAg positive (n = 20), anti-HBe positive patients with chronic hepatitis B (n = 10) and long-term asymptomatic HBsAg carriers with normal serum transaminases (n = 42). Assay of HBV-DNA by PCR was performed as described in detail elsewhere [4]. Furthermore, a 40 year old male of South-East Asian origin with a five year history of chronic hepatitis B before the diagnosis of primary hepatocellular carcinoma, was studied. Immunological characterization was performed using preS-specific monoclonal antibodies as described in detail elsewhere [2, 6]. Using the PCR assay the presence of deleted HBV-DNA sequences was studied [3]. The PCR product was subsequently cloned and sequenced using the dideoxy extension method (Sequenase Kit, US Biochemicals, Cleveland, USA).

Results

Assessment of the PCR assay

HBV-DNA was detectable in the serum of HBsAg positive HBV carriers using aliquots as small as 100 attoliter. The detection limit for cloned HBV-DNA was 100 attogram. Primer pairs from different regions of the amplified

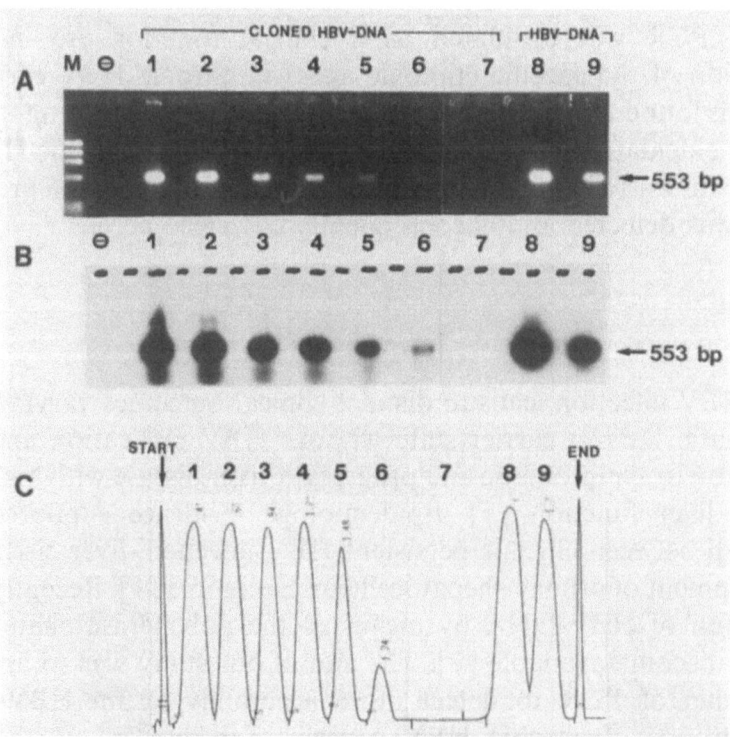

Fig. 1. Semiquantitation of PCR products using densitometry. Standardized cloned HBV-DNA, 10 ng (*1*), 100 pg (*2*), 1 pg (*3*), 100 fg (*4*), 1 fg (*5*), 100 ag (*6*), 1 ag (*7*); natural HBV-DNA from serum samples before (*8*) and after (*9*) interferon treatment. *M* reference molecular weight marker; *A* ethidium bromide staining; *B* Southern blot; *C* densitometric pattern

HBV genome resulted in different sensitivity. The detection of the amplified HBV-DNA by Southern blotting and subsequent scintillation counting or densitometry allowed a semi-quantitative assay (Fig. 1).

Clinical evaluation of the PCR assay

Using several primer pairs in parallel for optimal detection, all HBeAg positive HBsAg carriers, 80% of the anti-HBe positive symptomatic HBsAg carriers, and 57% of the asymptomatic anti-HBe positive HBsAg carriers were found to have HBV-DNA in the serum.

During anti-viral therapy HBV-DNA disappeared in the PCR assay in patients who became HBe negative but PCR detected a relapse earlier than did the conventional dot blot. PreS antigens were assayed in serum and liver samples of most chronic carriers. While most viremic carriers were strongly positive for preS1 and preS2 antigen, some HBV-DNA positive HBsAg carriers did not have detectable preS antigen and vice versa.

Characterization of a defective hepatitis B virus

Using PCR, the presence of deleted HBV forms was observed in serum, PBMC and liver tissue in an HBsAg and anti-HBeAg positive patient with

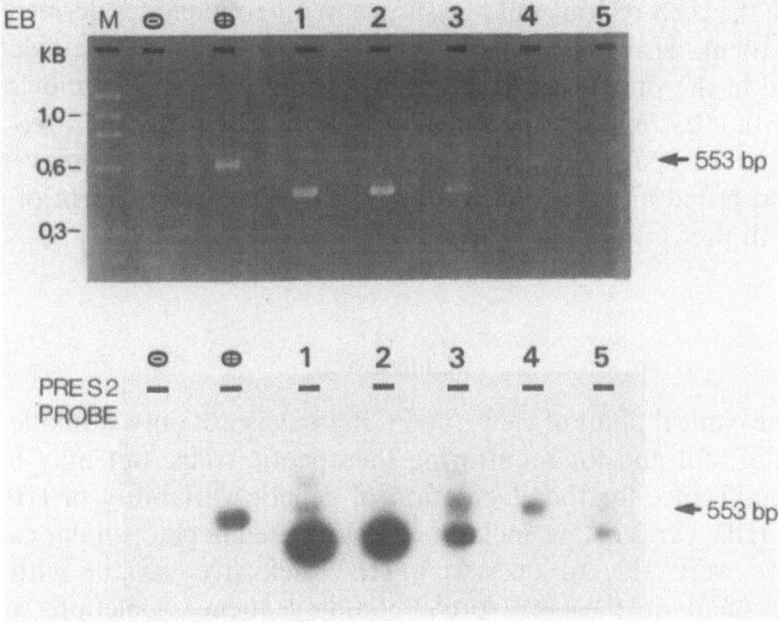

Fig. 2. Defective HBV-DNA observed after PCR analysis with preS primers and subsequent Southern blot using preS specific oligonucleotide probe. Negative control: PCR mix without template DNA. Positive control: 100 ng DNA extracted from HBsAg positive liver tumor tissue; *1* DNA extracted from serum, *2* DNA extracted from PBMC, *3* liver derived DNA, *4* normal viral particles, *5* defective viral particles, *EB* ethidium bromide staining

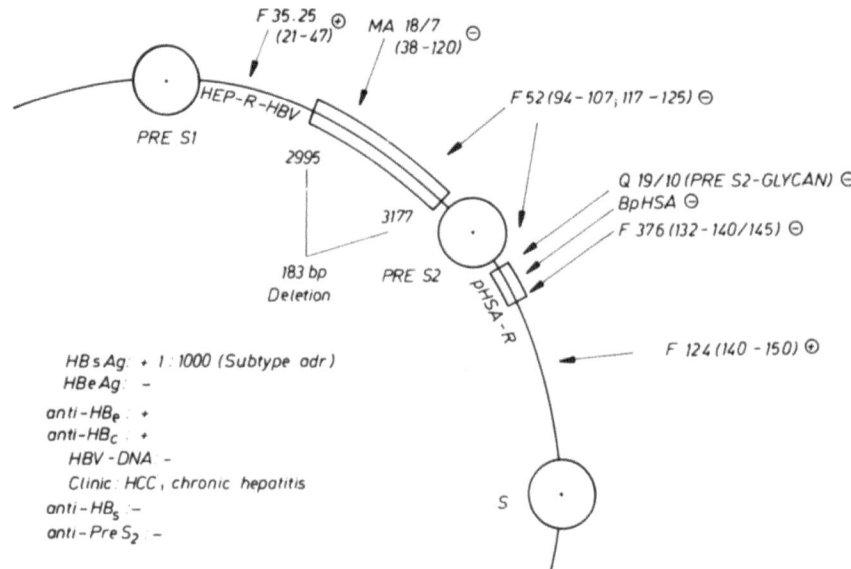

Fig. 3. Immunological and genetic characterization of a defective HBV with rearrangements in the preS gene sequence. *Hep-R-HBV* preS1 binding site of HBV to liver membranes; *pHSA* preS2 binding site of HBV for polymerized human serum albumin

liver cancer (Fig. 2). After cloning and sequencing, the genetic rearrangements of the preS region could be shown in different clones derived from the same patient. However, in 9/12 clones a 183 bp in-frame-deletion was recorded in the preS1 region (2995–3177) (Fig. 3). Using monoclonal anti-preS antibodies, immunological mapping showed a loss of preS epitopes located at the 3′-part of the preS1 and the 5′-part of preS2. Moreover, PBMC were also tested and only PCR alone showed the major form of defective HBV with preS1 183 bp deletion.

Conclusion

From the clinical point of view, the PCR assay is not only valuable as a new diagnostic tool and for monitoring therapeutic trials, but may be also of great significance for the elucidation of genetic variability of HBV in the chronic HBV carrier state, including HBV related hepatocellular carcinoma. Thus, we were able to observe a HBV defective variant with marked rearrangements in the preS1/preS2 coding sequence. Deletions and insertions could potentially lead to an impairment in the immune clearance of the virus by retaining a major hepatocyte binding site. Generation of such defective viruses during the course of HBV infection might therefore be an important factor in the persistence of an HBV chronic carrier state with development of hepatocellular carcinoma as an endstage complication.

Acknowledgements

This study was in part supported by grants from INSERM, ARC, DFG, DAAD, and NATO. G. Gerken was awarded by the Deutsche Gesellschaft für Verdauungs- und Stoffwechselkrankheiten, sponsered by ASCHE AG. We thank Wolfram H. Gerlich, Gießen, FRG, for providing us with the monoclonal antibodies MA18/7 and Q19/10. We are grateful to C. Förster-Schorr for excellent technical assistance.

References

1. Beasley RP, Hwang LY, Lin CC, Chien CS (1981) Hepatocellular carcinoma and HBV: a prospective study of 22707 men in Taiwan. Lancet ii: 1129–1133
2. Deepen R, Heermann KH, Uy A, Thomssen R, Gerlich WH (1990) Assay of preS epitopes and preS1 antibody in hepatitis B virus carriers and immune persons. Med Microbiol Immunol 179: 49–60
3. Gerken G, Kremsdorf D, Capel F, Petit MA, Dauguet C, Manns M, Meyer zum Büschenfelde K-H, Brechot C (1991) Hepatitis B defective virus with rearrangements in the preS gene during chronic HBV infection. Virology 183: 555–565
4. Gerken G, Paterlini P, Manns M, Housset C, Terre S, Dienes HP, Hess G, Gerlich WH, Berthelot P, Meyer zum Büschenfelde K-H, Brechot C (1991) Assay of hepatitis B virus DNA by polymerase chain reaction and its relationship to preS- and S- encoded viral surface proteins. Hepatology 13: 158–166
5. Hoofnagle JH, Shafritz DA, Popper H (1987) Chronic type B hepatitis and the healthy carrier state. Hepatology 7: 758–763
6. Petit MA, Zoulim F, Capel F, Dubanchet S, Dauguet C, Trepo C (1990) Variable expression of preS1 antigen in serum during chronic hepatitis B virus infection: an accurate marker for the level of hepatitis B replication. Hepatology 11: 809–814
7. Thiers V, Nakajima E, Kremsdorf D, Mack D, Schellekens H, Driss F, Goudeau A, Wands J, Sninsky J, Tiollais P, Brechot C (1988) Hepatitis B seronegative blood can transmit hepatitis B: polymerase chain reaction evidence of mutant viruses not currently detected. Lancet ii: 1273–1276

Authors' address: Dr. G. Gerken, INSERM Unité 75, CHU Necker, 156, Rue de Vaugirard, F-75730 Paris Cedex 15.

IV Serodiagnosis of hepatitis B

Arch Virol (1992) [Suppl] 4: 105–112
© Springer-Verlag 1992

PreS antigen expression and anti-preS response in hepatitis B virus infections: relationship to serum HBV-DNA, intrahepatic HBcAg, liver damage and specific T-cell response

M.-A. Petit[1]*, F. Capel[1], F. Zoulim[2], S. Dubanchet[1], I. Chemin[2], A. Penna[3], C. Ferrari[3], and C. Trépo[2]

[1] INSERM Unité 131, Clamart
[2] INSERM Unité 271, Lyon, France
[3] Cattedra Malattie Infettive, Università di Parma, Parma, Italy

Summary. The diagnostic value of preS antigens and anti-preS antibodies during Hepatitis B virus (HBV) infections have not yet been clearly elucidated. Therefore, the objectives of this study were: 1) to better understand the clinical significance of the expression of both preS1 and preS2 antigens (preS1Ag and preS2Ag) in the serum of chronic HBsAg carriers, and 2) to define the respective role of antibody responses to HBs-, preS2- and preS1-specific determinants in the course of acute hepatitis B (AH-B) infections with different outcomes.

Our data showed that the serum preS1Ag/HBsAg ratio correlated well with the level of viral replication (serum HBV-DNA as monitored by PCR assay and liver HBcAg), especially in anti-HBe positive chronic carriers. The complete eradication of virions required a persistent antibody response to conformation-dependent HBs-epitopes, temporally associated with a vigorous T cell response to nucleocapsid antigens. Recovery from hepatitis B can be achieved when there is no early antibody response to preS2- and preS1-proteins, which was found to be transient, concomitant with a flare-up of the liver disease, and preceding anti-HBs production.

Information on the patterns of preS antigens and their antibodies remained clouded because of the varying specificities and sensitivities of research methods used in studies to date. We have, therefore, developed original Polyclonal-Monoclonal RadioImmunoAssays (PAb-MAb RIAs) [5, 8] by using monoclonal antibodies (MAbs) having previously well-defined specificities [3, 4, 5, 7]. We could thus detect and quantify simultaneously the three distinct antigenicities of the HBV envelope, HBsAg, preS2Ag and preS1Ag, with the same sensitivity. In this way, the preS1Ag/HBsAg and preS2Ag/HBsAg ratios can be calculated to estimate the serum expression of both preS1Ag and preS2Ag in relation to total HBsAg activity during different stages of chronic HBV infection. For optimal management of the state of HBV replication in chronic viral infection, the detection of HBV-DNA in serum was monitored by the Polymerase Chain Reaction (PCR) assay [1].

We extended our work by investigating the clinical significance of antibody response to preS-specific determinants in patients with AH-B showing different outcomes in both natural course or response to α-interferon therapy. In a first attempt, we chose to use the Western Immuno-Blotting Assay (WIBA) to obtain a qualitative assessment of the nature of preS antibody responses [6]. Finally, the cell-mediated immune response to HBV antigens was also studied in several patients with self-limited AH-B [2] leading to a relevant finding which may help to clarify the mechanisms responsible for complete clearance of HBV.

PreS antigen expression in serum during chronic HBV infection: correlation with liver HBcAg and serum HBV-DNA detection by PCR

For studying the serum expression of preSAgs on circulating HBV/HBsAg particles during chronic HBV infection, we selected 39 HBsAg carriers belonging to three relevant groups, as shown in Fig. 1A. Nineteen patients were referred for chronic liver disease (CLD): 11 had HBeAg (group 1) and 8 had anti-HBe (group 2). Twenty patients (all with anti-HBe) were referred as healthy carriers (HCs) (group 3). All patients studied were found to be negative for antibodies to hepatitis delta virus. Liver biopsies were performed because they were clinically indicated in all patients with high serum ALT levels and persistent HBs antigenemia for at least 6 months (groups 1 and 2). For the detection of HBsAg, preS2Ag or preS1Ag in serum samples, the PAb-MAb RIAs were used, as described in detail elsewhere [8]. Briefly, polystyrene beads were coated with rabbit polyclonal IgG to HBsAg, and incubated with serial 10-fold dilutions of human sera. Then, murine MAbs to HBsAg (F39.20), preS2Ag (F124), or preS1Ag (F35.25) were added, and ^{125}I-labeled F(ab')2 fragment of anti-mouse Ig (Amersham) was used as tracer. Results were expressed as cpm of ^{125}I-labeled antibody bound, and as the preS2Ag/HBsAg and preS1Ag/HBsAg ratios of cpm $\times$ 100 calculated for each serum sample diluted tenfold.

The HBs-specific MAb F39.20 recognized a disulfide-bond dependent conformational epitope on the HBsAg molecule [4]. The preS2-specific MAb F124 recognized the sequential preS2 N-terminal region (120 to 126) [3], and the conformation-dependent pHSA-binding site (between amino acid residues 140–150) of the preS2-sequence (M.-A. Petit, unpublished results). The preS1-specific MAb F35.25 recognized the hepatocyte receptor-binding site (between amino acid residues 21 to 47) of the preS1-sequence, independently of d/y subtype changes [5, 7].

The results of tests for ALT, HBV-DNA and DNApolymerase (DNAp) in serum, the histological diagnosis, and the detection of HBcAg in liver by immunofluorescence (IF) for all patients are reported in Fig. 1. The serum expression of preSAgs in the different groups of chronic HBsAg carriers are presented in Fig. 2. PreS1Ag was found in most (90%) patients with CLD (group 1 and group 2), irrespective of the presence of HBV-DNA detected by slot-blot hybridization and the HBeAg/Anti-HBe status in sera. In contrast, preS1Ag was not detected in sera (diluted $10\times$) from almost any HCs (group 3). More significantly, the mean serum preS1Ag/HBsAg ratio is higher (34%) in group 1, which showed active viral replication, than in group 2 (18%), which demonstrated low-grade viral replication. Furthermore, a correlation between preS1Ag in serum and HBcAg in liver is observed in anti-HBe-positive chronic hepatitis B patients (group 2).

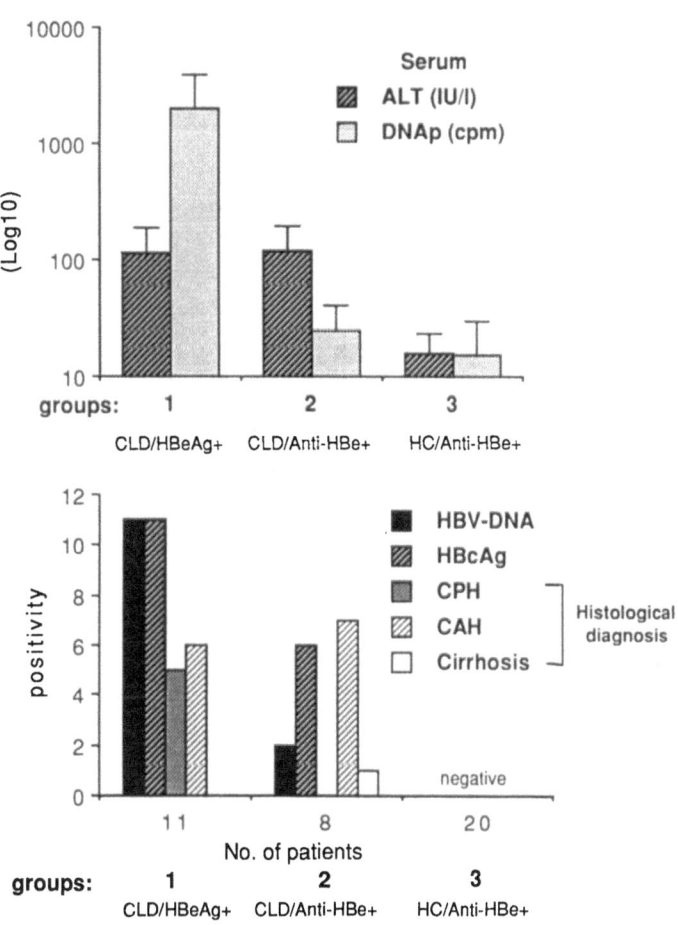

Fig. 1. Clinical details of the patients studied

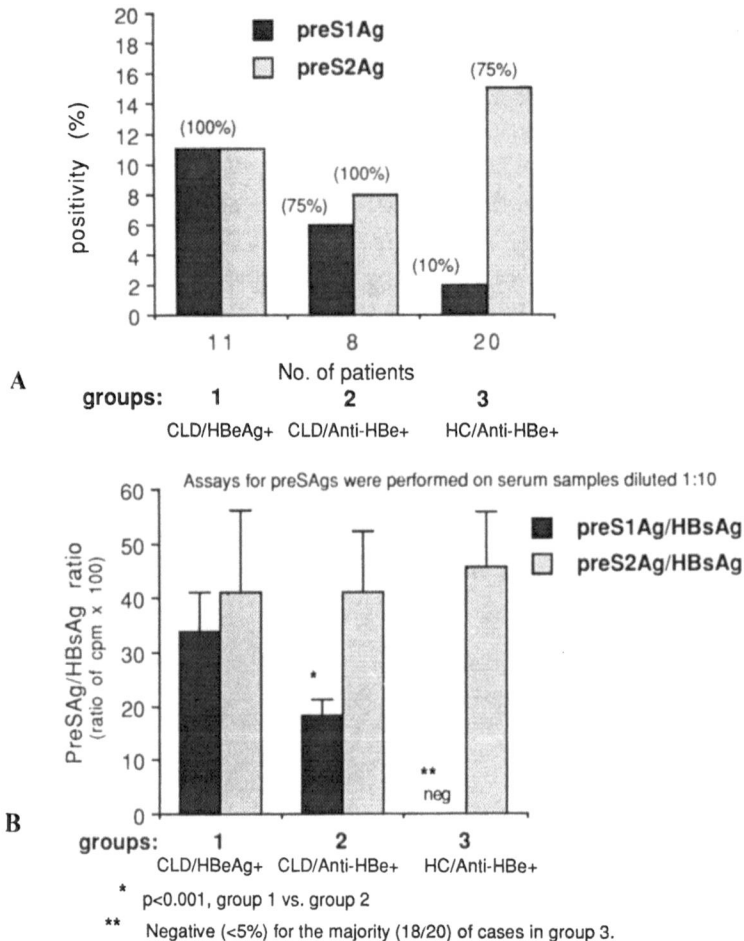

Fig. 2. PreS antigen expression in serum during chronic HBV infection. **A** Percentage of preS1Ag or preS2Ag positive patients. **B** Ratio of preSAg to HBsAg in the groups of chronically infected individuals

By using the same approach for detecting preS2Ag in the different groups of chronic HBsAg carriers, we showed that the preS2 proteins were expressed in all patients with chronic hepatitis B and in most HCs, irrespective of HBV replication markers such as HBV-DNA and HBeAg in the serum and HBcAg in the liver. This suggests that persistent HBs antigenemia is associated with a continuous synthesis of preS2 proteins as 22-nm defective viral particles in most chronic HBsAg carriers, even in nonviremic healthy blood donors.

Thus, the PAb-MAb RIA system developed by us for the simultaneous detection and quantification of HBsAg, preS2Ag and preS1Ag in human sera represents an improved approach for obtaining valuable information on the biology of HBV during the chronic HBsAg carrier state. Although the serum preS2Ag/HBsAg ratio appears to be of little relevance in patients with chronic HBV infection, the serum preS1Ag/HBsAg ratio seems to reflect the degree of viral replication, especially among anti-HBe carriers.

The presence of HBV-DNA in sera from 85 symptomatic and asymptomatic HBsAg carriers was studied by PCR with primers specific for the S and C regions in parallel with the detection of preS1Ag by our PAb-MAb RIA [1]. Whatever the clinical status of patients studied, the preS1Ag expression statistically correlated ($p < 0.0001$) with the presence of HBV-DNA detected by the PCR assay, where amplified DNA was analyzed by ethidium bromide (PCR-EB) [1]. This finding demonstrates that both methods (PAb-MAb RIA for preS1Ag and PCR-EB for HBV-DNA in serum) provide sensitive and accurate markers for HBV replication.

Clearance of HBV surface antigens and anti-preS response during acute hepatitis B (AH-B) infections

The clearance of HBsAg, preS2Ag and preS1Ag monitored by PAb-MAb RIAs, the anti-HBs production determined by RIA (Abbott Laboratories, North Chicago, IL) and the antibody response to preS2- and preS1-proteins detected by WIBA, were studied in 70 serum samples from 10 patients with acute self-limited AH-B (group 1), and in 39 serum samples from 3 patients with protracted (≥ 10 weeks but < 6 months) AH-B [9] (group 2). In group 2, one patient received placebo and the two others were treated with recombinant human α-interferon (rIFNα-2b, Intron A from Shering) for 24 weeks, with follow-up for 1 year. In group 1, seven patients with AH-B were analyzed for the proliferative response of peripheral blood lymphomononuclear cells (PBMC) to HBV envelope, core (HBcAg) and e (HBeAg) antigens [2].

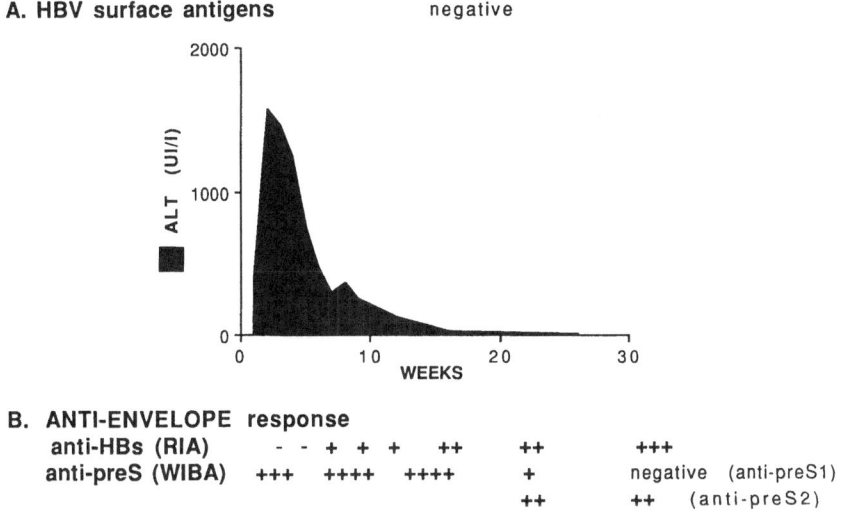

Fig. 3. Course of acute rapidly resolving hepatitis B

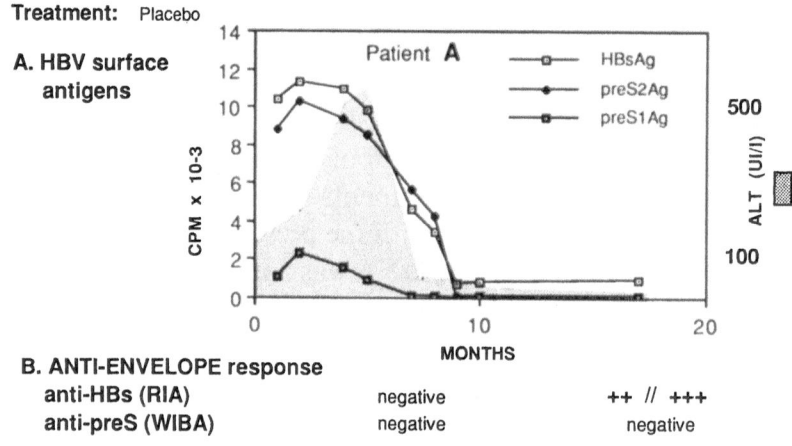

Fig. 4. Course of a protracted self-limited hepatitis B

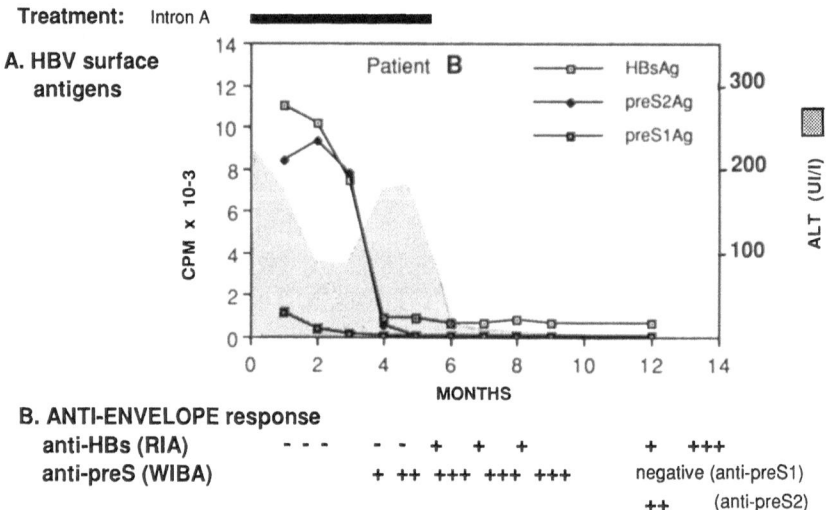

Fig. 5. Course of a protracted acute hepatitis B with resolution after interferon therapy

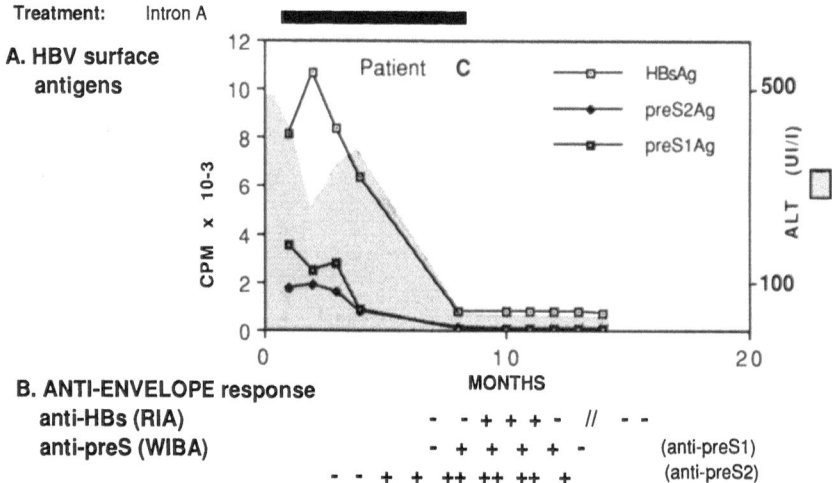

Fig. 6. Course of a protracted acute hepatitis B with resolution after interferon therapy

The results of proliferation assays indicated that the clearance of serum HBV surface antigens (HBsAg, preS2Ag and preS1Ag) was temporally associated with a strong T cell response to nucleocapsid antigens (HBcAg and HBeAg) in self-limited AH-B [2]. This intriguing observation suggests that the elimination of preS1Ag-positive complete virions may be facilitated by a vigorous nucleocapsid-specific T cell response.

Results obtained in a patient from group I with typical self-limited AH-B who rapidly progressed to recovery, are reported in Fig. 3. HBV surface antigens were cleared very early and found negative from the second serum sample on. Interestingly, the appearance of a strong antibody response to preS1- and preS2-proteins was found to be concomitant with a peak of ALT, reflecting an exacerbation of hepatitis. Then, the anti-preS1 response became negative while the anti-preS2 response decreased, together with a high anti-HBs response.

Figures 4 to 6 show results obtained in three patients with protracted AH-B. Patient A who received a placebo progressed spontaneously to remission of disease (ALT normalization), and seroconverted to anti-HBe and anti-HBs. Nevertheless, antibody response to preS2 or preS1 proteins was not detectable throughout follow-up. Patient B completely cleared HBV infection after α-interferon therapy, and exhibited typical seroconversion illness. The induced antibody response to preS-proteins was similar to that observed in self-limited AH-B as described above (Fig. 3), and was again found to be associated with a flare-up of liver disease. Patient C who cleared HBV surface antigens after α-interferon therapy did not, however, develop a sustained anti-HBs response. Thus, the transient anti-preS response may be not followed by anti-HBs seroconversion.

Our data indicate that complete eradication of HBV requires a strong sustained antibody response to disulfide bond-dependent epitopes on the HBsAg molecule, and that spontaneous recovery from hepatitis B can be achieved without prior antibody response to preS-proteins. Moreover, appearance of antibodies to both preS2- and preS1-proteins seems to be concomitant with a flare-up of the liver disease, suggesting a possible role of anti-preS response in the pathogenesis of HBV infection.

References

1. Chemin I, Baginski I, Petit M-A, Zoulim F, Pichoud C, Capel F, Hantz O, Trepo C (1991) Correlation between HBV DNA detection by PCR and preS1 antigenemia in symptomatic and asymptomatic HBV infections. J Med Virol 33: 51–57
2. Ferrari C, Penna A, Bertoletti A, Valli A, Antoni AD, Giuberti T, Cavalli A, Petit M-A, Fiaccadori F (1990) Cellular immune response to hepatitis B virus-encoded antigens in acute and chronic hepatitis B virus infection. J Immunol 145: 3442–3449
3. Neurath AR, Strick N, Kent SBH, Parker K, Courouce A-M, Riottot M-M, Petit M-A, Budkowska A, Girard M, Pillot J (1987) Antibodies to synthetic peptides from the preS1

and preS2 regions of one subtype of the hepatitis B virus (HBV) envelope protein recognize all HBV subtypes. Mol Immunol 24: 975–980

4. Petit M-A, Capel F, Riottot M-M, Dauguet C, Pillot J (1987) Antigenic mapping of the surface proteins of infectious hepatitis B virus particles. J Gen Virol 68: 2759–2767
5. Petit M-A, Dubanchet S, Capel F (1989) A monoclonal antibody specific for the hepatocyte receptor binding site on hepatitis B virus. Mol Immunol 26: 531–537
6. Petit M-A, Maillard P, Capel F, Pillot J (1986) Immunochemical structure of the hepatitis B surface antigen vaccine-II. Analysis of antibody responses in human sera against the envelope proteins. Mol Immunol 23: 511–523
7. Petit M-A, Strick N, Dubanchet S, Capel F, Neurath AR (1991) Inhibitory activity of monoclonal antibody F35.25 on the interaction between hepatocytes (HepG2 cells) and preS1-specific ligands. Mol Immunol 28: 517–521
8. Petit M-A, Zoulim F, Capel F, Dubanchet S, Dauguet C, Trepo C (1990) Variable expression of preS1 antigen in serum during chronic hepatitis B virus infection: an accurate marker for the level of hepatitis B virus replication. Hepatology 11: 809–814
9. Trepo C, Chemin I, Petit M-A, Chossegros P, Zoulim F, Chevallier P, Sepetjan M (1990) Possible prevention of chronic hepatitis B by early interferon therapy. J Hepatol 11: S95–S99

Authors' address: Dr. Marie-Anne Petit, INSERM Unité131 Immunopathology & Viral Immunology, 32, rue des Carnets, F-92140, Clamart, France.

Arch Virol (1992) [Suppl] 4: 113–115
© Springer-Verlag 1992

PCR analysis of HBV infected sera: relationship between expression of pre-S antigens and viral replication

A. R. Garbuglia[1], A. Manzin[2], A. Budkowska[3], G. Taliani[4], C. Delfini[1], and G. Carloni[5]

[1] Cell Biology, Istituto Superiore di Sanità, Rome
[2] Institute of Microbiology, University of Ancona, Italy
[3] Institut Pasteur, Paris, France
[4] Institute of Infectious Diseases, University of Rome
[5] Institute of Experimental Medicine, C.N.R., Rome, Italy

Summary. In order to determine the biological significance of the pre-S antigens in HBV infection, HBsAg sera were tested for the presence of pre-S1 and pre-S2. HBV DNA was detected by spot-hybridization and PCR. The data show a complete correlation between pre-S antigenemia and HBV DNA replication in anti-HBe positive cases. PCR but not spot-hybridization was adequately sensitive to also detect HBV DNA in roughly half of the preS negative sera as well. Thus PCR appears to be a valuable technique for detection of potentially infectious anti-HBe carriers.

*

Complete HBV virions, 22 nm spherical HBsAg particles and tubular forms all contain different amounts of pre-S surface antigens [1]. Pre-S expression is independent of symptomatology [2] and depends on the HBeAg/anti-HBe status of infected sera and on the stage of infection. Since the synthesis of pre-surface proteins increases during complete viral replication, the content of pre-S1 and pre-S2 sequences can be considered an important marker of infectivity.

A variable portion of anti-HBe subjects still exhibit HBV DNA in serum [3]. To investigate the biological significance of pre-S antigens with respect to viral replication, we studied 125 sera, 28 from HBeAg and 97 from anti-HBe positive HBsAg carriers. Pre-S1 and pre-S2 were determined by an enzyme immunoassay method using specific monoclonal antibodies, and HBV DNA was detected by spot-hybridization (SH) [3]. Pre-S1, pre-S2 and HBV DNA were present in 96.4%, 89.3% and 78.6% of HBeAg+ and in

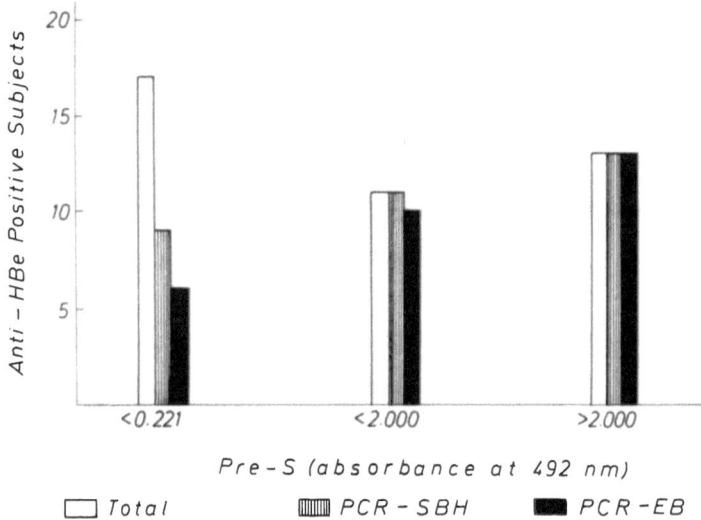

Fig. 1. Relationship between pre-S antigen activity in enzyme immune assay and detection of HBV DNA by PCR and Southern blot hybridization (*SBH*) or ethidium bromide (*EB*) staining

67.0%, 77.3% and 13.4% of anti-HBe + carriers, respectively. In addition, the sera that were found to be negative for HBV DNA by SH (6 HBeAg and 84 anti-HBe) were further investigated by PCR technique. Specific primers for HBV "S" and "C" regions were used for the amplification reaction. Samples were migrated in agarose gel (PCR-EB), blotted, and the filters hybridized with T4 polynucleotidekinase ^{32}P labeled specific probes (PCR-SBH). PCR analysis revealed that 4 out of 6 HBeAg positive sera and 33 out of 41 (80.5%) anti-HBe cases investigated, contained HBV specific amplified templates of S and C regions, absent in healthy control subjects. The PCR hybridization bands were in all cases fainter than those of the sera that were HBV DNA positive by SH. A strong correlation (100%) between pre-S antigenemia and HBV DNA replication, as detected by PCR-SBH, was found in 13 anti-HBe positive cases with very high titers of pre-S (pre-S value > 2.000) (see Fig. 1). However, PCR also detected the presence of circulating HBV DNA molecules in 9 out of 17 (52.9%) pre-S negative (pre-S value < 0.221) sera. These data suggest that, in anti-HBe subjects, strong viremia is associated with very high titers of pre-S antigens.

In conclusion, the PCR technique allows the detection of very small amounts of HBV (10–200 viral particles per ml serum), not detectable by SH, the sensitivity of which normally reaches about 10^6 viral particles per ml. PCR, therefore, helps identifying, with increased sensitivity, potentially infectious anti-HBe carriers, who exhibit a residual, although low, viremia. In addition, this technique allows us to better understand the significance of the pre-S antigen expression in these subjects.

References

1. Neurath AR, Kent SBH (1988) Adv Virus Res 34: 65–142
2. Budkowska A et al (1986) Hepatology 6: 360–368
3. Carloni G et al (1987) J Med Virol 21: 15–23

Authors' address: Dr. A. R. Garbuglia, Cell Biology, Istituto Superiore di Sanità, I-00161 Rome, Italy.

Arch Virol (1992) [Suppl] 4: 116–118
© Springer-Verlag 1992

Detection of HBV DNA by PCR in serum from an HBsAg negative blood donor implicated in cases of post-transfusion hepatitis B

H. Norder[1], B. Hammas[1], J. Larsen[2], K. Skaug[3], and L. O. Magnius[1]

[1] Department of Virology, The National Bacteriological Laboratory, Stockholm, Sweden
[2] Blood Bank and Department of Immunology, Ullevaal Hospital
[3] Microbiological Laboratory, Ullevaal Hospital, Oslo, Norway

Summary. An HBsAg negative blood donor, and three of her recipients, who developed HBsAg positive post-transfusion hepatitis B, were all positive for serum HBV DNA by polymerase chain reaction (PCR), and by subtype discriminating PCR were found to harbour HBV specifying *ayw*. Thus HBV DNA may be detected and sub-typed by PCR in infectious HBsAg negative individuals.

*

Various reports suggest that hepatitis B surface antigen (HBsAg) negative blood donors may transmit HBV infection. A Norwegian blood donor, negative for HBsAg and anti-HBs in serum, but positive for anti-HBc in high titer, had been implicated in three cases of overt post-transfusion hepatitis B (PTH-B) with demonstrable HBsAg in serum [1]. This blood donor and another from the same blood center, also with high-level anti-HBc as serological marker for HBV, were identified as positive for serum HBV DNA by a previously described polymerase chain reaction (PCR) [2].

A nested PCR for subtyping of HBV DNA was developed. Amplification within the S-gene with the primer pairs hep33-34 or hep3-33 directed to conserved regions was followed by another amplification step with the subtype specific oligonucleotide primer pairs, hep7-8 and hep3-20, for discriminating d/y, and hep3-30 and hep3-31 for discriminating w/r, respectively. The characteristics of these primers are given in the table.

Subtyping by PCR revealed that the three cases of PTH-B, and interestingly, both of the blood donors without detectable HBsAg in serum were harbouring HBV genomes specifying HBsAg of ayw subtype, despite the fact

Table 1. Genomic positions according to the nucleotide enumeration of Okamoto et al [3], sequences, annealing and extension temperatures for the utilized primer pairs

Primer designation	Genomic position	Sequence	Annealing temperature centigrade	Extension temperature centigrade
Hep3	748–736	CTCAAGCTTCATCATCCATATA	47°	66°
Hep4	637–649	CTTGGATCCTATGGGGAGTGG		
Hep7	339–357	CAACCTCTTGTCCTCCAAT	55°	74°
Hep8	536–518	CAGGAATCGTGCAGGTCTT		
Hep20	503–519	AGCACGGGACCATGCCG	47°	67°
Hep30	614–633	TCATCCTGGGCTTTCGGAAA	47°	69°
Hep31	614–633	TCATCCTGGGCTTTCGCAAG	55°	71°
Hep33	131–146	AGGACTGGGGACCCTG	47°	66°
Hep34	986–970	ACTTTCCAATCAATAGG		

that the majority of Scandinavian HBsAg carriers among blood donors have HBsAg of subtype adw in serum.

By utilizing subtype-specific PCR, we here provide positive epidemiological evidence for the association of HBsAg positive cases of PTH-B to a blood donor with high level anti-HBc as marker of HBV replication. In conclusion, HBV DNA may be detected by PCR in HBsAg negative individuals, infectious for HBV by blood donation, and the subtype of the HBV strain in such individuals might also be determined by this technique.

References

1. Larsen J, Hetland G, Skaug K (1990) Post-transfusion hepatitis B transmitted by blood from a surface-antigen negative hepatitis B carrier. Transfusion 30: 431–432
2. Norder H, Hammas B, Magnius LO (1990) Typing of hepatitis B virus genomes by a simplified polymerase chain reaction. J Med Virol 31: 215–221
3. Okamoto H, Imai M, Tsuda F, Sakugawa H, Sastrosoewignjo R, Miyakawa Y, Mayumi M (1988) Typing hepatitis B virus by homology in nucleotide sequence: Comparison of surface antigen subtypes. J Gen Virol 69: 2575–2583

Authors' address: Dr. Helene Norder, Department of Virology, The National Bacteriological Laboratory, Stockholm, Sweden.

Arch Virol (1992) [Suppl] 4: 119–121
© Springer-Verlag 1992

HBc and HBe specificity of monoclonal antibodies against complete and truncated HBc proteins from E. coli

E. Korec[1] and **W. H. Gerlich**[2]*

[1]Institute of Molecular Genetics, Prague, CSFR
[2]Department of Medical Microbiology, University of Göttingen, Federal Republic of Germany

Summary. We have prepared and used monoclonal antibodies against various populations of full-length and truncated hepatitis B core proteins in order to distinguish between epitopes of HBcAg and HBeAg. Our results show that various epitopes are specific for the different proteins. Certain epitopes, however, are ubiquitous to HBc/e proteins and these are probably exposed on the surface of HBc particles.

*

The 27 nm core particle of hepatitis B virus is serologically defined as a core antigen (HBcAg). HBe Antigen is related to HBcAg as it can be generated from HBcAg by denaturation. Human plasma-derived HBeAg is heterogeneous in size and is composed of polypeptides of about 15–20 kd. Carboxyterminal sequence analysis indicates that these HBc/e proteins are truncated at the carboxy terminus. An accurate distinction between epitopes of HBcAg and HBeAg is possible only by the use of monoclonal antibodies.

We have prepared sets of 4 monoclonal antibodies [mabs, Ref. 1] against recombinant full-length HBcAg (rHBc), 8 against carboxyterminal truncated HBc protein (tHBc), missing 33 amino acids [2], and 4 against SDS-denatured recombinant HBc (dHBc) proteins [3]. These antibodies were used for detection of epitopes which are specific for HBcAg or for

---

* Present address: Institute of Medical Virology, University of Giessen, Frankfurter Strasse 107, D-W-6300 Giessen, Federal Republic of Germany.

Table 1. Characterization of mabs which were induced by different HBc/e immunogens

Immunogen	Hybridoma line	Isotype	Reaction in ELISA with			
			rHBc	lHBc	tHBc	HBeAg
rHBc	18H5G1	IgG2A	+	+	+	−
rHBc	18C E11B12	IgG3	+	+	+	−
rHBc	18A E11B7	IgG1	+	+	+	−
rHBc	13:F9F1	IgG2A	+	+	+	−
tHBc	41F4	IgG2A	+	+	+	−
tHBc	41E2	IgG2A	+	+	+	−
tHBc	44H11B6	IgG2B	+	+	+	−
tHBc	46C12E6	IgG2A	+	+	+	−
tHBc	44A5C1	IgG2A	+	+	+	−
tHBc	44A7B9	IgG2B	+	+	+	−
tHBc	42B12	IgG2A	+	+	+	+
tHBc	44B12H10	IgG2B	+	+	+	+
dHBc	35H2	IgG2A	+	+	+	+
dHBc	A2C12	IgG2B	−	−	+	+
dHBc	34E5	IgG2A	−	−	+	+
dHBc	34E10	IgG2A	−	−	+	+

HBeAg using ELISA. The results are summarized in the table. All four mabs induced by rHBc reacted with epitopes specific for liver HBcAg (lHBc, Ref. 4) and tHBc, but these epitopes are not present on plasma-derived HBeAg. Six mabs induced against tHBc particles reacted with epitopes specific for rHBc and lHBc, 2 mabs (numbers 42B12 and 44B12H10) of this set reacted with common epitope(s) present in rHBc, lHBc and HBeAg. Three mabs against dHBc reacted with tHBc and HBeAg, but not with rHBc or lHBc. One monoclonal antibody reacted with all four antigens.

We conclude that a) rHBc and lHBc particles are indistinguishable by the 13 HBc reactive mabs used, b) tHBc has all epitopes which are present on rHBc or lHBc, c) tHBc has in addition epitope(s) which are absent on rHBc or lHBc but present on HBeAg. Generation of these epitopes is due to carboxyterminal truncation and not due to lack of capsid-like structures, because tHBc assembles to core-like particles [2]. d) Some epitopes are present on all kinds of HBc/e protein. They are probably sequential and surface exposed on HBc particles.

Acknowledgements

E. Korec was a recipient of an A. v. Humboldt stipendium.

References

1. Korec E, Korcova J, König J, Hlozanek I (1989) Detection of antibodies against hepatitis B core antigen using the avidin-biotin system. J Virol Method 24: 312–326
2. Melegari M, Bruss V, Gerlich WH (1991) The arginine-rich carboxyterminal domain is necessary for RNA packaging by hepatitis B core protein. In: Hollinger FB, Lemon SM, Margolis H (eds) Viral hepatitis and liver disease. Williams & Wilkins, Baltimore, pp 164–168
3. Korec E, Dostalova V, Korcova J, Mancal P, König J, Borisova G, Cibinogen V, Pumpen P, Gren E, Hlozanek I (1990) Monoclonal antibodies against hepatitis B e antigen: production, characterization, and use for diagnosis. J Virol Meth 28: 165–170
4. Gerlich WH, Goldmann U, Müller R, Stibbe W, Wolff W (1982) Specificity and localization of the hepatitis B virus associated protein kinase. J Virol 42: 761–766

Authors' address: Dr. E. Korec, Institute of Molecular Genetics, Flemingovo 2, CS-16637 Prague, CSFR.

Arch Virol (1992) [Suppl] 4: 122–123
© Springer-Verlag 1992

Measurement of anti-HBc IgM levels using the Amerlite anti-HBc IgM assay

J. A. Diment, J. Tyrrell, and **J. Brown**

Amerlite Diagnostics, Pollards Wood Laboratories, Amersham U.K.

Summary. The Amerlite anti-HBc IgM assay was evaluated as a tool for determination of antibody levels. With an assay time of 1 hour, the test showed a broad dose response range with high sensitivity and specificity. The software configuration of the analyser can be customised to suit specific research requirements.

*

Measurement of anti-HBc IgM levels can be of value, when considered along with other hepatitis markers, in identification of acute hepatitis B as well as differentiating states of chronic hepatitis B infection [1].

We have evaluated the potential of the Amerlite assay for measurement of antibody levels. We demonstrated a broad dose response range without detectable prozone effects. The protocol is such that results can be obtained in 1 hour.

Data obtained using the Paul Ehrlich reference serum and by measuring more than 500 normal sera showed that antibody levels between 10 and 900 U/ml can be determined. In the normal assay configuration the positive cut-off is equivalent to about 250 U/ml but this can be adjusted within the Amerlite Analyser software for specific clinical research purposes to the level required, for example to a level equivalent to 600 U/ml so as to flag only recent acute HBV infection, or to a level equivalent to 10 U/ml so as to flag essentially all detectable anti-HBc IgM levels. The dose response and result printout is such that medium (250–600 U/ml) levels can be readily distinguished from higher levels of antibody.

In addition the Analyser software can be configured in the research mode to directly printout the core IgM levels in units that correspond closely with the Paul Ehrlich reference calibrator. Regression analysis showed that the

PEI U/ml equals (Amerlite result × 176) −13, for an Amerlite result from 0.05 to 5.5.

Our results show that the high sensitivity and high specificity and wide dynamic range of the Amerlite assay make it useful for a range of applications.

References

1. Gerlich WH, Uy A, Lambrecht F, Thomssen R (1986) Cutoff levels of immunoglobulin M antibody against viral core antigen for differentiation of acute chronic and past hepatitis B viral infections. J Clin Micro 24: 288–293

Authors' address: Dr. J. A. Diment, Amerlite Diagnostics, Pollards Wood Laboratories, Amersham HP7 OHJ, U.K.

Arch Virol (1992) [Suppl] 4: 124–125
© Springer-Verlag 1992

Evaluation of a new enhanced luminescence immunoassay for confirming the presence of HBsAg in human serum or plasma

G. H. Spiller, A. Stalham, J. Holian, and **M. Jones**

Amersham International plc, Pollards Wood Laboratories, Chalfont St Giles, U.K.

Summary. An enhanced luminescence assay was developed to confirm the presence of HBsAg in weakly reactive sera. The test proved reliable in differentiating between truly reactive and falsely reactive samples.

*

With any highly sensitive assay for HBsAg a small number of HBsAg negative samples consistently give reactive results. Even where high levels of specificity are achieved, these results can have a significant effect on positive predictive value. For instance, in external trials of the Amerlite HBsAg assay, two of 3268 blood donor samples gave repeatably reactive results but neither was shown to be genuinely HBsAg positive. Such results are typical of HBsAg screening assays and illustrate the continuing requirement for reliable confirmatory tests. We have developed an assay based on enhanced luminescence to confirm the presence of HBsAg in human serum or plasma. The Amerlite HBsAg Confirmatory Assay uses the principle of specific antibody neutralisation. Test samples are pre-incubated with human antiHBs for one hour at 37 °C, prior to measurement in a one hour, single-incubation assay. A sample is confirmed positive if neutralisation gives a reduction in result of at least 50%, and the non-neutralised sample gives a result greater than the cut off. In a further external trial of the Amerlite HBsAg assay 500 blood donor samples were tested. One sample (0.2%) gave a repeatably weakly reactive result which could not be confirmed using the Amerlite Confirmatory assay. Of 132 samples known to be HBsAg positive (and reactive in the Amerlite HBsAg assay), all were confirmed positive by the Confirmatory assay. At Amersham the presence of HBsAg was confirmed in each of the 11 members of the Paris subtype panel. Fifteen samples were artificially created by deliberate gross haemolysis, addition of haemoglobin or sheep anti-mouse to give repeatably reactive results in the Amerlite HBsAg assay. None of these samples could be confirmed positive using the Confirmatory assay. These results support the use of the Amerlite HBsAg Confirmatory assay in differentiating between truly HBsAg positive specimens and those giving falsely reactive results.

Authors' address: Dr. G. H. Spiller, Amerlite Diagnostics, Pollards Woods Laboratories, Amersham HP7 OHJ, U.K.

V Hepatitis B surface proteins and vaccination

Arch Virol (1992) [Suppl] 4: 129–132
© Springer-Verlag 1992

Functions of hepatitis B surface proteins

W. H. Gerlich, K.-H. Heermann, and **Lu Xuanyong**

Department of Medical Microbiology, University of Göttingen, Federal Republic of
Germany

Summary. HBV surface proteins play a number of functional roles in cellular
infection, viral synthesis and in immune responses of the host. Three
coterminal proteins of differing sizes and three subdomains of the individual
molecules can be recognized. In this brief review, functions of the various
proteins and domains are described and their significance as potential
immunogens is discussed. Although it is apparent that the surface proteins
are involved in the development of persistent HBV infections, the underlying
mechanisms of their involvement remain unknown.

*

Hepatitis B surface (HBs) proteins are involved in i) the assembly and
secretion of hepatitis B virus (HBV), ii) in induction of immune tolerance in
the transiently or persistently infected host, iii) in attachment and penetration to liver cells of a new host, iv) and finally in development of protective
immunity by naturally infected or vaccinated persons.

There are three coterminal HBs proteins of different size: small (SHBs),
middle (MHBs), and large (LHBs). They represent three domains of the viral
surface: PreS1 which is present only in LHBs; preS2 which is present in LHBs
and MHBs, but in MHBs is glycosylated and in LHBs is not; and domain S
which is present in all three HBs proteins and forms the sole component of
SHBs. All three HBs proteins are present in the envelope of HBV particles,
and in the subviral HBs filaments and HBs spheres of 20 nm diameter.
However, the HBs spheres contain less LHBs than the other two particles.

SHBs with 226 amino acids (aa) is the major component of HBs particles
and of the HBV envelope. SHBs is required and sufficient for the assembly
and secretion of HBs spheres in animal cells. During biosynthesis, SHBs, and

very likely the S domains of MHBs and LHBs, cross the membrane of the ER at least two times but probably more often. Hydrophobic interactions and possibly intermolecular cross-linking of the numerous cysteins in the S domain lead to the budding of HBs particles to the ER lumen, from where they are secreted. Part of the naturally expressed S domain is glycosylated. HBs proteins which are expressed in transfected yeast cells form particles, but the S domains are not glycosylated and the particles are not secreted. For vaccine production they have to be extracted from the yeast cells.

The ability of SHBs to induce protective immunity is experimentally well established and confirmed by world-wide vaccination experience. The mechanism of how an anti-SHBs antibody protects is, however, not understood, since no function of SHBs in attachment or penetration has yet been found. Usually, antibodies neutralize viral infectivity by blocking attachment and penetration sites on the viral surface.

These functions are probably located in the preS domains of HBV. The preS1 domain of 119 aa binds by its sequence 21 to 47 (subtype adw) with low affinity to hepatocyte membranes and to monocytes. Binding may be indirect via the serum protein, apolipoprotein H. PreS1 peptide 21–47 has been found to induce neutralizing antibodies. Convalescents from hepatitis B develop such antibodies. Chronic carriers of HBV, however, may develop antibodies against preS1 sequence (94–117) but not against (21–47). These observations underline the importance of sequence 21–47 in the natural infection process. Possibly, this sequence binds to a novel IgA receptor of human liver, because this sequence has a partial homology to the Fc part of the alpha-chain, and binding of HBV to liver membranes is inhibited by IgA.

LHBs is required a) for the envelopment of core particles and subsequent maturation of HB virions, and b) for the morphogenesis of HBs filaments. The aminoend of preS1 carries an ER retention signal. Thus, pure LHBs can still form HBs filaments but these cannot be secreted. In transgenic LHBs mice this may even lead to storage disease and finally to hepatocellular carcinoma. Secretion of HBs filaments and of HB virions occurs only if SHBs and LHBs are coexpressed in ratios between 2:1 and 10:1.

The small preS2 domain of 55 aa carries a surprising number of functions. The glycan of preS2 has a novel liver-specific composition. It contains a mannose antenna and two complex antennas with some unusual linkages. This glycan binds specifically to a still unidentified lectin of the human hepatoma cell line HepG2. Since the natural preS2 glycan from human liver has a terminal sialic acid, this lectin is different from the mannose, galactose or asialoglycoprotein receptors of the liver. Binding of the glycan to the lectin can be inhibited by preS2-glycan specific antibodies. This lectin is the most convincing candidate for an organ-specific HBV receptor in human liver.

PreS2 binds, moreover, human serum albumin (HSA). Although HSA treated with glutaraldehyde binds much better than native HSA, a small subfraction of native HSA binds also in vivo and in vitro to the central part of the preS2 domain. The binding site overlaps with neutralizing epitopes. Possibly, a naturally modified HSA has both affinities to preS2 and hepatocyte membranes and would be an indirect receptor. It is interesting to note that patients with more severe liver disease or high HBV replication have usually free HSA binding sites, while asymptomatic HBsAg carriers have less preS2 domains which are, moreover, completely covered by HSA.

Penetration of HBV into cells has not yet been demonstrated in vitro, and infection of hepatocytes in vitro is, although possible, very ineffective. Penetration of viruses to the cytoplasm of cells is often mediated by a hydrophobic fusion peptide of 10 aa which is exposed only after proteolysis or after a pH-induced conformational change. Treatment of MHBs containing HBs spheres with proteases, such as trypsin or V8 protease, cleaves rapidly MHBs at the carboxyend of preS2 and induces nonspecific irreversible binding of HBs spheres to various cell types. Chymotrypsin cuts in MHBs, in contrast, at the aminoend of the S domain and removes a sequence which could be a fusion peptide. In fact, chymotrypsin treated HBs spheres do not bind to cells at all. The proteolysis-sensitive site of preS2 forms at the same time a very accessible, highly immunogenic region with great sequence variability. One could speculate that antibodies against this region could also neutralize infectivity.

HBs spheres are a good immunogen in the majority of vaccine recipients, but the amount of 10 to 20 μg SHBs protein per dose is much higher than the one required for e.g. a killed hepatitis A virus vaccine (0.1–0.3 μg/dose). The defect leading to non-responsiveness in certain healthy and in many immunodeficient recipients is insufficient presentation by monocytes and B lymphocytes, and/or absent recognition of HBs epitopes by T helper cells. SHBs is not easily processed by proteolysis, and this may lead to insufficient presentation by antigen presenting cells. Both, in mice and in man, preS1 has been found to be more immunogenic for T helper cells than preS2 or S. Since B-cells depend on stimulation by T helper cells, the whole humoral immune response including antibodies against S and preS2 is probably enhanced by the presence of preS1 T cell epitopes in HBs particles. According to the model of Milich, preS1 or other T helper epitopes have to be on the same particle as the B cell epitopes. As a consequence future hepatitis B vaccines should contain the preS1 and preS2 domains on SHBs spheres. Plasma-derived HBs spheres—isolated without proteolytic treatment—have been a source of such a vaccine, but for biological safety and production costs, recombinant vaccines are preferable.

Current recombinant SHBs vaccines rely on very few epitopes. Recently, development of escape mutants has been observed in vivo which contain one single aa exchange in the group-specific a-epitope of SHBs. Inclusion of several HBs subtypes, and of neutralizing preS1 and preS2 epitopes would probably prevent such a rapid development of escape mutants. Moreover, more attention should be paid to the protective effect of T-cell immunity.

While HBs spheres (even without preS domains) are usually satisfactory immunogens if applied as a vaccine, they are a miserable immunogen during natural infection. Even in resolving acute hepatitis B HBs-antigenemia reaches levels of 100 µg/ml or more for several weeks without an efficient antibody response. Furthermore, partial immune tolerance against HBs protein is the basis for persistent HBV infection. One potential factor in the induction of immune tolerance may be the adsorption of the host's serum proteins, like HSA, apolipoprotein H and others. Besides simple masking of neutralizing epitopes, the host protein may induce T suppressor cells after presentation by HBs specific B cells and, thus, inhibit anti-HBs production. Furthermore, HBV can infect the cells of the immune system and may disturb their biological function. At present, the role of HBs protein in establishment of acute and persistent HBV infection can only be described as a phenomenon, but its mechanisms are not understood.

Authors' present address: Dr. W. H. Gerlich, Institute of Medical Virology, University of Giessen, Frankfurter Strasse 107 D-W-6300 Giessen, Federal Republic of Germany.

Arch Virol (1992) [Suppl] 4: 133–136
© Springer-Verlag 1992

Deletion and insertion mutants of HBsAg particles

U. Machein, R. Nagel, R. Prange, A. Clemen, and **R. E. Streeck**

Institute for Medical Microbiology, University of Mainz, Federal Republic of Germany

Summary. We have found previously that hybrid 22-nm HBsAg particles can be created by insertion of short antigenic sequences into the HBV major envelope protein [1]. We have now performed a detailed deletion mutagenesis of the S gene of HBV encoding HBsAg. Deletion of the 51 C-terminal amino acids including most of the third and all of the fourth hydrophobic domain of the S protein did not affect particle assembly and secretion. However, secretion of 22-nm particles was abolished by minor deletions in the N-terminal region. Insertion and deletion/substitution mutants carrying a poliovirus epitope at the N-terminus and the preS1 region at the C-terminus have been characterized.

Introduction

During an HBV infection, large amounts of 22-nm empty envelope particles carrying the surface antigen (HBsAg) are secreted from the infected hepatocytes. Transfection of cell lines with the S gene encoding the 226-amino acid sequence of the major envelope protein yields similar particles which are secreted into the cell culture medium. Such 22-nm particles from serum, cell lines, or yeast are currently used as vaccine against hepatitis B.

We have previously shown by insertion of a poliovirus epitope into the major envelope protein that 22-nm particles can be used as efficient immunogenic carriers of small heterologous antigenic sequences [1]. However, assembly and secretion of such particles were strongly affected by longer inserts [2]. We have now constructed deletion mutants of the S gene to identify regions dispensable for envelope assembly and to create space for the insertion of longer sequences.

Results and discussion

In the small envelope protein of HBV two hydrophilic domains (30–80, 100–155) are separated by a strongly hydrophobic region which is essential for the transmembrane orientation [3]. The second hydrophilic domain carries the surface antigen. Leaving the essential hydrophobic and the antigenic region unchanged, we have deleted progressively larger sequences from the carboxyterminus (176–226) and shorter sequences of five to thirty amino acids at variable places near the aminoterminus (3–80). To determine whether the mutant proteins could still be assembled into particles and secreted, HBsAg was assayed in cellular lysates and in cellular supernatants using a monoclonal ELISA (Abbott) specific for HBsAg particles.

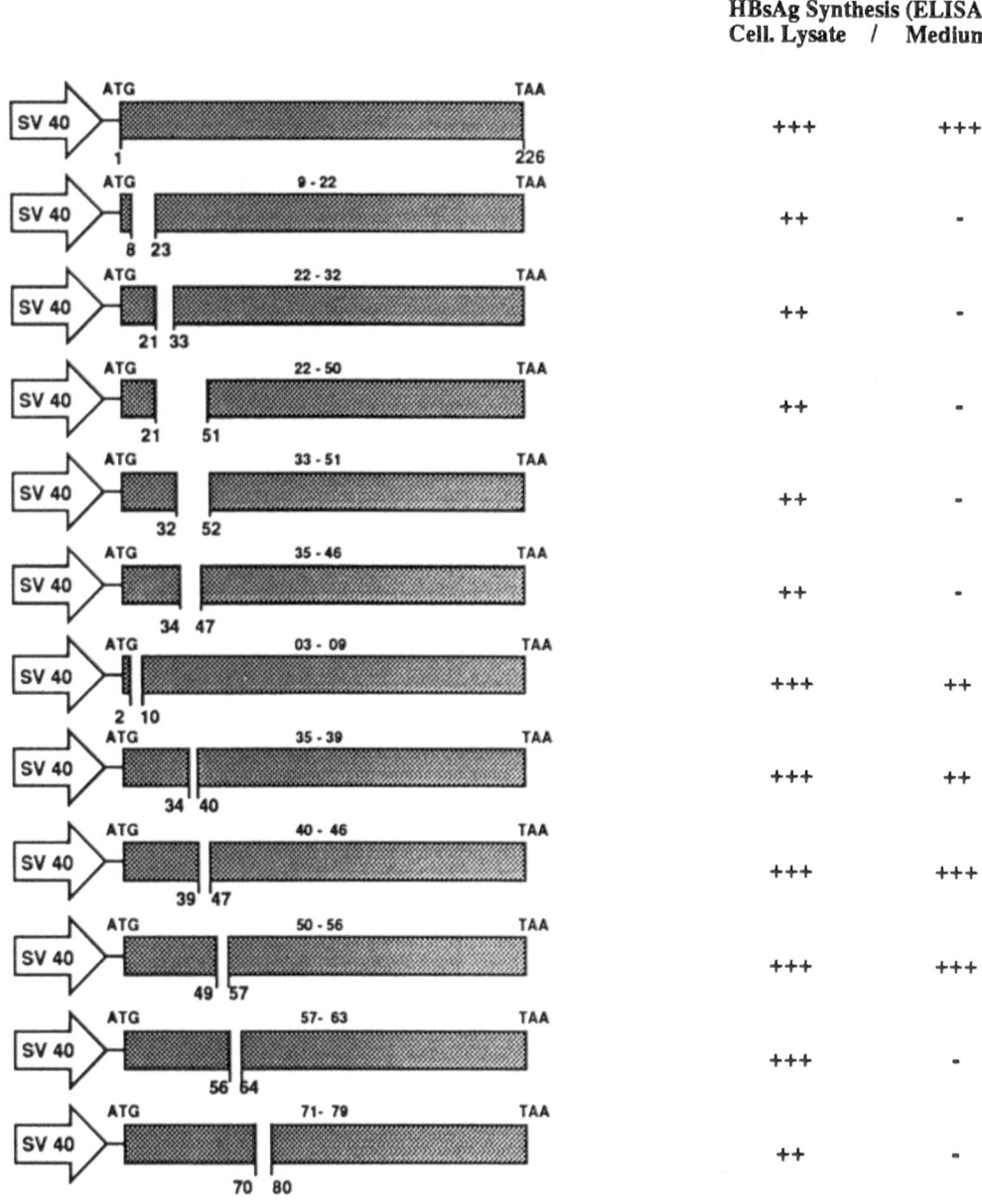

Fig. 1. Aminoterminal deletion mutants and their level of expression in HepG2 cells

Deletions were introduced into the S gene using either exonuclease treatment starting from unique restriction sites or oligonucleotide-directed mutagenesis. All mutants were verified by DNA sequence analysis and were assayed using a eucaryotic expression vector after transient expression in HepG2 cells (Fig. 1).

Deletions of five to eight amino acids at the very N-terminus (Δ3–9) and in the proximal part of the first hydrophilic domain (Δ35–39, Δ40–46, Δ50–56) had no effect on intracellular assembly and only a minor effect on secretion (Fig. 1). However, when deletions of the same size were introduced further downstream (Δ57–63, Δ71–79), the level of secretion was strongly reduced. These two regions are now being analysed by single point mutations to identify the critical amino acid residues essential for secretion.

When larger deletions were introduced into the first hydrophilic domain (Δ22–32, Δ33–51, Δ35–46) or into the N-terminal hydrophobic region (Δ9–22), HBsAg could not be detected in the cellular supernatants, but intracellular assembly was found. It will be interesting to learn whether intracellular HBsAg in the secretion-defective mutants corresponds to true 22-nm particles or to transmembrane aggregates.

Carboxyterminal deletions did not interfere with assembly and secretion of HBsAg particles. Even a large deletion of 51 amino acids covering nearly the entire C-terminal hydrophobic domain was compatible with the formation of 22-nm particles. This indicates the absence of essential topogenic sequences in this part of the molecule and suggests that large substitutions and insertions can be tolerated in this region.

Because of the increased immunogenicity of a hepatitis B vaccine which includes the preS1 region of the large surface protein, we substituted the deleted C-terminal 51 amino acids by either the entire preS1 sequence or its first 42 amino acids. Hybrid particles were detected intracellularly in HepG2 cells, but secretion was severely affected (Fig. 2). This demonstrates that a simple deletion/substitution is not sufficient for a high level expression of a heterologous antigen on HBsAg particles.

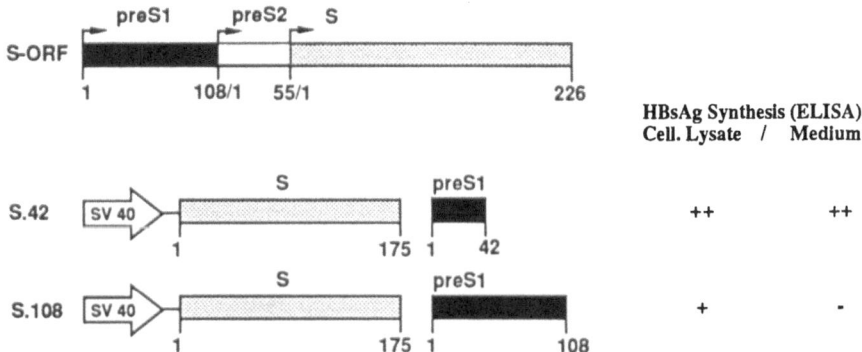

Fig. 2. Substitution of carboxyterminal wild type sequences with the entire preS1 region or its first 42 amino acids

We then considered the possibility that the junction with HBsAg at the C-terminal amino acid 175 of the deletion mutant could be unfavourable for the presentation of preS1 on the outside of 22-nm particles. Therefore all or part of the preS1 sequence was inserted next to amino acid 223. With these constructs, a much better secretion was observed. Hybrid HBsAg carrying the preS1 sequence 1–42 at the C-terminus were secreted with the same efficiency as unmodified HBsAg particles. The hybrid particles were fully reactive with monoclonal antibodies both to HBsAg and preS1.

Conclusion

Nearly all the deletion mutants were secreted with lower efficiency. The largest space for insertions can be created in the carboxyterminus. The substitution of wild-type with foreign sequences and the insertion of longer epitopes decreased the production of particles.

The insertion of parts of the preS1 domain into the carboxyterminus of the S gene is a novel possibility of producing hybrid middle or large surface proteins, which are secreted as particles from mammalian cells. Such particles could have a great potential for future HBV vaccines.

References

1. Delpeyroux F, Chenciner N, Lim A, Malpièce Y, Blondel B, Crainic R, van der Werf S, Streeck RE (1986) A poliovirus neutralization epitope expressed on hybrid hepatitis B surface antigen particles. Science 233: 472–475
2. Delpeyroux F, Chenciner N, Lambert M, Malpièce Y, Streeck RE (1987) Insertions in the hepatitis B surface antigen: Effect on assembly and secretion of 22-nm particles from mammalian cells. Mol Biol 195: 343–350
3. Eble BE, MacRae DR, Lingappa VR, Ganem D (1987) Multiple topogenic sequences determine the transmembrane orientation of hepatitis B surface antigen. Mol Cell Biol 7: 3591–3601

Authors' address: Dr. U. Machein, Institute for Medical Microbiology, University of Mainz, D-W-6500-Mainz, Federal Republic of Germany.

Arch Virol (1992) [Suppl] 4: 137–141
© Springer-Verlag 1992

Properties of a recombinant yeast-derived hepatitis B surface antigen containing S, preS2 and preS1 antigenic domains

J. Petre, T. Rutgers, and **P. Hauser**

SmithKline Beecham Biologicals, Rixensart, Belgium

Summary. Yeast cells have been engineered to express mixed HBsAg particles containing the S and a modified large (L*) protein. Their construction resulted in reduced protease sensitivity, reduced glycosylation and complete inactivation of the polymerized human albumin binding site. The particles exposed the S, preS1 and preS2 antigenic determinants and induced an immune response against the three domains. Highly purified preparations have been obtained and are presently being tested in human volunteers.

*

The envelope of hepatitis B virus (HBV) contains differently glycosylated forms of three proteins species: the small or major (S) protein, the middle (M) protein and the large (L) protein [7]. These forms incorporate the S, preS2 +S and preS1 +preS2 +S domains respectively. Although hepatitis B vaccines derived from human plasma or, more recently, from engineered yeast [2, 5] have shown that the S domain alone is sufficient to induce a highly protective immunity, experiments in animals have highlighted the potential benefits which might result from the inclusion of the preS domain in vaccines, namely the augmentation of the efficacy by broadening the protective immune response and the circumvention of non-responsiveness to the S determinant [10, 11]. Moreover, the preS domains are immunogenic in humans and elicit anti-preS responses during natural HBV infection, which often occur prior to any other anti-HBV response [1].

Therefore, we have first attempted to produce the middle and large protein particles in yeast, using the type of expression vector which was applied earlier to the expression of the major protein [6]. The large protein failed to assemble into typical HBsAg particles and the recovery was low.

The middle protein was expressed at satisfactory levels and formed spherical particles. These (M) particles were purified and, in mouse experiments, were shown to induce high levels of anti-S and anti-preS2 antibodies. In human tests, however, these particles gave disappointing results: they induced a largely depressed S response compared to yeast-derived HBsAg containing S only, while the preS2 response appeared delayed and transient.

We tried to determine the possible reasons for this failure, and identified the following factors which may affect the immunogenicity of the (M) particles in humans:

— the preS2 domain of the M protein is highly susceptible to proteolytic cleavage, which results in an antigen of poor stability;
— the preS2 domain of the M protein expressed in yeast is hyper-glycosylated: in addition to the expected mannose-rich N-glycosylation, it carries a number of abnormal O-glycosylations;
— the preS2 domain carries a human albumin binding site, which has been suspected of contributing to immune escape mechanisms [8];
— the (M) particles present the highly hydrophylic preS2 domain mostly at their surface, while the S determinants are largely masked. This may result in altered recognition and processing by the immune system.

Although it is not clear how all these mechanisms could actually contribute to a different behaviour of the antigen in mice and humans, we tried to incorporate these ideas into an idealized design of HBsAg particles containing the preS domain. Given the lack of self-assembly of the large protein in yeast, this design was also conceived as one way to incorporate the important regions of the preS1 domain shown to contain protective epitopes [12] into molecules which could be efficiently assembled in yeast.

Two concepts were used in this design: the modified large molecule, which we designate L*, and the mixed or composite particles, assembled from different molecular species (S and L*). The modified large molecule L* retains the potentially useful regions of the preS1 and preS2 domains, namely regions containing B and T cell epitopes (including those which were shown to be neutralizing or protective in chimpanzee challenge tests) and the preS1 associated hepatocyte binding site (Fig. 1). Most of the sensitive proteolytic sites and of the glycosylation sites have been deleted. In addition, deletion of the preS2 peptide 120–132 abolished the binding of the polymerized human albumin and removed a cytotoxic T-cell epitope with possible involvement in chronic evolution of HBV infection [3]. To form mixed particles, two different species, i.e. L* and S, were co-expressed in yeast. We have in this way succeeded to form particles containing about 25% of L* and 75% of S, which simultaneously expose preS1, preS2 and S determinants, more like HBV does.

The mixed particles containing S and L*, which we designate (S,L*), were purified from recombinant yeast using standard methodology. Electron

DESIGN OF A MODIFIED L PROTEIN

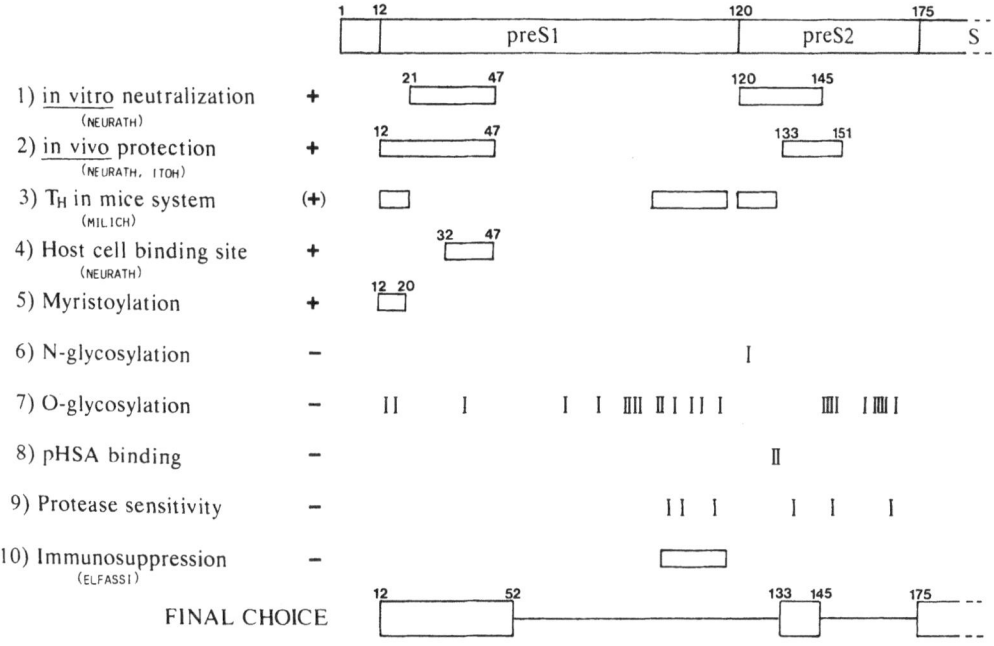

Fig. 1. Design of a modified L protein. The truncated L protein (L*) expressed in yeast, contains sequences 12–52 (*preS1*), 133–145 (*preS2*) and the whole S domain of the original L protein

Table 1. Antigenic characterization of (S,L*) particles

Reagent	Specificity	Properties	Binding to (S,L*) (ELISA)
HBsRF1 (H. Thomas)	S (common a determ.)	Virus neutralizing Conformational epitope	yes
S1.1 (SKB Bio)	PreS1 (12–32)	Binding to peptide	yes
F35.25 (M-A. Petit)	PreS1 (32–47)	Inhibition of virus binding	yes
S2.5 (SKB Bio)	PreS2 (133–145)	Binding to peptide	yes
pHSA	PreS2	Binding to pHSA receptor	no

microscopy showed spherical particles of about 25 nm, resembling recombinant (S) particles or natural plasma derived HBsAg particles. By electrophoresis, two major bands were seen in a roughly 1:4 ratio, corresponding to the L* (31 kDa) and S (24 kDa) species, respectively. In Western blot, the two species were stained by the S specific reagent HBS-1, while only the 31 kDa species was revealed with the preS1 specific monoclonal antibodies S1.1 and F35.25.

The particles were further characterized and compared to (M) and (S) particles by a series of ELISA assays with S, preS1 and preS2 specific monoclonal antibodies. (S,L*) particles were highly reactive with the monoclonal antibody HBs-RF1 shown to be virus-neutralizing, S-specific and conformation dependent [14, 9], indicating that the corresponding antigenic domain is correctly folded and exposed, as it is in (S) particles. The preS2 specific reagent S2-5 showed the strongest reaction with (M) particles, a reduced but significant reaction with (S,L*), and no reaction at all with (S) particles, in line with the respective preS2 content of these particles. The preS1 specific reagents S1.1 and F35.25, the latter recognizing the hepatocyte binding site [13], were both strongly bound by (S,L*) particles only. Finally, a strong binding to polymerized human albumin coated plates was observed with (M) particles, while (S) or (S,L*) particles were not bound at all.

Immunization of mice and monkeys with (S,L*) produced a strong S-specific immune response, similar to that induced by (S) particles. Compared to (M) particles, a weaker but significant response was detected against the preS2 peptide 120–145. This result may be a reflexion of the lower preS2 content of (S,L*) compared to (M) particles, and of the truncation of the preS2 domain, which is reduced to only 13 amino acids in (S,L*). A strong response was obtained against the preS1 peptides 12–32 and 32–47. Finally, sera of monkeys immunized with (S,L*) were able to inhibit the binding of F-35.25 to (S,L*) particles in a competition assay. This inhibition was maintained in the presence of excess (S) particles, indicating that it is due to competing preS antibody species with the same specificity as F35.25 rather than to hindrance of the binding by S-specific antibodies also present in the serum.

It was shown recently that the preS1 domain is highly immunogenic at the T cell level in humans [4] and several preS1-specific CD4+ cell clones were obtained from high responders immunized with a plasma-derived vaccine. (S,L*) particles strongly stimulated these clones in vitro (C. Ferrari, personal communication), indicating that the corresponding epitopes are correctly processed and recognized on (S,L*) and providing a firm basis for a contribution of the preS1 domain present in (S,L*) to an enhanced T cell help in humans.

In conclusion, we have shown that engineered yeast can produce HBsAg particles which express simultaneously the S, preS2 and preS1 antigenic domains. These particles have been obtained in highly purified form and are remarkably stable. They are an ideal candidate to further explore the eventual benefits of hepatitis B immunization with preparations containing preS antigenic determinants in populations which respond poorly to vaccines containing only the S domain of HBsAg.

References

1. Alberti A, Pontisso P (1987) Antibody to preS antigens in hepatitis B virus infection. Hepatology 7: 207–208
2. André FE, Safary A (1988) Clinical experience with a yeast-derived hepatitis B vaccine. In: Zuckerman AJ (ed) Viral hepatitis and liver disease. Liss, New York, pp 1025–1030
3. Barnaba V, Franco A, Alberti A, Benvenuto R, Balsano F (1990) Selective killing of hepatitis B envelope antigen-specific B cells by class I-restricted, exogenous antigen-specific T lymphocytes. Nature 345: 258–260
4. Ferrari C, Penna A, Bertoletti A, Cavalli A, Valli A, Schianchi C, Fiaccadori F (1989) The preS1 antigen of hepatitis B virus is highly immunogenic at the T cell level in man. J Clin Invest 84: 1314–1319
5. Gerety RJ (1988) Recombinant hepatitis B vaccines. In: Zuckerman AJ (ed) Viral hepatitis and liver disease. Liss, New York, pp 1017–1024
6. Harford N, Cabezon T, Crabeel M, Simoen E, Rutgers A, De Wilde M (1983) Expression of hepatitis B surface antigen in yeast. Dev Biol Stand 54: 125–130
7. Heermann KH, Goldmann V, Schwartz W, Seyffarth T, Baumgarten H, Gerlich WH (1984) Large surface proteins of hepatitis B virus containing the preS sequence. J Virol 52: 396–402
8. Hellström U, Sylvan S (1986) Human serum albumin and the enigma of chronic hepatitis type B. Scand J Immunol 23: 523–527
9. Iwarson S, Tabor E, Thomas HC, Goodall A, Waters J, Snoy P, Shih JWK, Gerety RJ (1985) Neutralization of hepatitis B virus infectivity by a murine monoclonal antibody. An experimental study in chimpanzee. J Med Virol 16: 89–96
10. Milich DR, Mc Lachlan A, Chisari FV, Kent BH, Thornton GB (1986) Immune response to the pre-S(1) region of the hepatitis B surface antigen (HBsAg): a pre-S(1) specific T cell response can bypass non responsiveness to the pre-S(2) and S regions of HBsAg. J Immunol 137: 315–322
11. Milich DR, Mc Namara MK, Mc Lachlan A, Thornton GB, Chisari FV (1985) Distinct H-2 linked regulation of T-cell responses to pre-S and S regions of the same hepatitis B surface antigen polypeptide allows circumvention of non responsiveness to the S region. Proc Natl Acad Sci USA 82: 8168–8172
12. Neurath AR, Seto B, Strick N (1989) Antibodies to synthetic peptides from the preS1 region of the hepatitis B virus (HBV) envelope (env) protein are virus-neutralizing and protective. Vaccine 7: 234–236
13. Petit M-A, Dubanchet S, Capel F (1989) A monoclonal antibody specific for the hepatocyte receptor binding site on hepatitis B virus. Mol Immunol 26: 531–537
14. Waters J, Pignatelli M, Galpin S, Ishihara K, Thomas HC (1986) Virus neutralizing antibodies to hepatitis B: the nature of an immunogenic epitope on the S gene peptide. J Gen Virol 67: 2467–2473

Authors' address: Dr. J. Petre, SmithKline Beecham Biologicals, Rue de l'Institut 89, B-1330 Rixensart, Belgium.

Arch Virol (1992) [Suppl] 4: 142–146
© Springer-Verlag 1992

Diverging policies for vaccination against hepatitis B

S. Iwarson

Department of Infectious Diseases, University of Göteborg, Östra Hospital, Göteborg, Sweden

Summary. Policies toward vaccination against hepatitis B vary globally according to local prevalence and the population of infected individuals. In the present report, vaccination plans, policies, risks, and experiences of both apparent successes and failures are described. Possible plans, including local vaccine production, are discussed in regard to problems of third-world countries with high HB prevalence. Recommendations are made for vaccination policies in various circumstances.

*

The use of hepatitis B vaccine usually has taken into consideration the geographic patterns of the prevalence of hepatitis B virus (HBV) infection. In areas of intermediate and high endemicity, infection is widespread and occurs more predominantly in infants and children. In these areas, immunization with hepatitis B vaccine has been recommended for all infants. In areas of low endemicity, however, the strategy for prevention of hepatitis B has been to immunize the groups (mostly adults) at high risk of infection.

Recommendation of HB vaccination in Europe and the US

The policies for recommendation of Hepatitis B vaccination differ from north to south in Europe due to different risks of HBV-exposure. In Scandinavia, vaccination is mainly recommended for health care workers with frequent blood contact while in Germany and in France vaccination is recommended for all health care workers with patient contact. These recommendations mirror the declining incidence of hepatitis B in Scandinavia.

Further south, as in Italy, all health care workers are considered a risk group and vaccination is recommended for all newly recruited workers and students. Recently it was suggested that Italy should follow the WHO-recommendation to immunize all newborns against hepatitis B since the overall prevalence of HBsAg-carriers range 2–4% in the country [1].

In 1989 the Greek Ministry of Health decided to implement a new vaccination programme. A pilot study showed that an increase in the acceptance rate of vaccination among health care personnel had taken place and now reached 92%. The Greek ministry intends to vaccinate medical and dental students, as well as student nurses, as soon as they have registered at the university. Vaccination will be implemented without screening, since HBV markers were detected in less than 8% of such students in a pilot study. In Greece as in most other western-European countries, hepatitis vaccination in health care workers is free of charge (Papaevangelou, G, personal communication).

Hepatitis B vaccine has been available in the United States since 1982. In spite of this the incidence of hepatitis B during the past 7–8 years has not decreased in the US.

The failure of the hepatitis B immunization program can be attributed in the majority of hepatitis B infections occurring in high-risk groups including intravenous drug abusers, promiscuous homosexual men and heterosexual persons with multiple partners.

Today, leading experts in the US seem to be convinced that to achieve the ultimate control of hepatitis B it will be necessary to incorporate hepatitis B vaccine in routine childhood immunization programs. This could be included at the same time as diphtheria, tetanus, pertussis and polio vaccines. In this way it should be possible to protect future members of high-risk "hard-to-reach" population groups in the US [2].

Postvaccination serologic testing and booster doses

Most European countries seem to agree with the WHO statement [3] that quantitative evaluation of the immune response after vaccination is desirable in order to identify inadequate responders and non-responders. Anti-HBs should ideally be measured one to three months after completion of the basic course of immunizations. To allow comparison of studies, anti-HBs should preferably be expressed as international units (IU/l).

According to the WHO-group, an individual with a peak of anti-HBs below 10 IU/l after the basic course of vaccination probably lacks protection against HBV infection. Individuals with peak anti-HBs levels of 10–100 IU/l (low responders) generally lack detectable anti-HBs within one year. Most western European centers seem to recommend booster doses when anti-HBs levels tend to fall below 10 IU/l or at least once 5–10 years after the initial course of vaccination. In several countries the initial level of anti-HBs following the basic immunizations is taken as a basis for the timing of booster doses. If 100 IU/l is not reached after three basic injections, a booster is often given within one year, while in those with a good response (> 100 IU/l) a booster dose after 5–10 years should be sufficient as suggested by Jilg et al. [3]. More sophisticated schedules with narrower intervals between boosters are being used in some central European countries.

The length of protection following vaccination remains, however, unknown. Partial immunity apparently persists also after anti-HBs has declined to undetectable levels. HBV infections have been reported in several adults who have responded to HB-vaccine, generally after antibody levels had become very low or undetectable, but very few of these infections have been clinically relevant.

Until more data are available, it seems logical to consider revaccination of individuals when anti-HBs levels tend to fall to or below 10 IU/l at least once in their lifetime. Immunocompromised individuals should, however, be revaccinated more frequently according to their anti-HBs responses.

In the US, the Public Health Service states (CDC 1990) that for adults and children with normal immune status, booster doses are not routinely recommended after vaccination, nor is routine serologic testing to assess antibody levels considered necessary for vaccine recipients. The possible need for booster doses after longer intervals will be assessed by the CDC as additional information becomes available. Obviously there is a discrepancy between official US recommendations and European policies concerning booster doses.

Post-exposure vaccination against hepatitis B

Vaccines are usually effective only if administered before exposure to an infectious agent. However, in infections with long incubation periods such as rabies, post-exposure vaccination may be at least partly effective. For post-exposure prophylaxis against hepatitis B (HB) vaccines are often used in combination with hepatitis B immune globulin (HBIG). Since hepatitis B usually has an incubation period of two or more months, rapid post-exposure vaccination without globulin should be effective, as well.

Hepatitis B vaccine is generally given at intervals of one month between the first two or three injections, followed by a booster injection 6–12 months after the first dose. Peak seroconversion rates and anti-HBs antibody titres occur after the booster dose, which is too late for post-exposure vaccination. Protective antibody titres must be achieved much earlier in these instances.

Wahl et al. [4] studied an accelerated vaccination schedule in medical students who received 10 μg of recombinant HB vaccine at 0, 2 and 6 weeks. Other students were given the same vaccine at 0, 1 and 6 months according to one recommended schedule for pre-exposure prophylaxis. Accelerated vaccination resulted in a significantly higher frequency of protective antibody titres (10 IU/l or above) one month after the first dose of vaccine (48% vs 4 %). All short-interval vaccinees had seroconverted within 2 months (i.e. 2 weeks after the third dose).

In a recent study of short-interval HB vaccination in chimpanzees exposed to HB virus, 4 chimpanzees received 10 µg of plasma-derived HB vaccine 4, 8, 48 and 72 hours, respectively, after intravenous injection of a HB virus inoculum [5]. A second and third vaccine dose was given to each animal at 2 and 6 weeks, and the chimpanzees were followed weekly for a year. Serial blood tests did not reveal any HBs-antigen, liver enzymes (ALT) remained normal and there were no histopathological changes in liver biopsy specimens.

Late appearance of serum anti-HBc antibodies was observed in one animal in which vaccination began 72 h after HB virus exposure. An unvaccinated control chimpanzee, which received the HB virus inoculum only, had symptoms of hepatitis B with raised alanine aminotransferase and HBs antigen in the serum.

These experimental data suggest that post-exposure vaccination with short intervals between vaccine injections can protect against hepatitis B infection, even when HBIG is not given. All chimpanzees in which post-exposure vaccination was started within 48 hours after HBV exposure were completely protected, while the chimpanzees which got the first vaccine injection after 72 hours had subclinical infection.

HB-vaccination in developing countries

Very few African countries have an HB-immunization programme in spite of very high prevalences of HBsAg carriers among pregnant women. The same accounts for certain Asian countries with high HBsAg-prevalences.

World-wide strategies for hepatitis B prevention in children will differ from area to area according to the epidemiology of HBV-infection. In East Asia high prevalences of HBsAg and HBeAg are seen in the serum of pregnant women and HBV-exposure of the child occurs at delivery.

The first hepatitis B vaccine injection may in these instances be given together with BCG-vaccine at birth. The second HB-vaccine dose being given 4–6 weeks later and the third at 10–11 weeks in accordance with suggestions by WHO. Ideally also these injections are combined with other vaccine injections like DTP.

In African countries also, spread of HBV seems to take place during the first few years of life and vertical transmission is less frequent than in East Asia. For that reason a somewhat slower HB-vaccination schedule may have the advantage of producing higher antibody levels for a longer period in young Africans. HB-vaccination of all children at birth, at 6 weeks and at 6–12 months may probably be a good alternative for many African countries.

Obviously, the main problem with HB-vaccination in third world countries is today's high vaccine costs. This problem will be at least partly overcome in the next few years when new HB-vaccines have entered the market. Local production of HB-vaccines has already been started in China and some other areas. When HBsAg-prevalences of several percent are seen in mothers, which is the fact in Africa and Asia, there is no reason to screen mothers for HBsAg. Furthermore, also in Asia horizontal spread of HBV occurs frequently, which speaks in favor of immunization of all Asian as well as African newborns.

In low-prevalence areas like most of Europe, vaccination only of children of mothers with HBsAg and anti-HBe in serum may be a reasonable choice while high-risk children of mothers with HBsAg and HBeAg in serum should receive HBIG in addition to vaccination until more is known about the protective effect of a more rapid vaccination procedure.

References

1. Angelillo B, Da Villa G, Fara GM, Piazza M, Pasquini P, Profeta ML, Toti L: Perspectives for mass vaccination against hepatitis B in Italy. The 1990 International Symposium on Viral Hepatitis and Liver Disease, Houston, USA (Abstract)
2. Margolis H, Alter M, Krugman S (1991) Strategies for controlling hepatitis B in the United States. In: Hollinger FB, Lemon SM, Margolis H (eds) Viral hepatitis and liver disease. Williams & Wilkins, Baltimore, pp 720–722
3. International Group (1988) Immunization against hepatitis B. Lancet i: 875–876
4. Jilg W (1989) Impfung gegen Hepatitis B. Impstoffe, Wirkungweise, Impfschutzdauer, Widerimpfung. Dtsch Med Wochenschr 114: 596–598
5. Wahl M, Hermodsson S, Iwarson S (1988) Hepatitis B vaccination with short dose intervals—a possible alternative for post-exposure prophylaxis? Infection 16: 229–232
6. Iwarson S, Wahl M, Ruttimann E, Snoy P, Seto B, Gerety RJ (1988) Successful postexposure vaccination against hepatitis B in chimpanzees. J Med Virol 25: 433–439
7. Iwarson S (1989) Post-exposure prophylaxis for hepatitis B: Active or passive? Lancet i: 146–148

Authors' address: Prof. Dr. S. Iwarson, Department of Infectious Diseases, University of Göteborg, Östra Hospital, S-416 85 Göteborg, Sweden.

Arch Virol (1992) [Suppl] 4: 147–153
© Springer-Verlag 1992

Immunogenicity and safety of a recombinant hepatitis B vaccine produced in mammalian cells and containing the S and the preS2 sequences

M. P. Corradi[1], C. Tata[2], P. Marchegiano[2], E. Villa[2], M. De Palma[3], G. Trianni[1], L. Fuiano[4], P. Rompianesi[2], T. Scacchetti[2]

[1] Direzione Sanitaria, [2] Cattedra di Gastroenterologia, and [3] Servizio Immunotrasfusionale, Ospedale Policlinico Modena
[4] Istituto Merieux Italia, Roma, Italy

Summary. A group of 273 health care workers, at risk of HBV infection, underwent vaccination with recombinant HBsAg produced in mammalian cells and containing protein sequences coded by both the S and pre-S2 regions (Genhevac B). Preliminary results show that a very early pre-S2 response occurred which may be useful in post-exposure prophylaxis. This observation, in addition to reduced influence by the vaccination protocol, provides grounds for optimism in spite of the fact that the efficiency spectrum of this vaccine was not superior to that of recombinant vaccines produced in yeast.

*

Hepatitis B represents one of the most important diseases in terms of morbidity and mortality. The need to eradicate the disease has lead in recent years to research and testing of vaccines that are capable of stopping recirculation of the virus.

Several studies [1, 2] underline the functional importance of proteins coded by the preS regions of the viral genome. This suggests that these proteins could play an important role in the assembly of viral components, or alternatively could be somehow involved in the ability shown by the virus to specifically infect liver cells.

If we assume that peptides coded by the preS1 and preS2 regions are involved in the anchorage of the virus to the hepatocyte, they should be a good target for a vaccine; this is supported by the evidence that antibodies reacting with intact virus [3, 4] or with peptides coded by the preS2 [5, 6] or preS1 [7] regions, as well as antibodies blocking the binding between HBV and pHSA [8] have been described in the first period of convalescence after acute B hepatitis. The presence of these antibodies in serum after recovery would indicate their protective valency.

Aim of this study was to evaluate preliminary results on a population of health care workers, at risk for HBV infection, undergoing vaccination with Genhevac B, a recombinant vaccine produced in mammalian cells containing the protein sequences coded by both the S and preS2 regions.

Subjects

A group of 273 health care workers (age range 18–40), who were negative at HBV marker evaluation, underwent anti-HBV vaccination from November 1988 to January 1990. Subjects were randomly allocated into 4 groups: A, B, C, D.

Vaccine

The vaccine, produced by Pasteur Vaccins (Paris, France), was provided by the Merieux Institute (Rome, Italy) as single dose syringes containing recombinant HBsAg (rHBsAg), produced in Chinese hamster ovarian cells transfected with genetic sequence coding for HBsAg (subtype ay), aluminum hydroxide up to a maximum of 1.25 mg aluminum and 0.5 ml buffer.

Administrations

Preparations contained different vaccine doses according to different batches: Individuals in groups A and B were administered 20 μg whereas groups C and D received 10 μg doses. Vaccine administration (intramuscular in the deltoid region) was performed according to a 4 dose schedule in groups A and C (French protocol) and according to a 3 dose schedule in groups B and D (American protocol).

Results and discussion

The results of the present study confirm the safety of the vaccine. Local side effects were mild and transient and general side effects were rare.

Transaminase levels did not change during the study, even with the higher dose (20 μg). This excludes the occurrence of significant cytolysis due to production of anti-serum albumin or anti-idiotype preS2 [9], despite administration of high amounts of preS2.

The main peculiarity of the vaccine utilized, is the content of large amounts of protein M (20% of total protein content of the vaccine). Data obtained in mice raised the hypothesis that the presence of the preS2 region might amplify the immune response, compared to other vaccines not containing the preS2, also in humans; on the other hand, other authors suggested that contemporary presentation of both self (albumin) and non-self (HBsAg) antigens to the cells of the immune system could have diminished the antibody response to HBsAg [9].

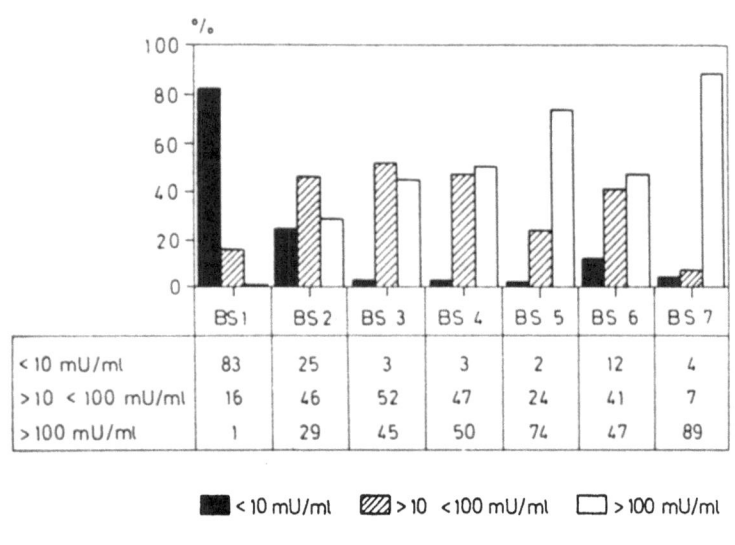

	BS 1	BS 2	BS 3	BS 4	BS 5	BS 6	BS 7
< 10 mU/ml	83	25	3	3	2	12	4
>10 < 100 mU/ml	16	46	52	47	24	41	7
>100 mU/ml	1	29	45	50	74	47	89

■ < 10 mU/ml   ▨ >10 <100 mU/ml   ▢ >100 mU/ml

Time course of HBsAb titers   A

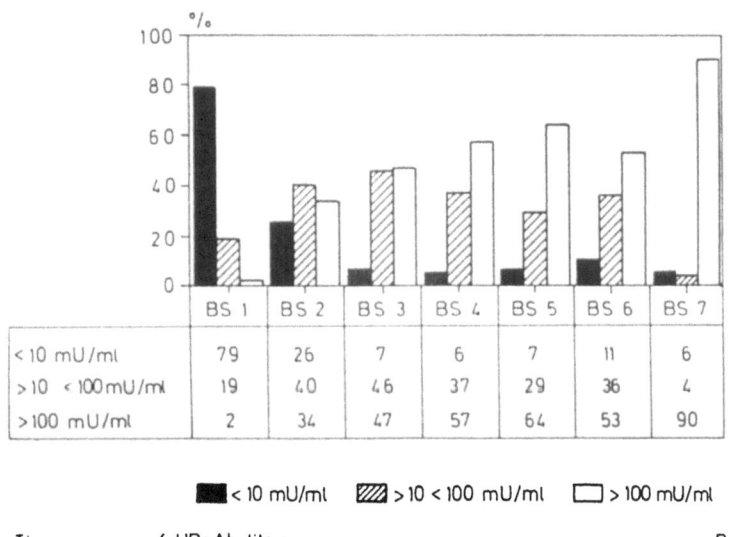

	BS 1	BS 2	BS 3	BS 4	BS 5	BS 6	BS 7
< 10 mU/ml	79	26	7	6	7	11	6
>10 < 100 mU/ml	19	40	46	37	29	36	4
>100 mU/ml	2	34	47	57	64	53	90

■ < 10 mU/ml   ▨ >10 <100 mU/ml   ▢ >100 mU/ml

Time course of HBsAb titers   B

Fig. 1. A Group A Seroconversion rate; **B** Group C Seroconversion rate

Our data do not appear to completely support either hypothesis; a good response to vaccine administration takes place (Figs. 1–2). The percentage of anti-HBsAg seroconversion in the 4 groups is maximal one month after the booster dose (92–98%), in agreement with previously published data regarding yeast-derived recombinant vaccines (10–13). The time course of antibody titers in the different groups appears to be slightly influenced by the administration schedule (Figs. 3–4).

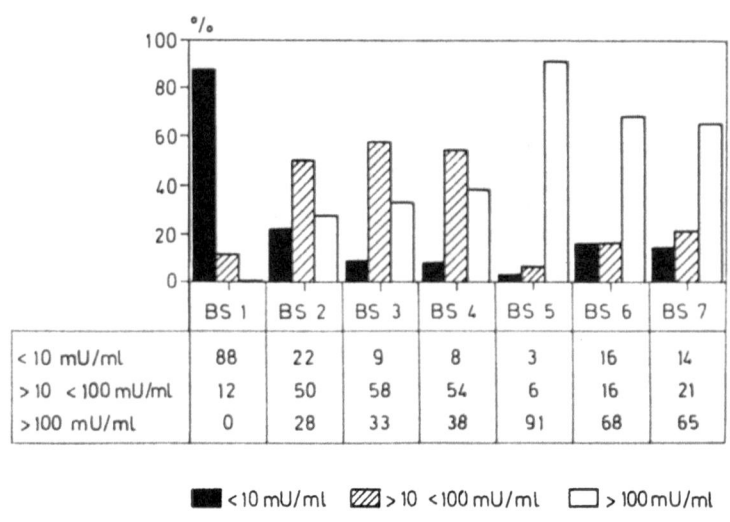

Time course of HBsAb titers   A

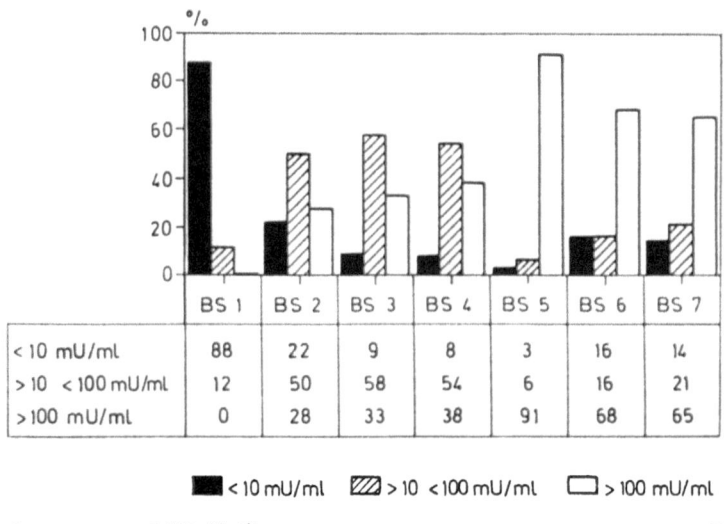

Time course of HBsAb titers   B

Fig. 2. A Group B Seroconversion rate; **B** Group D Seroconversion rate

Concerning preS2 antibody response, we still only have preliminary results. However, some considerations can be made: preS2 antibody production is clearly dose-dependent; in fact, only groups A and B, who received 20 µg vaccine, show a good antibody production. In contrast, groups C and D show lower antibody titers, never exceeding the 30 mU/ml titer, which is conventionally considered protective. This probably reflects different sensitivity to the factors underlying the immune response to the preS2 coded

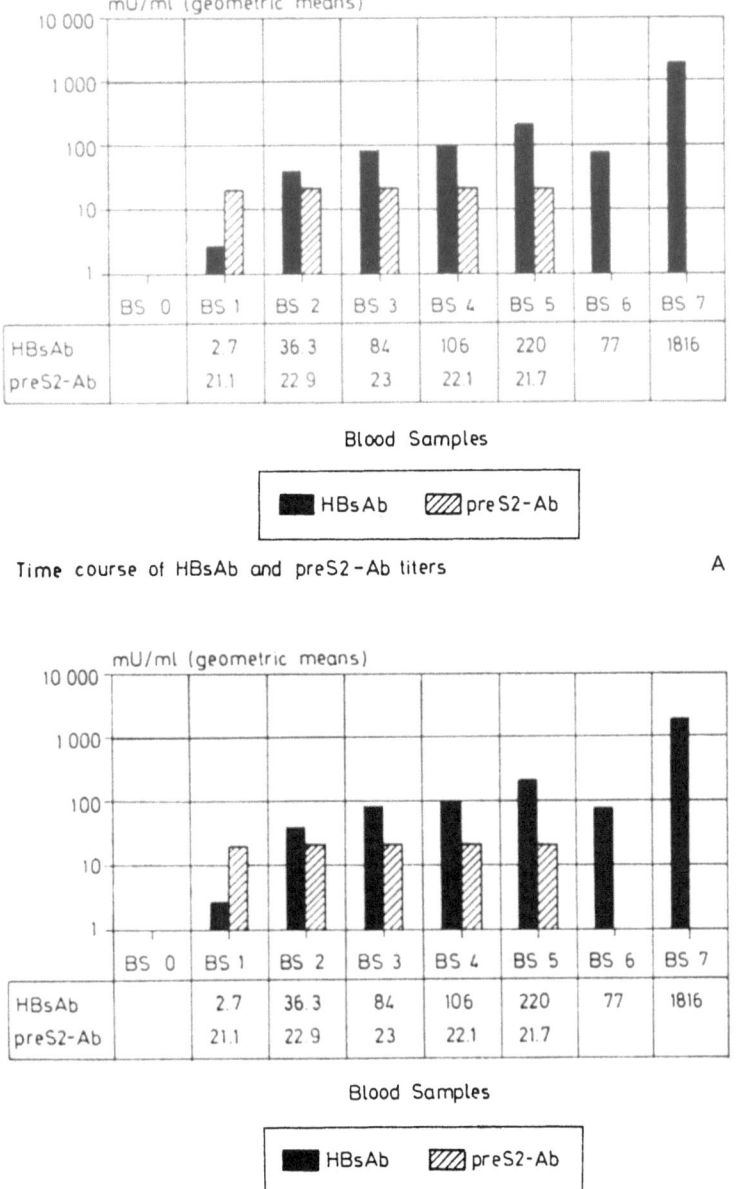

Fig. 3. A Group A Antibody response; B Group C Antibody response

protein, which seems to be partly overcome by increasing the administered dose.

Groups A and B show significant anti-preS2 response as soon as 60 days after the first administration, whereas comparable anti-HBsAg response develops at a later time. This data, although preliminary, suggest that the administration schedule could affect the time course of antibody titers.

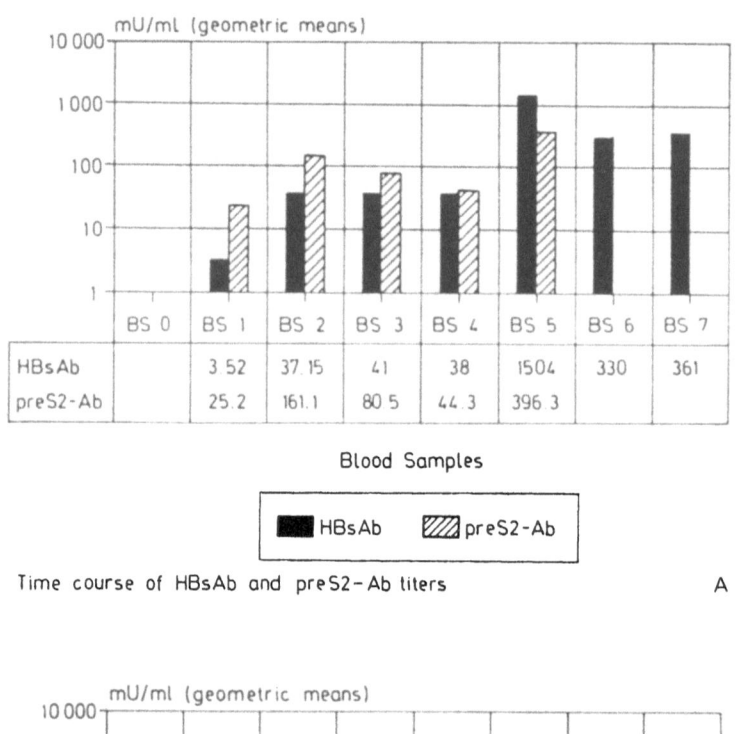

Time course of HBsAb and preS2-Ab titers   A

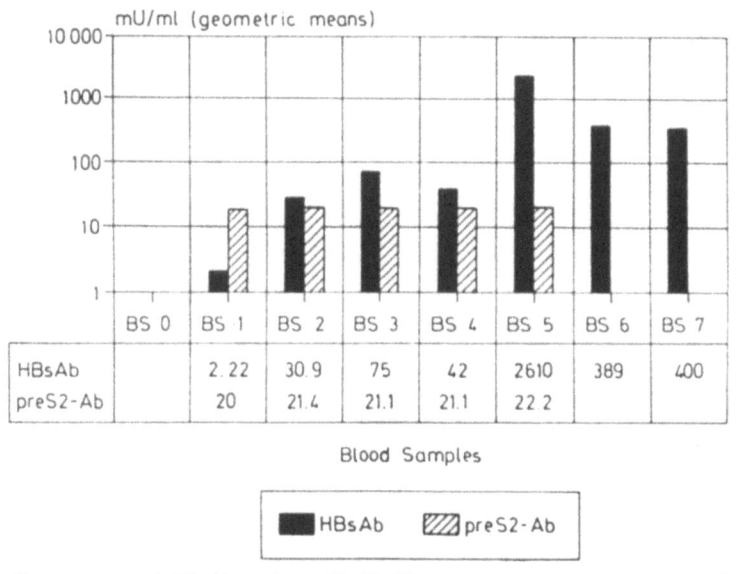

Time course of HBsAb and preS2-Ab titers   B

Fig. 4. A Group B Antibody response; **B** Group D Antibody response

In conclusion, this study shows that the immunogenicity of mammalian rHBsAg vaccines is good and in general comparable to that of yeast-produced recombinant and blood-derived vaccines. Preparations containing high dosages of protein M (at least 20% of total vaccine protein) are required in order to obtain a good anti-preS2 response. At these doses, however, anti-HBsAg response is also less influenced by the vaccination protocol and this may allow to carry out shorter and less expensive vaccine programs.

Although the present study could not show a definite improvement of the efficacy spectrum of these vaccines, compared to recombinant vaccines produced in yeasts, the precocity of the anti-preS2 response could constitute a valuable advantage, suggesting that this vaccine might be a useful tool for post-exposure prophylaxis to HBV infection.

References

1. Pugh JC et al (1987) J Virol 61: 1384
2. Shaeffer E et al (1986) J Virol 57: 137
3. Alberti A et al (1978) Br Med J 2: 1056
4. Alberti A et al (1984) Hepatology 4: 220
5. Budkowska A et al (1986) Hepatology 6: 360
6. Okamoto H et al (1986) Hepatology 6: 354
7. Klinkert MQ et al (1986) J Virol 58: 522
8. Pontisso P et al (1986) J Hepatol 3: 939
9. Hellström U et al (1986) Scand J Immunol 23: 523
10. Andre FE et al (1988) Viral hepatitis and liver disease. Liss, New York, p 1025
11. Zajac BA et al (1986) J Infec 13 [Suppl A] 39
12. Jilg W et al (1986) J Infec 13 [Suppl A] 47
13. Yamamoto S et al (1986) J Infec 13 [Suppl A] 53

Authors' address: Dr. M. P. Conradi, Direzione Sanitaria, Ospedale Policlinico Modena, I-411000-Modena, Italy.

Arch Virol (1992) [Suppl] 4: 154–155
© Springer-Verlag 1992

Kinetics of anti-HBs after hepatitis B vaccination: a comparison of two recombinant and one plasma-derived vaccines

M. Gesemann[1], S. Schröder[1], N. Scheiermann[2], and C. Maurer[2]

[1] Institut für Medizinische Virologie und Immunologie, Universitätsklinikum, Essen
[2] Institut für Laboratoriumsmedizin, Städtisches Krankenhaus, Heilbronn, Federal Republic of Germany

Summary. Geometric mean titers were determined for three groups of medical students who had been vaccinated against hepatitis B with two different yeast-derived recombinant vaccines and one plasma-derived vaccine. The antibody kinetics for the three groups were similar over a period of 4 years. A formula for prediction of titers from the post-booster anti-HBs concentration is provided.

*

To assess the long-term immunogenicity of hepatitis B vaccines, 2 groups of medical students were each immunized with one of two recombinant, yeast-derived vaccines (YDV) (group 1, n = 169: Engerix B™, SmithKline Biologicals [1], 4 vaccinations at months 0, 1, 2 and 11; group II, n = 32: CC 2572-P-101, Cilag [2], 3 vaccinations at months 0, 1, and 6) and with the 20 µg formulation of a plasma-derived vaccine (PDV) (HB-Vax, MSD, n = 45 and n = 18, resp.). Individual anti-HBs concentrations were measured at intervals for 49 and 30 months (resp.) after the last (booster) dose.

One month after the booster vaccination, all of 147 vaccinees tested in group I (schedule 0–1–2–11 months) and 94% of the 50 subjects in group II (schedule 0–1–6 months) had anti-HBs levels ≥ 10 IU/l. After completion of the vaccination courses, no significant differences in antibody titers were observed between YDV and PDV in their respective vaccine groups. However, geometric mean titers (GMT) were significantly higher after 4 vaccinations (group I, GMT = 21330 IU/l) than in group II (GMT = 2013 IU/l).

In both groups, GMTs decreased rapidly by a factor of 5 to 10 during the first 12 months after the booster vaccination and by a factor of 2 in the 2nd year, but 3 and 4 years after the booster (group I), GMTs decreased only by factor 1.3 to 1.5 per year.

These antibody kinetics represent an increase of antibody half-lives with time which is similar for YDV and PDV and may best be described by a logarithmic function

$$y = ax^{-k} \text{ with}$$

y = actual anti-HBs concentration;
a = anti-HBs concentration 1 month after the last vaccination;
x = time (months) since last vaccination;
k   (mostly $0.5 < k < 1.0$) = constant depending on the vaccination schedule.

To predict the actual titer from the post-booster anti-HBs concentration, the formula "$y = a/x$" is suggested. Whether this formula may be applied to extended periods beyond 4 years after the booster vaccination will have to be determined by long-term follow-up studies.

References

1. Scheiermann N, Gesemann KM, Kreuzfelder E, Paar D (1987) Effects of a recombinant yeast-derived hepatitis B vaccine in healthy adults. Postgrad Med J 63 [Suppl] 2: 115–119
2. Gesemann M, Scheiermann N, Friedmann N, Mirman I (1988) Reactogenicity and immunogenicity of a third yeast-derived hepatitis B vaccine (abstract) Antiviral Res 9: 143

Authors' address: Dr. M. Gesemann, Institut für Medizinische Virologie und Immuno-logie, Universitätsklinikum, D-W-4300-Essen, Federal Republic of Germany.

Arch Virol (1992) [Suppl] 4: 156–159
© Springer-Verlag 1992

Enhanced luminescent assays for hepatitis markers: Assessment of post vaccine responses

E. H. Boxall

Regional Virus Laboratory, Birmingham, West Midlands, England, U.K.

Summary. A range of solid phase immunoassays have been developed using enhanced chemiluminescence to provide a signal which can be measured in a qualitative or quantitative manner. The "Amerlite" anti-HBs assay has been used routinely to test more than 3000 post-vaccine anti-HBs levels. The results show that 5% of vaccinees are non-responders, but that more than 30% produce antibody levels greater than 1000 mIU/ml.

Introduction

A range of solid phase immunoassays have been developed using enhanced chemiluminescence, a light generating system based on the oxidation of a chemiluminescent substrate (luminol) by horse-radish peroxidase. By using a chemical enhancer, the light signal can be greatly enhanced enabling the development of sensitive immunoassays.

Assays for Hepatitis A and Hepatitis B markers (HBsAg, antiHBs, antiHBc, antiHBc IgM and antiHAV IgM) have been assessed and have proved to have the sensitivity and specificity of radio immunoassay, but without the radiation hazards.

In this report the use of the "Amerlite" anti HBs assay in quantitating Hepatitis B vaccine responses has been evaluated.

Method

The Amerlite antiHBs assay was used according to the manufacturers protocol.

Opaque microtitre wells are used and the light is detected and measured using an analyser whose software includes data reduction facilities for both qualitative and quantitative assays.

Amerlite anti-HBs can be used quantitatively using a panel of antiHBs standards of 0, 10, 50, 250, 500 and 1000 mIU/ml to generate a standard curve.

Results

The Amerlite assay and a radioimmunoassay for antiHBs (Ausab, Abbott Laboratories) were compared. The correlation between the two assays is very good in the range up to 300 mIU/ml. However, the range of Amerlite assay is much wider and can generate values up to 1,000 mIU/ml *without dilution*.

This assay is in routine use to measure the responses to Hepatitis B vaccines. The following analysis is on a total of 3,296 individual sera from vaccinated health care staff (Fig. 1).

Of the vaccinees 10% are non-responders with titres below 5 mIU/ml, 36% produce titres in the range 11–200 mIU/ml, but more than half of those vaccinated (54%) produce titres greater· than 200 mIU/ml i.e. outside the range of most enzyme immunoassays. Using Amerlite it can be seen that many people (20%) are capable of producing titres in excess of 1,000 mIU/ml. (Actual values were established using a 1/10 dilution of sample.)

The immune response in males and females of different ages is shown (Fig. 2). As previously reported, there is a fall in immune response with increasing age which is more marked in men than women.

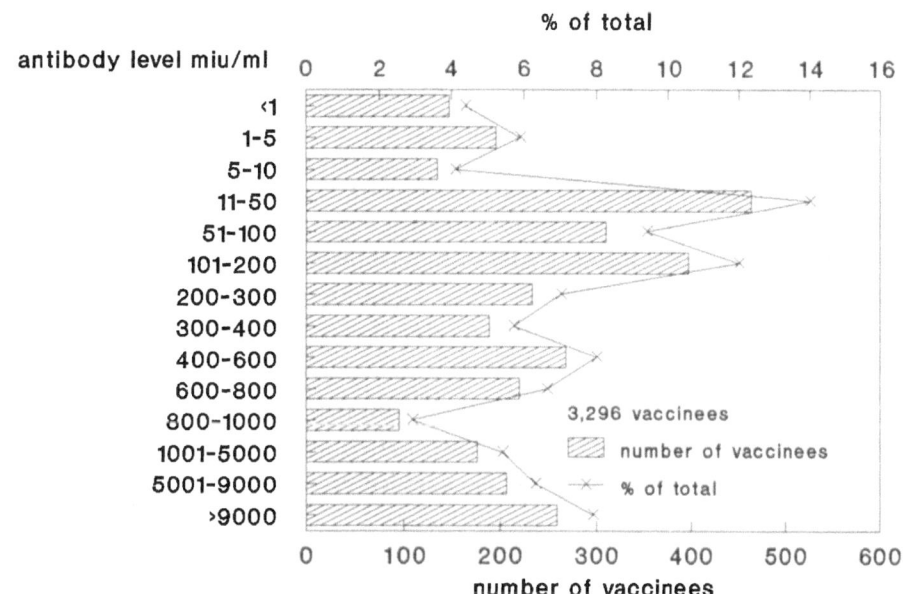

Fig. 1. Post vaccine responses: quantitation by Amerlite antiHBs assay.

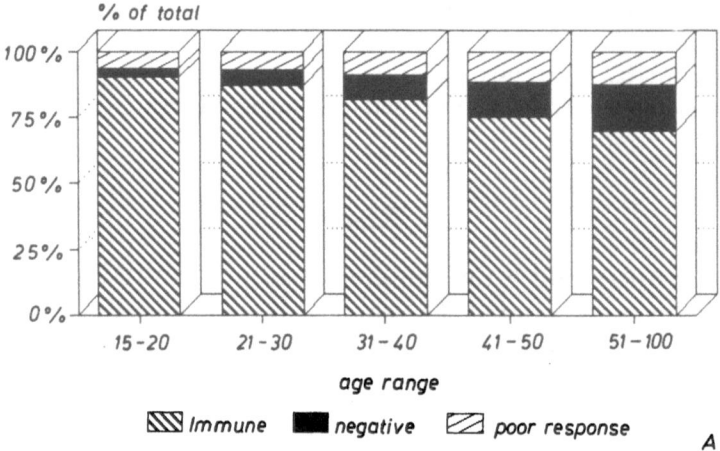

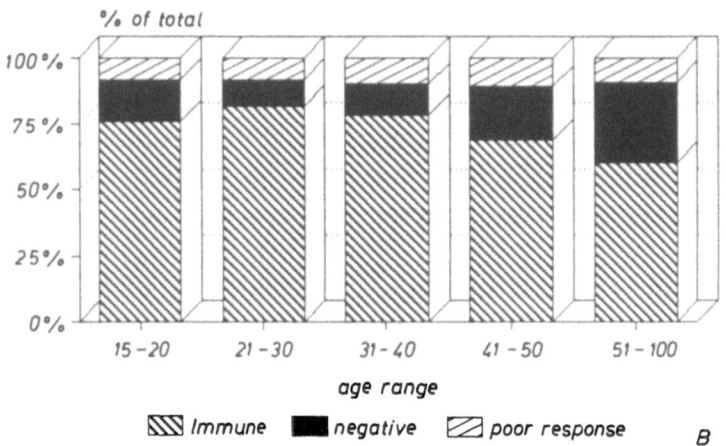

Fig. 2. A Responses to hepatitis B vaccine, age ranges-females; **B** Responses to hepatitis B vaccine, age ranges—males

Conclusion

The accurate quantitation of immune status is of importance in planning revaccination intervals and will continue to be of importance as other vaccines are developed and a wider range of individuals are immunized. The quantitation of antiHBs over a wide dynamic range is of importance in health care planning, for example to decide on revaccination intervals, to assess the immune status of contacts, and to advise health care staff about their risk. The enhanced luminescent method provides an efficient method of quantitation over a wide dynamic range: such systems will become increasingly important in infectious disease serology.

References

1. Atherton CJ, Boxall EH (1986) A sensitive screening test of the simultaneous detection of hepatitis B surface antigen and antibody. J Virol Met 13: 245–253

2. Lane RS (1981) Hepatitis B surface antigen testing: The blood products laboratory radio immunoassay (BPL/RIA) system. Med Lab Sci 38: 323–329
3. Wheeley SM, Boxall EH, Tarlow MJ, Gatrad AR, Anderson J, Bissenden J, Chin KC, Mayne A (1990) Hepatitis B Vaccine in the prevention of perinatally transmitted hepatitis B virus infection: J Med Virol 30: 113–116

Author's address: Dr. Elizabeth H. Boxall, Regional Virus Laboratory, Birmingham B9 5ST, West Midlands, England, U.K.

VI Characterization of hepatitis C virus

Arch Virol (1992) [Suppl] 4: 163–171
© Springer-Verlag 1992

Comparative molecular biology of flaviviruses and hepatitis C virus

F. X. Heinz

Institute of Virology, University of Vienna, Vienna, Austria

Summary. Currently available sequence information suggests that the genome organization of hepatitis C virus is similar to that of flaviviruses. A positive-stranded genomic RNA contains a single long open reading frame (ORF) which is flanked by 5′ and 3′ noncoding sequences. This RNA codes for structural proteins at the 5′ end (starting with the capsid protein) and a set of nonstructural proteins in the remainder of the genome. The latter provide essential virus-specific functions for the viral life cycle, such as protease, helicase, and RNA replicase activities. The sequence motifs characteristic of the corresponding functional protein domains are separated by similar spacings in the nonstructural regions of hepatitis C virus and flaviviruses. The structural region of the hepatitis C virus appears to consist of a capsid protein which is larger than that of flaviviruses and two putative envelope proteins which are presumably different in molecular weight and much more heavily glycosylated than their counterparts in flaviviruses. A study group of the International Committee on the Taxonomy of viruses proposes to include hepatitis C virus as a genus into the family 'flaviviridae'.

Introduction

The family flaviviridae was recently established as a separate family from the togaviridae due to fundamental differences in genome structure and replication cycle. To date, about 70 different but serologically related flaviviruses have been identified. Most of these are arthropod-borne, being transmitted to their vertebrate hosts by chronically infected mosquitoes or ticks. Several flaviviruses are important human pathogens. The highest annual disease rates for flaviviruses are caused by yellow fever (YF), Japanese Encephalitis, Dengue, and tick-borne encephalitis (TBE) virus [for review see ref. 16].

Flaviviruses are spherical, enveloped, positive-stranded RNA viruses with a diameter of about 50 nm. Mature particles contain three distinct structural proteins: capsid (C); envelope (E); and membrane (M). The single-stranded genomic RNA is about 11,000 nucleotides in length and carries a cap structure at its 5' end [for review see ref. 2]. It was generally believed that flavivirus RNAs are not polyadenylated but recent evidence indicates the presence of poly A tracts in certain strains of TBE virus [14]. The viral RNA contains a single long open reading frame of about 10,000 nucleotides which is flanked by 5' and 3'-noncoding sequences of 95 to 132 and 385 to 585 nucleotides, respectively. The 3'- noncoding sequence of some TBE viruses seems to be much shorter (114 bases). The genomic RNA serves as the only viral messenger and translation of the ORF gives rise to a polyprotein which is proteolytically processed into at least 11 different proteins. The 5' quarter of the genome encodes the structural proteins (C-prM(M)-E), whereas a set of nonstructural proteins is encoded by the rest of the genome (NS1 – NS2A – NS2B – NS3 – NS4A – NS4B – NS5) [Fig. 1A]. Characterization of the cleavage sites by amino-terminal and carboxy-terminal sequence analyses as well as the intracellular localization of individual proteins provide evidence that at least three proteases (both cellular and viral) are involved in the

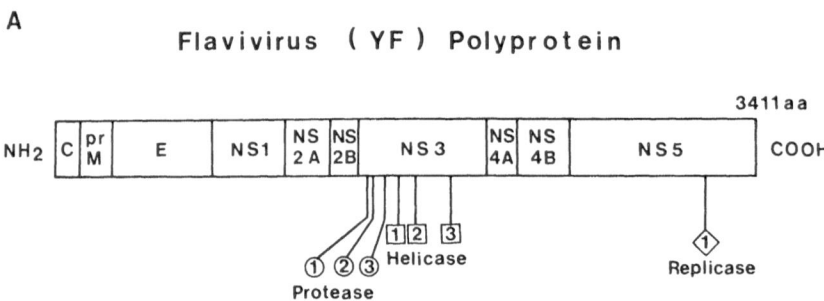

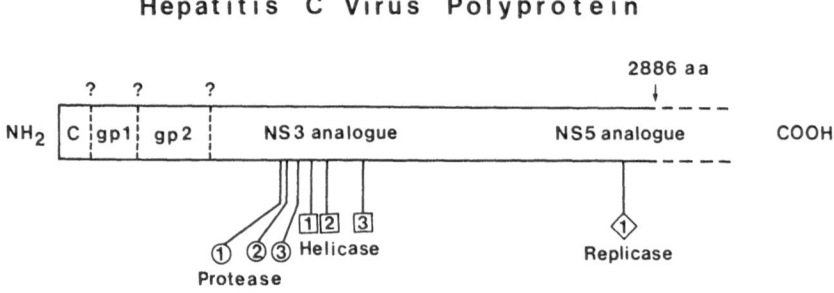

Fig. 1. Polyproteins of flaviviruses (**A**) and hepatitis C virus (**B**)

Protease domain: ① H   ② D   ③ GXSGXP

Helicase domain: 1 GXGKT   2 DEAH (DECH in hepatitis-C)

3 GRXGR

Replicase: ◇₁ GDD

cotranslational and posttranslational processing of the polyprotein: The cellular enzyme signalase, a viral protease which is associated with the nonstructural protein NS3 (see below) and a cellular Golgi protease.

Recently an ingenious approach was used to clone and sequence the genome of hepatitis C virus [3], which was previously suspected of bearing resemblance to viruses of the togavirus family. A large part of the genomic RNA sequence was elucidated and this for the first time allowed sequence comparisons to be made with other known viruses [10, 15]. The sequence was further extended to the 5'- noncoding region [17, 21] thus revealing an RNA genome of about 10 kilobases which contained a single long open reading frame coding for at least 2886 amino acids (Fig. 1B). The analysis of this sequence (discussed in more detail below) suggests a genome organization similar to that of flaviviruses. A study group of the 'International Committee on the Taxonomy of Viruses' therefore proposes to include hepatitis C virus as well as the related pestiviruses as individual genera in the family flaviviridae.

Structural proteins

The structural proteins of flaviviruses are encoded at the 5' end of the genomic RNA in the order C -prM(M)-E. While the C protein is a cytoplasmic protein, both prM and E are amphiphilic membrane proteins which carry carboxyterminal membrane anchors. The polyprotein must therefore provide sorting signals, which allow the generation of one cytoplasmic protein and two separate identically oriented membrane proteins upon translation from a single messenger RNA. This is accomplished by a specific set of internal signal sequences [22] and stop transfer sequences, which direct the nascent polypeptide chain into the lumen of the ER and anchor individual proteins to the ER membrane in their correct orientation [23].

The whole process is believed to proceed as follows (Fig. 2): An internal signal sequence at the carboxyterminus of protein C causes transport of the nascent protein through the ER membrane. The signal sequence is cleaved off by signalase, which thus generates the N-terminus of the prM protein. The C protein is presumed to be liberated from the membrane, probably by the action of a viral protease. The transfer of prM is stopped by a hydrophobic sequence which is immediately followed by a second internal signal sequence for the cross-membrane transport of protein E. Again, this signal sequence is cleaved off by signalase. According to this scheme, both prM and E are anchored to the membrane by two hydrophobic stretches of amino acids. The last of these functions as a further internal signal sequence for the first nonstructural protein NS1. Virus assembly apparently occurs at intracellular sites. First, immature virions containing the protein C, prM and E are formed [20]. The proteolytic cleavage of prM into M, presumably by a

F. X. Heinz

SYNTHESIS OF FLAVIVIRUS STRUCTURAL PROTEINS

RNA

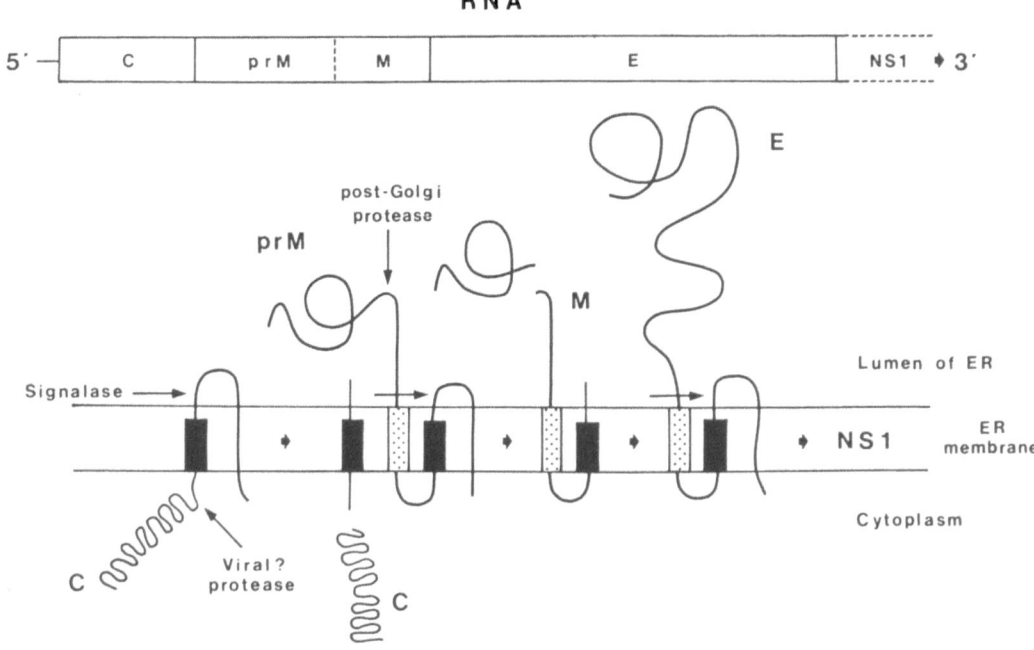

Fig. 2. Synthesis of the flavivirus structural proteins. Translocation through the ER membrane and membrane integration is accomplished by consecutive internal signal sequences ▌ and stop transfer sequences ▓

post-Golgi protease [19], apparently represents a late event in virus maturation and yields fully infectious virions.

Based on this background the following considerations suggest that the 5′-terminal part of the hepatitis C virus RNA codes for a capsid protein and probably for two membrane-associated glycoproteins (Fig. 3). The N-terminus of the polyprotein is enriched in basic amino acids, characteristic of proteins associating with nucleic acids. It is therefore likely that approximately the first 160 amino acids code for the capsid protein. The following sequence contains a total of 16 potential N-glycosylation sites. For glycosylation, this part of the protein must be transported into the lumen of the ER. This could be accomplished by a hydrophobic stretch of amino acids between residues 170 to 190, which may function as an internal signal sequence. Following five N-glycosylation sites we encounter a structural element (residues 347 to 390) similar to those described above for flaviviruses: A stop transfer sequence consisting of hydrophobic amino acids and a positively charged amino acid (Lys) is followed by a second potential internal signal sequence. This could give rise to a second glycoprotein containing 11 N-glycosylation sites. Although the final identification of these proteins will require N- and C-terminal sequence analysis, the analogy to flaviviruses

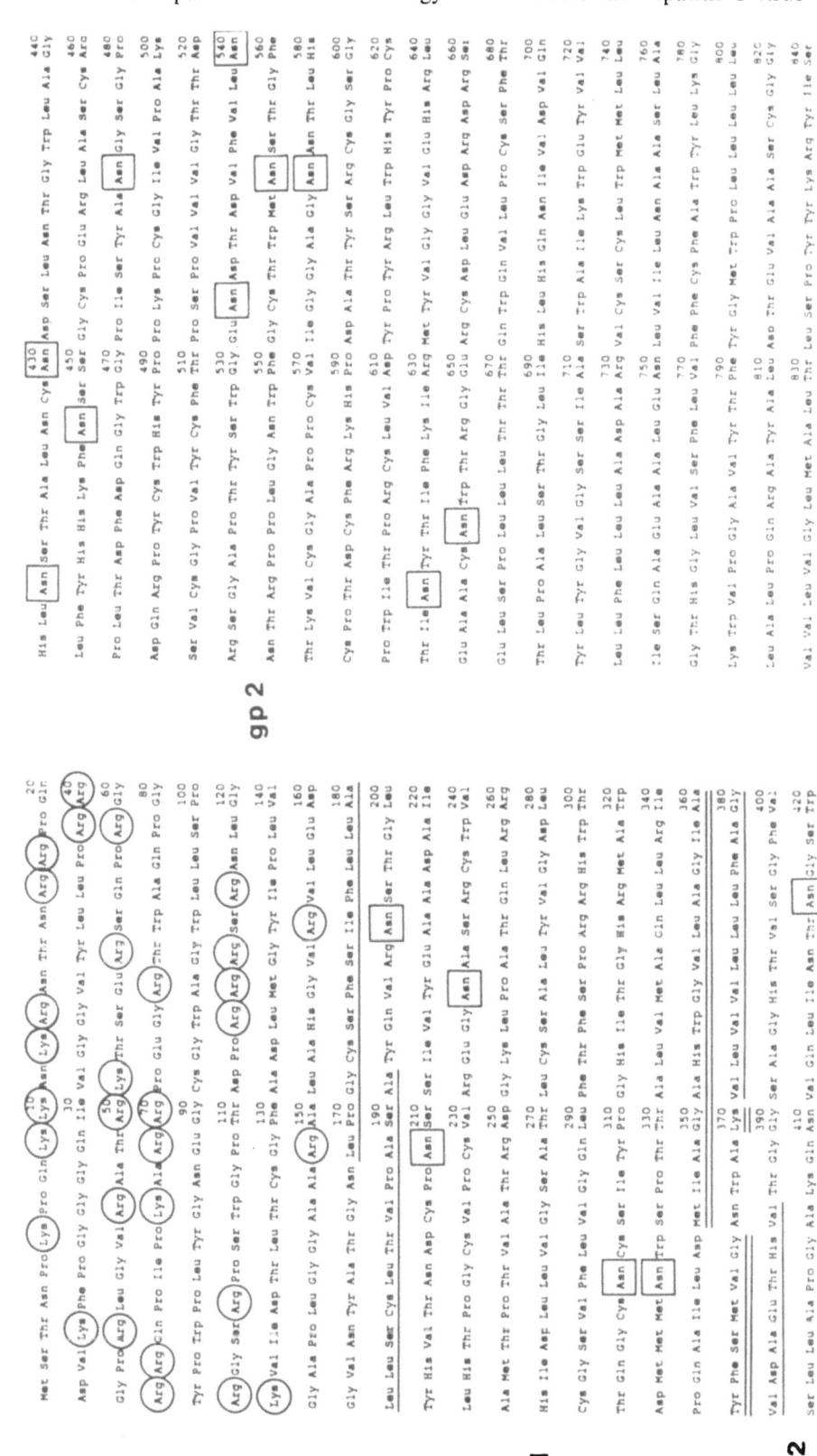

Fig. 3. Amino acid sequence of the hepatitis C virus structural region (the sequence shown is that of the isolate HCV UC as described in [10] and [21]). Positively charged amino acids in the putative capsid protein are encircled. Potential N-glycosylation sites are marked by rectangles. Putative internal signal sequences are underlined, a possible stop transfer sequence including a positively charged lysine is indicated by a double line

makes it possible to predict that this part of the hepatitis C virus ORF encodes a capsid protein and two heavily glycosylated membrane proteins. It remains to be shown, however, whether both glycoproteins are structural components of the virion.

Nonstructural proteins

For the replication of flaviviruses in infected cells, several essential virus-specific functions are provided by nonstructural proteins. These include viral proteases, RNA dependent RNA-polymerases, and probably helicases, similar to all positive-stranded RNA viruses of plants and animals [6]. Although NS1 has been implicated as being involved in virus assembly, its function is presently unknown. The same also holds true for the proteins NS2A/2B and NS4A/4B. Functional activities, however, were ascribed to NS3 and NS5, either by sequence homologies with other proteins of known function, or directly by the use of functional assays.

NS3 apparently represents a bifunctional protein which carries an N-terminal protease domain related to cellular trypsin-like serine proteases [1, 8], and C-terminal sequence motifs characteristic of helicases [7, 9, 12]. The protease activity in the N-terminal domain of NS3 was recently verified using in vitro transcription and translation of Dengue type 2 cDNA constructs [18]. Sequence comparisons and structural modelling suggest amino acids that are crucial for catalytic activity and substrate binding. Specifically H 53, D 77, and S 138 (YF virus numbering) are believed to represent the catalytic triad of the flavivirus protease. The same residues are also found with practically identical spacings in the hepatitis C virus polyprotein at positions 1083, 1107, and 1165. The serine is part of the conserved motif GXSGXP which is a general characteristic of the nucleophilic site of serine proteases [8].

In addition to this N-terminal protease domain, NS3 contains sequence motifs characteristic of nucleoside-triphosphate binding (NTPase) or helicase enzymes which are also found in other positive strand RNA viruses of animals and plants [7, 9, 12]. The synthesis of proteins with helicase activity is probably necessary for the unwinding of "Replicative Form" RNA molecules in the course of genome replication. The positions of the most highly conserved sequence elements are indicated in Fig. 1. These are GXGKT, DEAH, and GRXGR for flaviviruses and are found in NS3 of YF virus at residues 201 to 205, 289 to 292, and 463 to 467, respectively. The homologous motifs GSGKS, DECH, and GRTGR are found at residues 1233 to 1237, 1316 to 1319, and 1489 to 1493 in the hepatitis C virus polyprotein.

The largest nonstructural protein (NS5) shows the highest degree of conservation among all flavivirus-specific proteins and is thought to represent the RNA-dependent RNA-polymerase, although direct proof is still lacking. The RNA-polymerase was functionally identified for poliovirus, [4, 13] and by sequence comparison characteristic sequence motifs were found in the homologous proteins of all other known positive stranded RNA viruses of animals and plants [11, 6]. The invariant GDD sequence is found at positions 666 to 668 in the YF virus NS5 and at positions 2737 to 2739 in the hepatitis C virus polyprotein (Fig. 1).

Concluding remarks

From extensive computer-assisted sequence comparisons using different data banks it became clear that hepatitis C virus is not closely related to any of the RNA viruses thus far described, despite restricted amino acid similarities with pestiviruses, flaviviruses and certain plant viruses [15]. However, sequence motifs characteristic of nonstructural proteins of all positive-stranded RNA viruses of animals and plants (protease, helicase, RNA dependent RNA polymerase) can readily be identified in the hepatitis C virus sequence. It is striking to note, that not only the spacings between individual elements of a given functional domain but also between each of the domains for protease, helicase and polymerase activity are almost identical in flaviviruses and hepatitis C virus. In the latter case, however, this whole nonstructural sequence block seems to be shifted closer to the N-terminus of the polyprotein (Fig. 1). Assessment of the recently published 5′-terminal sequences of the hepatitis C virus genome suggests considerable differences to the structural region of flaviviruses. Specifically, the hepatitis C virus capsid protein is considerably larger and the two putative membrane proteins are potentially much more heavily glycosylated than their flavivirus counterparts. Nevertheless, the gene order of both the structural and the nonstructural proteins seems to be similar, thus justifying the inclusion of hepatitis C virus into the family flaviviridae.

It is currently believed that the evolution of positive stranded RNA viruses occurs by two major mechanisms: Divergency from a common ancestor and interviral recombination, by which sets of genes may be exchanged among different viruses [6]. This has led to the concept of modular evolution, i.e. new viral RNA genomes may be generated by the mixing and joining of gene modules [5, 24]. Further studies, especially on the location of individual proteins and the processing of the polyprotein, will be necessary to gain more insight into the biology of hepatitis C virus and its evolutionary relationship to other viruses. It is nevertheless exciting and satisfying to see how much information on the molecular aspects of virus cell interactions can be obtained by the comparative analysis of sequence data alone.

Acknowledgements

I am grateful to Dr. Michael Roggendorf and Dr. Peter Highfield for helpful discussions and information provided for the preparation of my lecture in Siena and this manuscript.

References

1. Bazan JF, Fletterick RJ (1989) Detection of a trypsin like serine protease domain in flaviviruses and pestiviruses. Virology 171: 637–639
2. Chambers TJ, Hahn CS, Galler R, Rice CM (1990) Flavivirus genome organization, expression and replication. Ann Rev Microbiol 44: 649–688
3. Choo QL, Kuo G, Weiner AJ, Overby LR, Bradley DW, Houghton M (1989) Isolation of a cDNA clone derived from a blood-borne non-A, non-B hepatitis genome. Science 244: 359–362
4. Flanegan JB, Baltimore D (1977) Poliovirus-specific primer-dependent RNA polymerase able to copy poly(A). Proc Natl Acad Sci USA 74: 2677–2680
5. Gibbs A (1987) Molecular evolution of viruses: 'trees', 'clocks' and 'modules'. J Cell Sci 7: 319–337
6. Goldbach R (1990) Genome similarities between positive-strand viruses from plants and animals. In: Brinton MA, Heinz FX (eds) New aspects of positive-strand RNA viruses. American Society for Microbiology, Washington, DC, pp 3–11
7. Gorbalanya AE, Koonin EV, Donchenko AP, Blinov VM (1988) A novel superfamily of nucleoside triphosphate binding motif containing proteins which are probably involved in duplex unwinding in DNA and RNA replication and recombination. FEBS Lett 235: 16–24
8. Gorbalanya AE, Donchenko AP, Koonin EV, Blinov VM (1989) N-terminal domains of putative helicases of flavi- and pestiviruses may be serine proteases. Nucleic Acids Res 17: 3889–3897
9. Hodgman TC (1988) A new superfamily of replicative proteins. Nature 333: 22–23
10. Houghton M, Choo QL, Kuo G (1988) NANBV diagnostics and vaccines. Europ Pat Appl Nr 88310922.5; Publ No. 318216
11. Kamer G, Arges P (1989) Primary structural comparison of RNA-dependent polymerases from plant, animal and bacterial viruses. Nucleic Acids Res 12: 7269–7282
12. Lain S, Riechmann JL, Martin MT, Garcia JA (1989) Homologous potyvirus and flavivirus proteins belonging to a superfamily of helicase-like proteins. Gene 82: 357–362
13. Lundquist RE, Ehrenfeld E, Maizel JV (1974) Isolation of a viral polypeptide associated with poliovirus RNA polymerase. Proc Natl Acad Sci USA 71: 4773–4777
14. Mandl CW, Heinz FX, Kunz C (1991) Presence of poly(A) in a flavivirus: Significant differences between the 3′ noncoding regions of the genomic RNAs of Tick-borne encephalitis virus strains.
15. Miller RH, Purcell RH (1990) Hepatitis C virus shares amino acid sequence similarity with pestiviruses and flaviviruses as well as members of two plant virus supergroups. Proc Natl Acad Sci USA 87: 2057–2061
16. Monath T (1990) Flaviviruses. In: Fields BN, Knipe DM (eds) Virology, 2nd edn. Raven Press, New York, pp 763–814
17. Okamoto H, Okada S, Sugiyama Y, Yotsumoto S, Tanaka T, Yoshizawa H, Tsuda F, Miyokawa Y, Mayumi M (1990) The 5′-terminal sequence of the hepatitis C virus genome. Japan J Exp Med 60: 167–177
18. Preugschat F, Yao C-W, Strauss JH (1990) In vitro processing of dengue virus type 2 nonstructural proteins NS2A, NS2B, and NS3. J Virol 64: 4364–4374

19. Randolph VB, Winkler B, Stollar V (1990) Acidotropic amines inhibit proteolytic processing of flavivirus prM protein. Virology 174: 450–458
20. Shapiro D, Brandt W, Russell PK (1972) Change involving a viral membrane glycoprotein during morphogenesis of group B Arboviruses. Virology 50: 906–911
21. Takeuchi K, Kubo Y, Boonmar S, Watanabe Y, Katayama T, Choo QL, Kuo G, Houghton M, Saito I, Miyamura T (1991). The putative nucleocapsid and envelope protein genes of hepatitis C virus determined by comparison of the nucleotide sequences of two isolates derived from an experimentally infected chimpanzee and healthy human carriers. J Gen Virol 71: 3027–3033
22. Von Heijne G (1984) How signal sequences maintain cleavage specificity. J Mol Biol 173: 243–251
23. Wengler G, Castle E, Leidner U, Nowak T, Wengler G (1985) Sequence analysis of the membrane protein V3 of the flavivirus West Nile virus and of its gene. Virology 147: 264–274
24. Zimmern D (1987) Evolution of RNA viruses. In: Holland J, Domingo E, Ahlquist P (eds) RNA genetics. CRC Press, Boca Raton, FL, pp 211–240

Author's address: Dr. F. X. Heinz, Institute of Virology, University of Vienna, Kinderspitalgasse 15, A-1095 Wien, Austria.

Arch Virol (1992) [Suppl] 4: 172–178
© Springer-Verlag 1992

Hepatitis C viral RNA in serum of patients with chronic non-A, non-B hepatitis: detection by the polymerase chain reaction using multiple primer sets*

K. Cristiano[1], **A. M. Di Bisceglie**[2], **J. H. Hoofnagle**[2], and **S. M. Feinstone**[1]

[1] Laboratory of Hepatitis Research, Division of Virology, Center for Biologics Evaluation and Research, Food and Drug Administration, Bethesda, MD
[2] Liver Diseases Section, National Institutes of Diabets and Digestive and Kidney Diseases, National Institutes of Health, Bethesda, MD, USA

Summary. The recently introduced antibody test for hepatitis C virus (HCV) infection has proven to have certain limitations. Since HCV itself is usually present in clinical specimens at very low titers, a useful assay for the virus must have very high sensitivity. We have developed a simple, highly sensitive assay for HCV RNA based on the polymerase chain reaction (PCR). In this test, RNA extracted from HCV infected serum or plasma is used as the template for double PCR with nested primers. Sensitivity studies demonstrate that this assay is able to detect HCV at or beyond the sensitivity level of chimpanzee infectivity. We tested, with several sets of nested primers, 40 patients with chronic non-A, non-B hepatitis (36 seropositive and 4 seronegative) and found that 35/40 were PCR positive including all 4 seronegative patients. Normal human plasma and plasma from hepatitis B infected patients did not react in this test. This assay has proven to be valuable for determining the presence of HCV in various samples; furthermore, it offers the possibility of diagnosis of HCV infection in seronegative patients.

Introduction

Hepatitis C virus is considered one of the agents of the formerly termed parenterally transmitted non-A, non-B hepatitis (NANBH). Recently, a group at Chiron Corporation revealed that the viral genome is a positive

---

* Informed consent was obtained from all patients according to the guidelines for human experiments of the U.S. Department of Health and Human Services.

sense, single stranded RNA of approximately 10 kb in length with a single long open reading frame [3]. The same group also developed an ELISA test specific for HCV antibodies that is now commercially available and that uses an expressed viral peptide termed C-100 as the antigen [9]. This test appears to be incapable of avoiding false positives, therefore, requiring confirmatory tests (such as the RIBA-HCV or the neutralization test) that are constantly being improved in order to give reliable results. It should also be noticed that since the mean time to seroconversion is about 5 months after exposure to the virus [1], this antibody test lacks the ability to diagnose recently infected patients. Furthermore, it may not correlate well with the presence of infectious virus. Since HCV is usually present in clinical specimens at very low titers and an appropriate tissue culture system has not yet been established, other types of assays with very high sensitivity must be used in order to detect virus or virus-related products.

We have developed a simple, highly sensitive assay for hepatitis C viral RNA based on the polymerase chain reaction (PCR). Using this assay we have been able to detect HCV RNA in clinical samples such as anti-HCV negative acute and chronic patients as well as anti-HCV positive chronic patients, and in samples from experimentally infected chimpanzees. Such an assay should be valuable whenever it is important to detect the presence of infectious HCV.

Methods

Subjects. In order to establish the assay, we used RNA extracted from liver of a chimpanzee acutely infected with the H strain [6]. To determine the sensitivity of the assay we made serial dilutions of several samples whose infectivity titer had been previously determined in chimpanzees. We also studied 40 patients (31 male, 38 caucasian) with chronic NANBH who were being evaluated at the National Institutes of Health for inclusion in a trial of alpha interferon therapy [5]. All patients had elevations of serum aminotransferase values for at least 12 months, and all had a history of previous parenteral exposure to blood or blood products. Their clinical, serum biochemical and histopathological characteristics are shown in Table 1. Plasma or serum was collected at the time of entry into the therapeutic trial and stored at $-70\,^{\circ}$C until tested. All serum or plasma samples were tested for anti-HVC by commercially available enzyme-linked immunoassay (Ortho HCV ELISA). As control samples, we tested serum from 2 patients with chronic hepatitis B and from 7 normal individuals under the same conditions.

RNA extraction. 200 µl of serum or plasma samples were extracted by mixing each with 550 ul of extraction buffer (4.4 M guanidinium isothiocyanate, 0.5% sodium lauryl sarkosate [2], and 5 mM Tris-HCl pH 8.0), followed by one phenol-chloroform extraction and one with chloroform alone. The total nucleic acid extracted by this procedure from 200 µl of serum or plasma, was resuspended in 200 µl of RNase-free water and 4 µl of RNasin (40,000 µ/ml) and used as the template in the PCR reaction.

PCR method. The PCR primers were synthesized according to the published HCV sequence [8]. A first screening of the samples was carried out with one set of nested primers (Table 2, no. 2). The PCR negative samples were subsequently retested with several sets of nested primers that are located in different regions of the HCV genome (Table 2).

Table 1. Clinical characteristics of 40 patients with chronic non-A, non-B hepatitis

Mean age (yrs)	44.2 (range 20–69)
Mean duration hepatitis (yrs)	3.7 (range 1.1–22)
Mean serum ALT (u/L)	246 (range 63–835)
Mean serum AST (u/L)	125 (range 40–436)
Source of hepatitis (%):	
Transfusion	63%
IV Drug Abuse	25%
Occupational exposure	13%
Liver histopathology (%):	
Chronic persistent hepatitis	8%
Chronic active hepatitis	78%
Cirrhosis	15%

The first PCR reaction was combined with the reverse transcription step in the same tube containing a 100 μl reaction mixture made up as follows: 25 μl of serum or plasma nucleic acid extract, 0.5 mM of each outer primer, 200 mM of each of the 4 deoxynucleotide triphosphates, 2.5 units of Taq DNA polymerase (Perkin-Elmer Cetus), 10 units of avian myeloblastosis virus reverse transcriptase and [1x] Taq buffer consisting of 10 mM Tris-HCl pH 8.3, 50 mM KCl, 1.5 mM $MgCl_2$ and 0.01% gelatin. The thermocycler (Perkin-Elmer Cetus) was programmed to first incubate the samples for 20 minutes at 43 °C for the initial reverse transcription step and then to carry out 35 cycles consisting each of 94 °C for 1 minute, 45 °C for 1.5 minutes and 72 °C for 3 minutes. For the second reaction, either 1 or 10 μl were removed from the first reaction, and were added to a tube containing the second set of primers and all the other reagents but no reverse transcriptase. PCR was then carried out for another 35 cycles as described for the first reaction with the only difference that the extension step at 72 °C lasted 1.5 minutes instead of 3. The PCR products were analyzed by electrophoresis on 1% agarose gels containing 0.5 μg/ml ethidium bromide. The gels were viewed and photographed on a UV light box.

Results

PCR amplification of HCV RNA from infected chimpanzee liver resulted in a product of the expected size, even when 1 μl of the extract (containing 2.6 μg of total RNA) was diluted up to 10^4. This specimen was therefore used as a positive control in all further experiments conducted on serum samples.

Studies of the sensitivity of this assay showed that the PCR technique was able to detect HCV RNA at or beyond the level of chimpanzee infectivity. Thus the H plasma that had been shown infectious at a dilution of 10^{-6}/ml but not at 10^{-7} in chimpanzees, had a PCR titer of $10^{7.5}/25$ μl (Table 3). Sera from 2 clinical samples titered $10^{1.5}$ and $10^{2.5}/25$ μl (Table 3). This finding is in keeping with the generally low infectivity titers of HCV in blood, as has previously been reported [6]. The PCR product synthesized from HCV

Table 2. Nested primer sets used for amplification of HCV RNA

Outer primers No.	Sense	Position	Sequence (5'--3')	Inner primers Sense	Position	Sequence (5'--3')	Second PCR product size
1	+	781–804	CGTCCTGGGCCATTAAGTGGGAGT	+	901–924	ACCTCGTAATAACTTAATGCAGCAT	569 bp
	–	1524–1501	CCCGAAGAGGAGTGAGATGGTTAT	–	1464–1441	TGACCATTTGCACGTAATGGCCTC	
2[a]	+	2127–2147	TTCTACTTGAAAGGCTCCTCG	+	2199–2219	EcoR1-GCGGGGTGTGCACCCGTGGAGTG	606 bp
	–	2840–2821	TGAATGACAGAAGATGAGAT	–	2804–2784	Hind III-TACTTCGAGGGGGATAGCCTT	
3	+	2407–2427	TAGTACTCAACCCCTCTGTTG	+	2442–2462	TTTGGTGCTTACATGTCCAAG	363 bp
	–	2840–2821	TGAATGACAGAAGATGAGAT	–	2804–2784	Hind III-TACTTCGAGGGGGATAGCCTT	
4	+	4742–4764	CTGTGGAGCTGAGATCACTGGAC	+	4760–4802	CGATGAGGATCGTCGGTCCTAGG	623 bp
	–	5414–5430	CAAGCGGATCGAAGGAG	–	5362–5383	ACCCTGGCTAGATGTTGCCGCCCA	
5	+	6661–6684	GCGTGGCCATCAAGTCCCTCACCG	+	6728–6751	GAACTGCGGCTATCGCAGGTGCCG	532 bp
	–	7273–7299	ACTAGATGGAGGTTAGTAAGTTTCTGA	–	7235–7259	ACGCTCTAGATGCCCCGGACGATGA	

[a] Set of primers used for the initial screening of the 40 chronic NANBH patients.

Table 3. Sensitivity of the PCR assay

Sample	Chimp infectivity titer	PCR titer
H strain (human plasma)[a]	$10^{6.5}$ CID/ml	$10^7/25\ \mu l$
F strain (human plasma)[a]	$\geqslant 10^0,\ \leqslant 10^2$ CID/ml	$10^1/25\ \mu l$
Acute chimp liver	Unknown	$10^4/1\ \mu l$
Chronic patient (K.)	Unknown	$10^1/25\ \mu l$
Acute patient (H.F.)	Unknown	$10^2/25\ \mu l$

[a] See reference no. 6

RNA was not detectable after the first PCR reaction in most clinical samples, the H serum being the only exception because of its high titer.

A group of 40 patients with chronic NANBH were tested for HCV RNA by PCR. Of these, 36 had anti-HCV by the ELISA assay and 4 were seronegative. In preliminary studies, 16 of the seropositive patients and none of the seronegative patients had HCV RNA in their plasma. Patients with HBV infections and patients without liver disease did not react in the test. The first screening was carried out with a set of nested primers whose sequences fall in the predicted protease/helicase coding region of the HCV genome that has some sequence conservation among flaviviruses. Since we knew that our test had very high sensitivity and that there is a very high degree of sequence variability in HCV isolates, all the PCR negative patients were subsequently retested with several sets of nested primers located in different positions between the 5′ and the 3′ ends of the HCV genome. During this process, we selected the sets of primers that proved to be capable of detecting HCV RNA more frequently than the others (Table 2). These primers produced PCR products between approximately 400 and 600 bp. The use of several sets of primers greatly improved the sensitivity of the PCR test by increasing the number of anti-HCV positive, PCR-positive patients from 16 to 31/36. Moreover, with these modifications to our procedure, we were able to diagnose HCV infection in all the four seronegative patients. In several instances where we had PCR failures and subsequently sequenced the DNA across the region covered by the primers, we found enough divergence in the target sequence compared to the Chiron sequence from which the primers were made, to account for the lack of amplification (data not shown).

In order to simplify the use of multiple primer sets we tried a co-amplification approach. In these experiments, still based on double PCR followed by electrophoresis, three primer sets were successfully combined together in the same PCR reaction mix, showing the ability to co-amplify multiple target regions of the HCV genome. These primer sets were selected in order to obtain three PCR products not overlapping and with distinctive sizes. This procedure that needs further investigation, has the important advantage of making this HCV PCR test less expensive and less time consuming and therefore more suitable for routine analysis.

Discussion

Due to a high degree of sequence variability, it has been found necessary to test clinical samples with several sets of nested primers. However, it appears that the use of multiple primer sets from diverse regions of the genome in one and the same reaction will yield similar results. We have recently cloned and sequenced the 5' end of the genome of the H strain of HCV (Wychowski, in preparation), whereby the sequence of the non-coding region (approximately the first 350 nucleotides) was found to be more than 98% identical to the sequence that has recently been published [10]. It should be possible to make primers based on this sequence that will be able to amplify most if not all HCV RNA's. Such primers are currently being tested and may substitute for the multiple primer approach described in this paper.

Recent reports demonstrate a close relationship between PCR results and infectivity of blood donors [7]. Since the PCR assay is a measure of the viral RNA, it should correlate well with infectious virions as it is not likely that there is free viral RNA in the plasma. We have determined that the PCR assay does, in fact, correlate with infectivity in a general way, but it is important to determine if it can be used as a reliable measure of infectivity.

References

1. Alter HJ, Purcell RH, Shih JW, Melpoder JC, Houghton M, Choo QL, Kuo G (1989) Detection of antibodies to hepatitis C virus in prospectively followed transfusion recipients with acute and chronic non-A, non-B hepatitis. N Engl J Med 321: 1494–1500
2. Chirwin J, Przybyla AE, MacDonald RJ, Rutter WJ (1979) Isolation of biologically active ribonucleic acid from sources enriched in ribonuclease. Biochemistry 18: 5294–5299
3. Choo QL, Kuo G, Wiener AJ, Overby LR, Bradley DW, Houghton M (1989) Isolation of a cDNA clone derived from a blood-borne non-A, non-B viral hepatitis genome. Science 244: 359–362
4. Cristiano K, Baker B, Di Bisceglie AM, Feinstone SM (1991) Detection of hepatitis C viral RNA by the polymerase chain reaction. In: Hollinger FB, Lemon SM, Margolis H (eds) Viral hepatitis and liver disease. Williams & Wilkins, Baltimore, pp 374–376
5. Di Bisceglie AM, Martin P, Kassianides C, Lisker-Melman M, Murray L, Waggoner J, Goodman Z, Banks S, Hoofnagle JH (1989) Recombinant interferon alpha therapy for chronic hepatitis C. N Engl J Med 321: 1506–1510
6. Feinstone SM, Alter HJ, Dienes HP, Shimizu Y, Popper H, Blackmore D, Sly D, London WT, Purcell RH (1981) Non-A, non-B hepatitis in chimpanzees and marmosets. J Infect Dis 144: 588–598
7. Garson JA, Tedder RS, Briggs M, Tuke P, Glazebrook JA, Trute A, Parker D, Barbara JAJ, Contreras M, Aloysius S (1990) Detection of hepatitis C viral sequences in blood donations by "nested" polymerase chain reaction and prediction of infectivity. Lancet 335: 1419–1422

8. Houghton M, Choo QL, Kuo G (18 November 1988). European Patent Application number 88310922.5. Publication number 0318216

9. Kuo G, Choo QL, Alter HJ, Gitnick GL, Redeker AG, Purcell RH, Miyamura T, Dienstag JL, Alter MJ, Stevens CE, Tegtmeier GE, Bonino F, Colombo M, Lee WS, Kuo C, Berger K, Shuster JR, Overby LR, Bradley DW, Houghton M (1989) An assay for circulating antibodies to a major etiologic virus of human non-A, non-B hepatitis. Science 244: 362–364

10. Okamoto H, Okada S, Sugiyama S, Yotsumoto S, Tanaka T, Yoshizawa H, Tsuda F, Miyakawa Y, Mayumi M (1990) The 5'-terminal sequence of the hepatitis C virus genome. Jpn J Exp Med 60: 167–177

Authors' address: Dr. Karen Cristiano, Instituto Superiore di Sanità, Laboratorio di Virologia, Reparto Virus Epatitici, Viale Regina Elena 299, I-00161 Roma, Italy.

Arch Virol (1992) [Suppl] 4: 179–183
© Springer-Verlag 1992

Detection and characterization of hepatitis C virus sequence in the serum of a patient with chronic HCV infection

E. Schreier[1], K. Fuchs[2], M. Höhne[1], M. Motz[3], R. Zachoval[4], J. Esteban[5], S. Dittmann[1], F. Deinhardt[2], and M. Roggendorf[2]

[1] Robert Koch-Institute of the Federal Health Office, Berlin
[2] Max-von-Pettenkofer-Institute, University of Munich
[3] Mikrogen, Munich
[4] Klinikum Großhadern, University of Munich, Federal Republic of Germany
[5] Ciudad Santaria, Vall d'Hebron, Barcelona, Spain

Summary. A cDNA fragment corresponding to the nonstructural gene region of Hepatitis C virus was cloned and sequenced. cDNA was obtained by reverse transcription of viral RNA extracted from serum of a German patient with chronic post transfusion hepatitis. "Nested" PCR resulted in a cDNA fragment of 345 nt. The sequence showed a homology of 96% to the American prototype HCV.

*

Recently, the polymerase chain reaction (PCR) has been developed for detection of viral RNA of Hepatitis C virus (HCV) in serum samples [1, 2]. We have amplified, cloned and sequenced a part of the HCV genome from a serum sample of a German patient (MOJ) with chronic post transfusion hepatitis.

Three sets of primer pair sequences which correspond to the NS3/NS4 region of HCV, synthesized according to the prototype HCV sequence (ptHCV) [3] were used for "nested" PCR technique (Fig. 1, Table 1).

RNA isolation from 500 μl serum was done by a 3 h incubation at 45 °C in 50 mM Tris/HCl pH 8.0, 100 mM NaCl, 10 mM EDTA, 4% SDS, 3 mg/ml Proteinase K and subsequent extraction of aqueous phase with phenol/TE, phenol/chloroform and chloroform. One-fifth of the RNA preparation was reverse transcribed into cDNA with the synthetic antisense primer MRMM3 using 20 units of AMV reverse transcriptase (Boehringer/Mannheim, FRG).

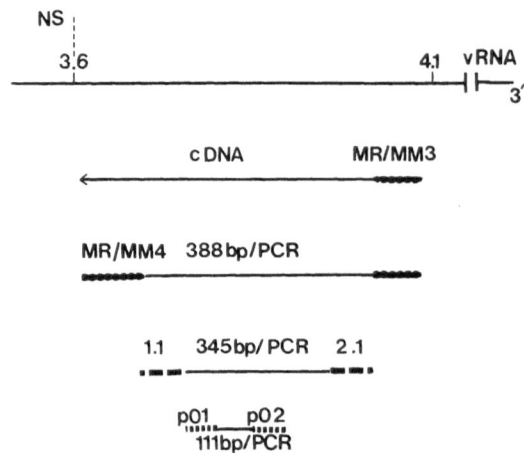

Fig. 1. Scheme of the amplification of a cDNA fragment from the nonstructural region of HCV by "nested" PCR

Table 1. Primer sequences and expected product size after PCR: Nucleotide positions are numbered according to the ptHCV sequence [3]. Restriction sites (EcoRI/HindIII) introduced to facilitate subsequent cloning of PCR products are underlined

Primer	Sequence (5'→3')	Nucleotide positions	Product size
MRMM3 antisense	GGCTGGTGACAGCAGCTGT	4050–4032	
MRMM4 sense	GCCGCGTATTGCCTGTCA	3663–3680	388 bp
2.1 antisense	AGCAAGCTTTAGGCGGGGTTACCAGG	4010–3996	
1.1 sense	ACTGAATTCCTGCGTGGTCATAGTGG	3686–3702	345 bp
pO2 antisense	AAGCTTACGGTAAGT	3802–3793	
pO1 sense	GAATTCGCAGGGTCG	3703–3711	111 bp

In accordance with the method originally described by Saiki et al. [4], the cDNA was amplified adding the sense primer MRMM4 using a Gene Amp DNA amplification reagent kit (Perkin-Elmer Cetus, CT, USA). After an initial 2 min denaturation at 92 °C, 35 cycles of PCR were carried out on all samples as follows: denaturation for 1 min at 92 °C, annealing of primers for 2 min at 37°, and extension for 3 min at 72 °C. The amplified cDNA

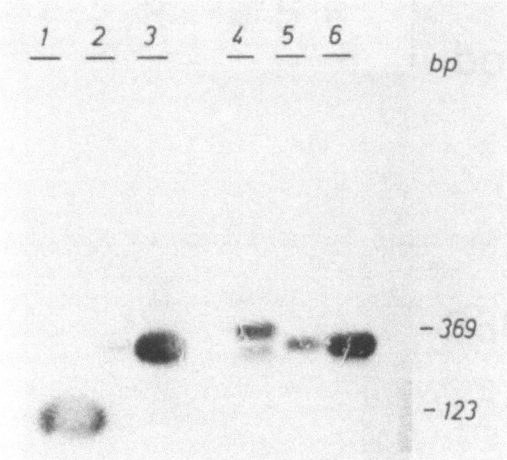

Fig. 2. Southern blot of PCR products of cDNA from serum of patient MOJ. Hybridization was done with the digoxigenin labelled synthetic DNA probe of 100 nt. Lane *1* "nested" PCR products with primer set p01/p02; *2* PCR products with primers 1.1/1.2; *3* "nested" PCR products with primer set 1.1/1.2; *4* PCR product with primer set MRMM3/MRMM4; *5* and *6* Constructs of PCR using the cloned HCV fragment as template, lane 5 contains a 1:10 dilution of template concentration used in lane 6

fragments were analyzed on a 1.8% agarose gel. Bands were visualized by ethidium bromide staining.

The specificity of the PCR products was demonstrated by hybridization with a digoxigenin labelled probe of 100 nucleotides generated from synthetic oligodesoxynucleotides of the same region (nt 3703–3802 of ptHCV sequence) and cloned in pGem-2. The probe has no sequence overlap to primers 1.1 and 2.1 (Fig. 1). Figure 2 shows the Southern blot of the expected PCR products of 345 and 111 nt hybridized with the synthetic probe.

The amplified "nested" PCR fragment of about 345 bp was excised from the agarose gel, purified, and cloned into the EcoRI/HindIII site of pUC8 [5].

The HCV was determined by the dideoxynucleotide chain termination method [6] on alkaline denaturated recombinant plasmid DNA [7] of three independent clones. The sequence of 293 nucleotides of our HCV isolate of the NS3/NS4 region is given in Fig. 3. We compared this sequence of the cDNA fragment with the ptHCV sequence [3] and with sequences of 5 isolates from Japanese patients with chronic Hepatitis Non-A, Non-B [8] (Fig. 3). The sequence of the cDNA fragment from our isolate MOJ shows extensive homology to pT HCV. Only 12 different nucleotides (indicating 96% homology) and 4 amino acid differences [A-T (2x), T-V, V-I] were observed. However, to the 5 Japanese isolates our HCV sequence showed only about 76% homology at the nucleotide level and about 80% homology at the amino acid level. These results indicate that our isolate is highly conserved in the NS3/NS4 gene region as compared to the ptHCV isolate.

```
a 3703 GCAGGATCGTCTTGTCCGGGAAGCCGGCAATTATACCTGACAGGGAAGTCCTCTACCGGG
b      -----G---------------------------C------------------------A-
c                                                             -A--
d                                                             ----
e                                                             -AA-
f                                                             -C--
g                                                             -A--

a 3763 AGTTTGATGAGATGGAAGAGTGCTCTCAGCACTTACCGTACATCGAGCAAGGGATGATGC
b      ---C--------------------------------------------------------
c      ----------------G------G-CTCA---C-T--T--------A-----A---CA--
d      ----C-----------G------G--TCA---C-C--T--------A--G--A---CG--
e      ----C--C--A-----------G-CTCA---C-T--T--------A--G--A---CA--
f      ----C-----A-----------TG-CTCA---C-C--T--------A-----A---CA--
g      ----------------G------G-CTCA---C-T--T--------A-----A---CA--

a 3823 TCGCCGAGCAGTTCAAGCAGAAGGCCCTCGGCCTCCTGCAGACCGCGTCCCGTCAGGCAG
b      ------------------------------------------------------------
c      ----------A--------------G-----G--G-----A--A--C---AAG--A--G-
d      ----------A--------------G-----G--GT-------A--CA--AAG--A--G-
e      ----------A-----------A--G-----TT-G-------A--TA--AAG--A--G-
f      ------A--A--------------G-----GT-G-----A--G--CA--AAG--A--G-
g      ----------A--------------G-----G--G-----A--G--C---AAG--A--G-

a 3883 AGGTTATCACCCCTACTGTCCAGACCAACTGGCAAAAACTCGAGGTCTTCTGGGCGAAGC
b      -------G-----G----------------------------AC--------------
c      ---C-GCTG-T--CGTG--AG--T----G------GCC--T----C--------------
d      ---C-GCAG-T--CGTG---G--T-T--A----GGGCT--T---AC--------------
e      ---C-GCAG----CGTG--GG--T----G----GGGCC--T---ACT------------
f      ---C-GC-G-T--CGTG--GG--T----G----GT-CC--T----C--------------
g      ---C-GCTG-T--CGTG--AG--T----G------GCC--T----C--------------

a 3943 ACATGTGGAACTTCATCAGTGGGATACAATACTTGGCAGGCCTGTCAACGTTG
b      -T------------------------------G---T-------C--
c      ------------------C-------G--TC-A------T----C--CC--
d      ---------T--------C-------G---C-A------T----C--TC--
e      ---------T--------C-------A-------T-A--C--T---
f      ---------T--------C--------C-A------T----C--TC--
g      ---------TC-------C-------G--TC-A------T----C--CC--
```

Fig. 3. cDNA sequences of the putative nonstructural gene region of HCV: nucleotide positions are numbered according to the ptHCV sequence [3]; *a* sequence from our isolate, *b* ptHCV sequence; *c–g* sequences from 5 Japanese isolates [8]

Further studies on the structural and nonstructural genes are necessary to demonstrate that our isolate is closely related to the ptHCV. The low homology to the Japanese isolates described by Kaneko et al. [8] confirms findings from Okamoto et al. [9] and Takeuchi et al. [10] that there are two types of isolates in Japan, one closely and one distantly related to ptHCV.

References

1. Garson JA, Tedder RS, Briggs M, Tuke P, Glazerbook JA, Trute A, Parker D, Barbara JAJ, Contreras M, Aloysius S (1990) Detection of hepatitis C viral sequences in blood

donations by "nested" polymerase chain reaction and prediction of infectivity. Lancet 335: 1419–1422

2. Weiner AJ, Kuo G, Bradley DW, Bonino F, Saracco G, Lee C, Rosenblatt J, Choo QL, Houghton M (1990) Detection of hepatitis C viral sequences in non-A, non-B hepatitis. Lancet 335: 1–3

3. Houghton M, Choo QL, Kuo G (1989) NANBV diagnostics and vaccines. Chiron Corporation, European Patent Application No. 318216

4. Saiki RK, Gelfand DH, Stoffel S, Scharf SJ, Higuchi R, Horn GT, Mullis KB, Erlich HA (1988) Primer-directed enzymatic amplification of DNA with a thermostable DNA polymerase. Science 239: 487–491

5. Vieira J, Messing J (1982) The pUC plasmid, a M13mp7-derived system for insertion mutagenesis and sequencing with synthetic universal primers. Gene 19: 259–268

6. Sanger F, Nicklen S, Coulson AR (1977) DNA sequencing with chain-terminating inhibitors. Proc Natl Acad Sci USA 74: 5463–5467

7. Chen EJ, Seeburg PH (1985) Supercoil sequencing: a fast simple method for sequencing plasmid DNA. DNA 4: 165–170

8. Kaneko S, Kunok, Yanagi M, Unoura M, Hattori N, Murakami S, Kobayashi K (1991) Sequence analysis of Hepatitis C virus genome isolated from 5 patients with chronic non-A, non-B hepatitis. In: Hollinger FB, Lemon SM, Margolis H (eds) Viral hepatitis and liver disease. Williams & Wilkins, Baltimore, pp 364–367

9. Okamoto H, Okada S, Sugiyama Y, Yotsumoto S, Tanak T, Yoshizawa H, Tsuda F, Miyakawa Y, Mayumi M (1990) The 5′-terminal sequence of the Hepatitis C virus genome. Japan J Exp Med 60: 167–177

10. Takeuchi K, Boonmar S, Kubo Y, Katayama T, Harada H, Ohbayashi A, Choo QL, Kuo G, Houghton M, Saito I, Miyamura T (1990) Hepatitis C viral cDNA clones isolated from a healthy carrier donor implicated in post-transfusion non-A, non-B hepatitis. Gene 91: 287–291

Authors' present address: Prof. M. Roggendorf, Institute of Medical Virology and Immunology, University Clinics, Hufelandstrasse 55, D-W-4300 Essen, Federal Republic of Germany.

Arch Virol (1992) [Suppl] 4: 184–185
© Springer-Verlag 1992

Sequence analysis of PCR amplified hepatitis C virus cDNA from French non-A, non-B hepatitis patients

J. Li, S. Tong, L. Vitvitski, and **C. Trépo**

INSERM U-271, Lyon, France

Summary. Using nested PCR and hybridization techniques, it was found that the predominant HCV strain in France is related to the U.S. strain rather than to Japanese isolates.

*

Hepatitis C virus (HCV) is a plus strand RNA virus with a 10 kb genome. Significant sequence divergence has been found between the prototype U.S. HCV strain and Japanese HCV isolates [1–3]. Sequence data of HCV strains circulating in other parts of the world would be extremely useful in estimating the variability of the viral genome as well as in the efficient detection of HCV infection by PCR and design of effective HCV vaccines. We have recently succeeded in detection of HCV infection in France by the nested PCR. PCR products were cloned into M13 vector and sequenced. A 407 bp sequence in the NS3 region was determined for 4 such cases. The nucleotide sequence in 3 of the 4 cases showed 97–98% homology to the prototype HCV strain, while the remaining case had a sequence moderately (92%) related to the Japanese strain [1]. At the amino acid level, the first 3 isolates were 97–98% homologous to the U.S. strain while the 4th isolate was 97% homologous to the Japanese strain. Hybridization experiments involving more samples suggested that the predominant HCV strain circulating in France is closely related to the prototype U.S. strain rather than to the Japanese isolate. It will be of interest to learn whether there is a similar prevalence of the prototype-related HCV strains in other European countries.

References

1. Kubo Y, et al (1989) Nucleic Acids Res 17: 10367–10372

2. Enomoto, et al (1990) Biochem Biophys Res Commun 170: 1021–1025
3. Takeuchi K, et al (1990) Gene 91: 287–291

Authors' address: Dr. Ji-su Li, INSERM U271, 151 Cours A. Thomas F-69003 Lyon, France.

Arch Virol (1992) [Suppl] 4: 186–190
© Springer-Verlag 1992

Antigenicity of synthetic peptides derived from C100 protein of hepatitis C virus

P. Neri[1], **A. Bonci**[1], **G. Campoccia**[2], **G. Fanetti**[2], **L. Lozzi**[1], **M. Scarselli**[1], and **P. Soldani**[1]

[1] Dipartimento di Biologia Molecolare, Universitá di Siena, [2] Servizio di Immunoematologia, Siena, Italy

Summary. Synthetic octapeptides spanning the 119–147 region of the Hepatitis C Virus (HCV) C100 protein were tested on HCV positive sera. The 138–145 region proved to be antigenic and possibly able to avoid undesired cross-reactions.

*

Seroepidemiological studies suggest that HCV is a major cause of non-A, non-B hepatitis (NANBH) [2]. Although a test based on the single HCV non-structural C100 protein is at present available it is necessary to prevent any cross-reactivity of the antigen used with unrelated viral proteins [3].

The amino acid sequence of C100 protein was divided into overlapping octapeptides (Fig. 1) and compared with all the viral sequences available in a protein database (National Biomedic Research Foundation, NBRF), using the SCAN routine of the P.I.R. (Protein Identification Resource, NBRF) software package. The 119–147 region, corresponding to an epitopic area in the hydrophobicity profile, showed sequence homologies with the following viral products: Simian Immunodeficiency Virus (gag polyprotein residues 7–11), Epstein-Barr Virus (ECRF protein, residues 57–61), Varicella-Zoster Virus (gene 22 and 32 products residues 841–845 and 136–143 respectively) (Fig. 2a). No matches with other viral proteins were found in the regions shown in Fig. 2b.

Twenty-two octapeptides spanning the 119–147 region were concurrently synthesized on activated polyethylene rods using the 9-fluorenyl-methyloxicarbonyl (Fmoc)/t-butyl protecting group combination and the highly activated pentafluorophenyl esters (Epitope Scanning Kit, CRB,

```
119 V L S G K P A I
      L S G K P A I I
        S G K P A I I P
          G K P A I I P D
            K P A I I P D R
              P A I I P D R E
                A I I P D R E V
                  I I P D R E V L
                    I P D R E V L Y
                      P D R E V L Y R
                        D R E V L Y R E
                          R E V L Y R E F
                            E V L Y R E F D
                              V L Y R E F D E
                                L Y R E F D E M
                                  Y R E F D E M E
                                    R E F D E M E E
                                      E F D E M E E C
                                        F D E M E E C S
                                          D E M E E C S Q
                                            E M E E C S Q H
                                              M E E C S Q H L¹⁴⁷
```

Fig. 1. Sequence of the overlapping octapeptides in the 119–147 region of HCV C100 protein

a

¹¹⁹VLSGK P A I
⁷VLSGK K A D   Gag polyprotein Simian Immunodeficiency Virus

¹²²GKPAI I P D
¹⁶⁰⁶GKPAI N V Q   Genome Polyprotein Murray Valley Enchephalitis Virus

¹²⁴PAIIP D R E
¹⁴¹PAIIP T E E   Probable early E13 21K protein Mastadenovirus

¹²⁶PAIIPDRE V L
¹³⁶PAIIPDRE Q P   Gene 32 Protein Varicella-Zoster Virus

¹³⁵REFDE M E E
⁸⁴¹REFDE L S R   Gene 22 Protein Varicella-Zoster Virus

¹²⁸PDREV L Y R
⁵⁷PDREV A H L   Hypothetical ECRF4 Protein Epstein Barr Virus

¹⁴⁴SQHLP Y E
⁵⁷SQHLP E L   Haemagglutinin praecursor   Influenza A Virus
                     (strain A/Duck/Alberta)

b

¹³¹EVLYREFD

¹³⁷FDEMEECS

DEMEECSQ

EMEECSQH¹⁴⁶

Fig. 2. a Identically matching pentapeptides in HCV C100 protein (119–147 fragment) and other viral proteins, revealed with the P.I.R. program package. **b** Fragments in the 119–147 region showing no homologies with other viral proteins

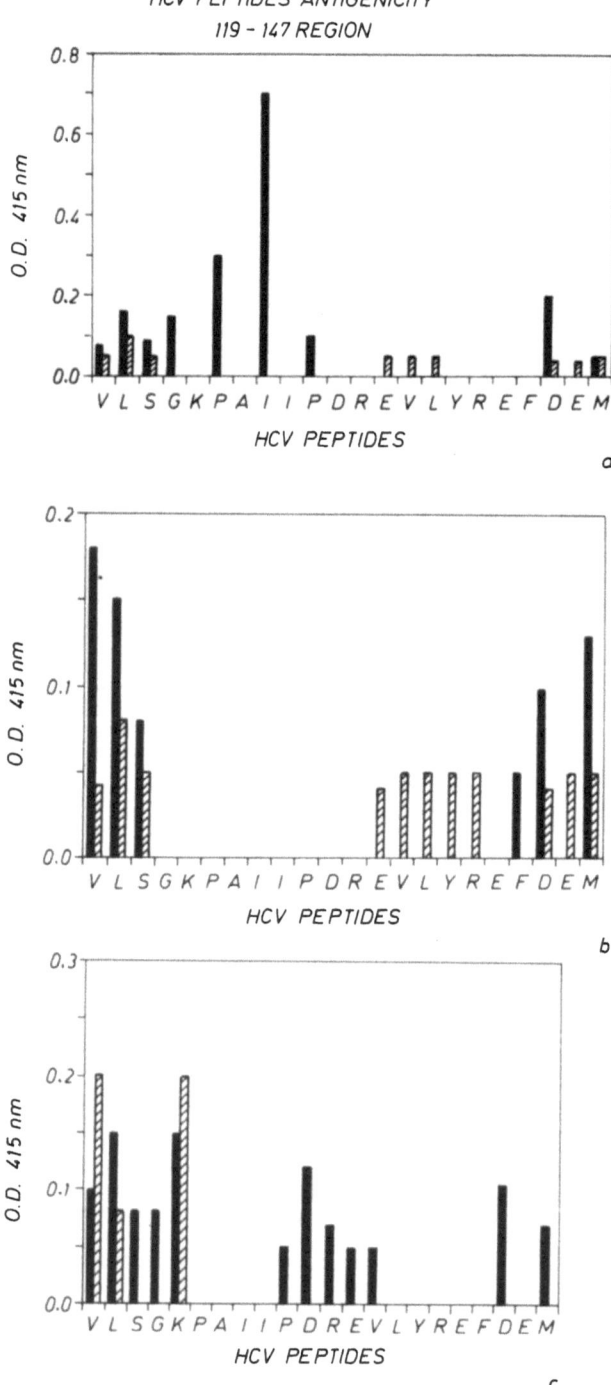

Fig. 3. Reactivity in ELISA of the overlapping octapeptides. The letters indicate the first aminoterminal amino acid shown in Fig. 1. Peptides were reacted with HCV positive blood donors sera (HCV2+, HCV3+, HCV4+) and with an HCV negative serum (HCV4−). Each value was the mean of two determinations. Antibody binding activity for each peptide is shown as a vertical line proportional to the absorbance value obtained after subtracting the background absorbance value. **a** HCV2+ ■, HCV4− ▨; **b** HCV3+ ■, HCV4− ▨; **c** HCV4+ ■; HCV4− ▨

Cambridge, UK). Side chain deprotection was obtained by the cleavage mixture of trifluoroacetic acid/phenol/ethanedithiol (95:2, 5:2,5) [1].

Support-coupled peptides were tested for their ability to bind antibodies both from HCV positive blood donors sera (Fig. 3 a, b, c) and from a patient with chronic non-A, non-B Hepatitis (Fig. 4); one peptide (^{138}DEMEECSQ^{145}) repeatedly proved to be antigenic. A serum positive for Varicella-Zoster virus was also tested on the same peptides and was found to bind a region other than the 138–145 (Fig. 5).

The fragment 138–145, therefore, seems to be able to avoid serological cross-reactions with unrelated antiviral antibodies and could be useful as diagnostic epitope.

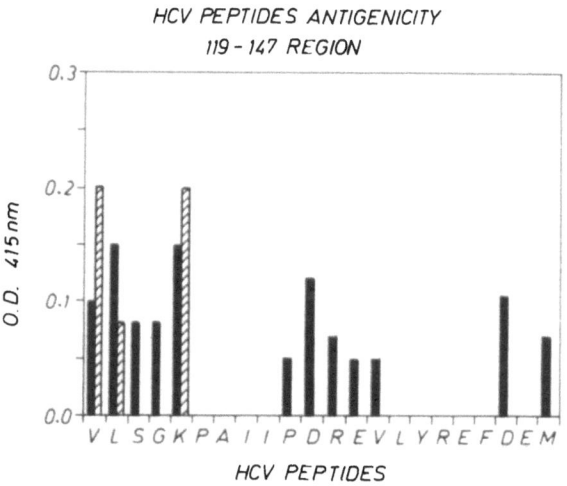

Fig. 4. Antigenicity of the synthetic peptides reacted with a serum from a chronic HCV positive patient (HCV5 + ■) or negative patient (▨)

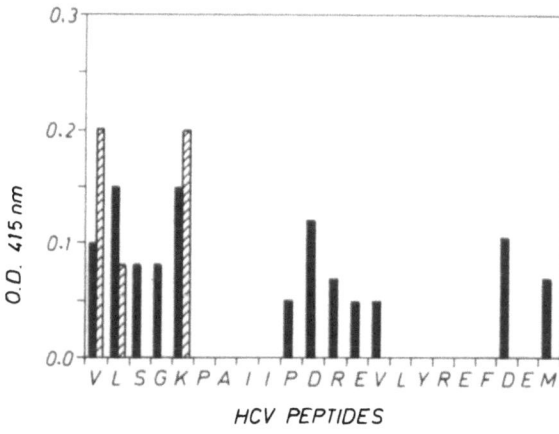

Fig. 5. Results of ELISA assays performed with Varicella-Zoster Virus positive/HCV negative (VHZV + ■) and HCV positive (HCV5 + ▨) sera

The potential use of this synthetic peptide to replace the whole antigen in HCV serological assays is currently under investigation.

Acknowledgements

The authors wish to thank Miss Silvia Scali, Miss Cecilia Fusi and Mr Stefano Bindi for their skilful technical assistance.

References

1. Geysen HM, Rodda SJ, Mason TJ, Tribbick J, Schoofs PG (1987) Strategies for epitope analysis using peptide synthesis. J Immunol Meth 102: 259–274
2. Kuo G, Choo Q-L, Alter HJ, Gitnick GL, Redeker AG, Purcell RH, Miyamura T, Dienstag JL, Alter MJ, Stevens CE, Tegtemeier GE, Bonino F, Colombo M, Lee W-S, Kuo C, Berger K, Shuster JR, Overby LR, Bradley DW, Houghton M (1989) An assay for circulating antibodies to a major etiologic virus of human non-A, non-B hepatitis. Science 244: 362–364
3. McFarlane IG, Smith HM, Johnson PJ, Bray GP, Vergani D, William R (1990) Hepatitis C virus antibodies in chronic active hepatitis: pathogenetic factor or false-positive result? Lancet 335: 754–757

Authors' address: Dr. P. Neri, Dipartimento di Biologia Molecolare, Università di Siena, I-53100 Siena, Italy.

Arch Virol (1992) [Suppl] 4: 191–195
© Springer-Verlag 1992

Localization of hepatitis C virus antigen(s) by immunohistochemistry on fixed-embedded liver tissue

D. Infantolino[1], M. Chiaramonte[2], A. R. Zanetti[3], R. R. Lesniewski[4], F. Bonino[5]

[1] Department of Pathology, Community Hospital, Castelfranco Veneto (TV)
[2] Department of Gastroenterology, University of Padua
[3] Institute of Virology, University of Milan, Italy
[4] Abbott Diagnostic Division, Abbott Laboratories, North Chicago, IL, USA
[5] Division of Gastroenterology, San Giovanni Battista Molinette Hospital, Turin, Italy

Summary. Using two sources of primary antibodies, we immunohistochemically stained hepatitis C virus-related antigen(s) on fixed-embedded liver specimens. These antigens were localized in the cytoplasm of hepatocytes. The results obtained serologically correlated well with immunohistochemistry.

Introduction

Hepatitis C virus (HCV) has recently been identified as a major cause of post-transfusion non-A, non-B (PT-NANB) hepatitis [1, 2]. A recombinant antigen (c100-3), representing a nonstructural protein of HCV, is currently being used as a target antigen in EIA and RIA to detect an anti-HCV (anti-c100-3) specific antibody. The presence of sequences of HCV-RNA have been observed in serum as well as in liver tissue of anti-HCV positive patients [3]. As previously reported [4], we detected intrahepatic hepatitis C virus related antigen(s) by immunohistochemistry. Here we report our preliminary immunohistochemical data, obtained on fixed and embedded liver biopsies, from patients with chronic hepatitis, using both human and rabbit anti-HCV antisera.

Material and methods

Antisera

We used two sources of anti-HCV primary antibodies: human and rabbit antisera.

Human Serum. IgG (final concentration 7.2 ug/ml) were obtained from a high-titre anti-HCV positive serum after precipitation, dialysis and protein A Sepharose chromatography.

Rabbit serum. Rabbit anti-HCV serum (r6096) was produced using a synthetic peptide (sp42), 42 amino acids in length, containing HCV sequences of the immunoreactive clone 5.1.1, immunogen. Anti-HCV (c100-3) titer of r6096 was determined to be greater than 1:100,000 by EIA, while pre-immune r6096 serum was nonreactive.

Patients

Thirty-one liver biopsies from pedigreed patients, with post-transfusion or cryptogenic chronic liver disease (26 anti-HCV positive and 5 anti-HCV negative), were examined. As control we selected 24 liver biopsies from patients with anti-HCV negative chronic liver disease, of well defined aetiology (14 Primary Biliary Cirrhosis, 5 Alcoholic Liver Disease, 5 HBsAg positive).

Immunohistochemistry

Liver specimens. For this study formalin-fixed, paraffin-embedded liver biopsies were used. Before immunostaining, liver sections were treated with 0.5% digitonin/methyl alcohol, to increase tissue permeability [5] (modified) and then with 9% hydrogen peroxide/methyl alcohol 1:1 to inhibit endogenous peroxidase and peroxidase-like enzyme activities, for 30′ respectively. Parallel sections of each biopsy were then stained by the modified peroxidase–antiperoxidase techniques (PAP) as described below [6], using both human and rabbit primary antibodies. Peroxidase reaction was revealed using diaminobenzidine solution and hydrogen peroxide.

Human serum. The sections were incubated overnight at 4 °C in normal swine serum (NSS) containing 0.1% monoclonal anti-blood group antibodies (Dakopatts, Glostrupp, Denmark). They were then incubated with human anti-HCV IgG, diluted (0.05 ug/ml) in 0.05M Tris Buffer Saline pH 7.4 (TBS) containing 0.1% bovine serum albumin (BSA) and 0.1% NSS for 30′ at room temperature (r.t.). The sections were incubated in rabbit anti-human, swine anti-rabbit (both diluted 1:200) and pap-rabbit (1:100), for 30′ at r.t respectively, using an immunoperoxidase technique with four stages. The third and fourth stages (swine anti-rabbit and pap-rabbit) were repeated twice to amplify the signal.

Rabbit serum. Overnight incubation at 4 °C in NSS was carried out. R6096 antiserum and preimmune r6096 serum were diluted 1:1600 with Tris/BSA/NSS and then incubated overnight at 4 °C. Conventional Pap technique, using swine anti-rabbit (diluted 1:200) and pap-rabbit (1:100), for 30′ each at r.t, was performed. Bridge and pap steps were repeated twice to amplify the signal.

Evaluation of the results. All the preparations were read blindly. We considered as "positive" those specimens with well-defined cytoplasmic staining of the hepatocytes.

Specificity. Positive liver specimens were retested as follows: i) omitting the incubation with primary antibodies or substituting them with NSS, using both methods; ii) carrying out the peroxidase reaction, in the absence of antibodies to rule out endogenous enzyme activities, using both methods; iii) substituting immune r6096 with preimmune r6096 serum in the second method. To rule out cross-reactions with other viral antigens, hepatic and extra-hepatic infected tissues (such as HPV in both human uterine portio and dog buccal mucosae;

HD in both human and woodchuck liver; HSV in human buccal mucosae; CMV in human lung) were tested with r6096 antiserum.

To exclude cross-reactions with endogenous human antigens, normal tissues (such as skin, brain, ganglia, breast, intestine, lung, placenta, fetal liver) and neoplastic tissues (such as myoma, angioma, melanoma, schwannoma, lymphoma, squamous and adenocarcinoma) were stained with r6096 antiserum.

Results and conclusions

Both human and rabbit antisera reacted with antigen(s) in the cytoplasm of hepatocytes, but unfortunately a high background staining was observed using human antiserum and the signal to noise ratio was frequently unsuitable for a correct evaluation of the staining. However, the results closely resembled that obtained with rabbit antiserum.

Table I summarizes the results obtained with rabbit r6096 antiserum. R6096 reacted with 24 liver specimens obtained from 26 anti-HCV sero-positive, as well with 3 from 5 anti-HCV sero-negative patients with nonA, nonB chronic hepatitis. Out of 24 anti-HCV sero-negative control patients, 18 gave clear cut negative results in tissue, while the remaining 6 gave tissue staining (4 primary biliary cirrhosis, 1 alcoholic liver disease and one from an HBsAg+ve/antiHBe+ patient with very active chronic hepatitis).

The immunoreactions showed a cytoplasmic drop-like (Fig. 1) or sand-like pattern (Fig. 2). Topographically, clusters or trabeculae of positive cells were distributed throughout the lobules with a prevalently focal pattern.

The specificity of the staining was confirmed by negative results obtained in positive specimens excluding primary antibodies, or using normal swine serum instead of anti-HCV antisera, or by means of preimmunization rabbit serum in the place of immune r6096. Endogenous peroxidase activity was also excluded by performing the histochemical reaction without the im-

Table 1. Results of HCV immunostaining in 31 liver biopsies from patients with PT or cryptogenic nonA, nonB chronic hepatitis

Liver	Serum	
	anti-C100-3 +ve	anti-C100-3 −ve
HCAg (anti-sp42) +ve	24	3 (6)
HCAg (anti-sp42) −ve	2	2 (18)
	26	5 (24)

In brackets are reported the results obtained in 24 control specimens from patients anti-HCV negative with chronic liver diseases of other well-defined aetiology

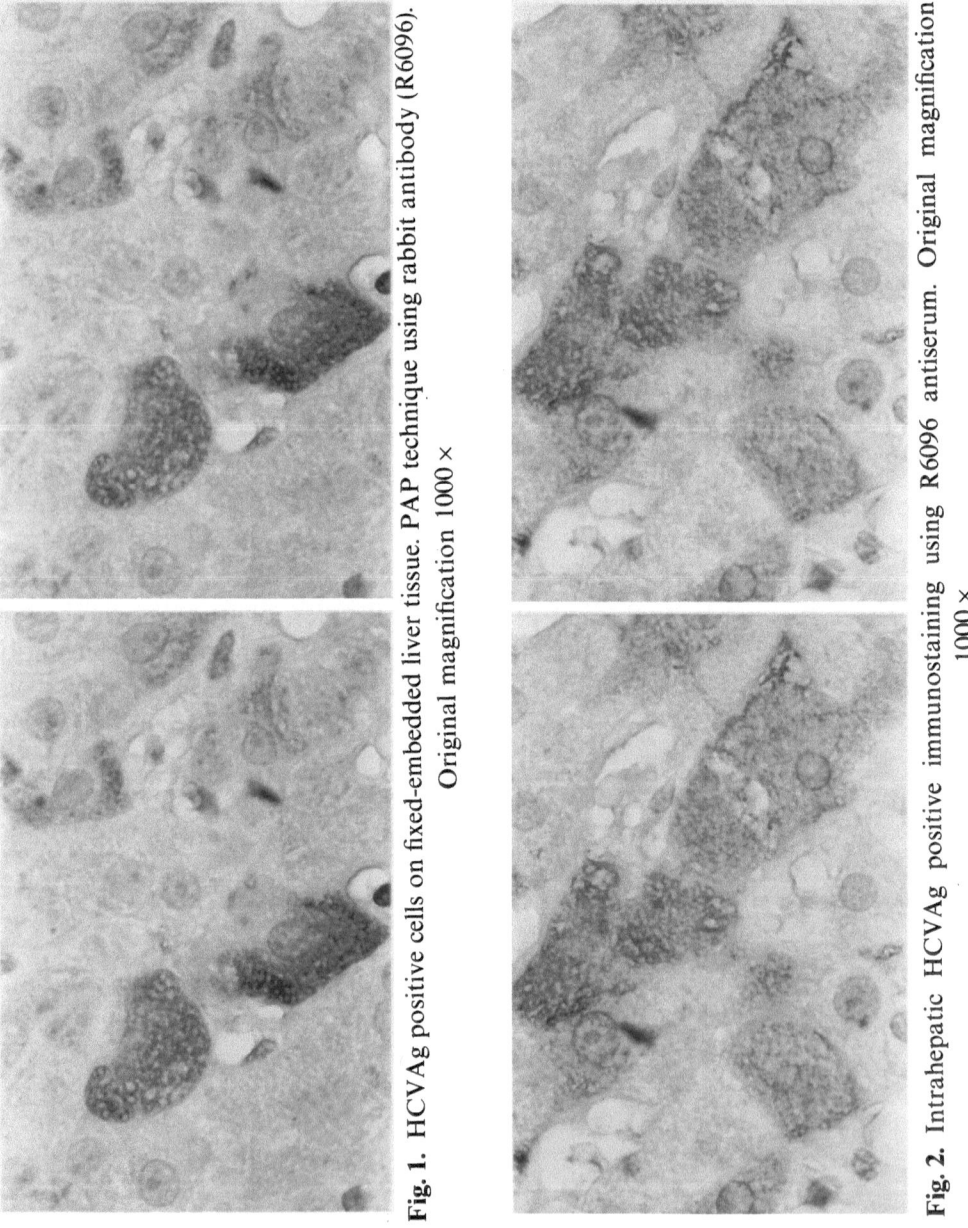

Fig. 1. HCVAg positive cells on fixed-embedded liver tissue. PAP technique using rabbit antibody (R6096). Original magnification 1000 ×

Fig. 2. Intrahepatic HCVAg positive immunostaining using R6096 antiserum. Original magnification 1000 ×

munological procedure. R6096 antiserum gives a weak cross-reaction staining in endothelial cells, reticular and other cells of mesenchymal origin, even in negative liver specimens. This staining observed both in hepatic and extrahepatic specimens, in normal and in neoplastic vascular areas may suggest a cross-reactivity between viral and host epitopes or the presence of antibodies unrelated to HCV, produced during the animal immunization. In conclusion, these preliminary results give evidence that hepatitis C virus antigen (HCAg) is detectable in fixed-embedded liver sections. Although the discussed cross-reactions limit the specificity of our rabbit antibody, the high concordance obtained between serology and immunohistochemistry provides evidence that HCAg detection bears clinical significance. The specificity is at least comparable to that of currently available serological tests.

References

1. Kuo G, Choo QL, Alter HJ, Gitnick GL, Redecker AG, Purcell RH, Miyamura T (1989) An assay for circulating antibodies to a major etiologic virus of non-A, non-B viral hepatitis. Science 244: 362–364
2. Choo QL, Kuo G, Weiner AJ, Overby LR, Bradley DW, Houghton M (1989) Isolation of cDNA clone derived from a blood-borne non-A, non-B viral hepatitis genome. Science 244: 359–362
3. Weiner AJ, Kuo G, Bradley DW, Bonino F, Saracco G, Lee C, Rosenblatt J, Choo QL, Houghton M (1990) Detection of hepatitis C viral sequences in non-A, non-B hepatitis. Lancet 335: 1–3
4. Infantolino D, Bonino F, Zanetti AR, Lesniewski RR, Barbazza R, Chiaramonte M (1990) Localization of hepatitis C virus (HCV) antigen(s) by immunohistochemistry on fixed-embedded liver tissue. Ital J Gastroenterol 22: 198–199
5. Infantolino D, Pinarello A, Cccato R, Baibazza R (1989) HBV DNA by in situ hybridization. A method to improve sensitivity on formalin-fixed, paraffin-embedded liver biopsies. Liver 9: 360–366
6. Sternberger LA, Hardy PH jr, Cuculis JJ, Meyer HG (1970) The unlabeled antibody enzyme method of immunohistochemistry. Preparation and properties of soluble antigen-antibody complex (horseradish peroxidase anti-horseradish peroxidase) and its use in the identification of Spirochetes. J Histochem Cytochem 18: 315–333

Authors' address: Prof. Dr. M. Chiaramonte, Department of Gastroenterology, Policlinico Universitario, Via Giustiniani, 2. I-35100 Padova, Italy.

Arch Virol (1992) [Suppl] 4: 196–198
© Springer-Verlag 1992

Identification of HCV-associated antigen(s) in hepatocytes

K. Krawczynski

Hepatitis Branch, Centers for Disease Control, Atlanta, GA, USA

Summary. HCV-associated antigens (HCAg) were localized morphologically using immunofluorescence methods. Fluorescence was found in the cytoplasm of individual liver cells or in groups of cells. Nuclear fluorescence was not observed. Blocking and absorption studies suggest that HCAg is related to nucleocapsid or envelope proteins.

*

Attempts to identify liver antigen(s) associated with posttransfusion hepatitis virus infection have not been adequately documented either immunologically or morphologically. In this study, hepatitis C virus (HCV) and disease-associated antigen (HCAg) were detected immunohistochemically in hepatocytes in patients with chronic hepatitis C and in chimpanzees experimentally infected with HCV isolates.

IgG fractions of chimpanzee and human sera were used as fluorescein isothiocyanate (FITC)-labeled probes to identify HCAg in liver. Immunologic studies were carried out to establish reactivity of immunoglobulins (PRC.2126, JEN.A) used for identification of HCAg in hepatocytes with HCV recombinant proteins. Enzyme immunoassays revealed that PRC.2126 and JEN.A immunoglobulins reacted with recombinant proteins expressed by clones F3 and F6 (encoded with NS3 and NS5, respectively). PRC.2126 reacted also with HCV nonstructural protein expressed by the F5 clone (equivalent of C100-3, NS4 [1, 2]).

In HCAg positive liver biopsies, the antigen was found in the entire cytoplasm of individual hepatocytes or in groups of liver cells. Liver cell nuclei never contained HCAg. HCAg fluorescence had a very fine granular, powder-like pattern with superimposed larger granules of distinct and brilliant fluorescence. HCAg was detected in nine tested chimpanzees with acute HCV hepatitis before and shortly after ALT elevation.

In control studies, FITC-labeled anti-HCAg did not react with chimpanzee liver biopsy specimens obtained either before HCV inoculation or during convalescence. HCAg-negative liver biopsy specimens were obtained from chimpanzees (n = 11) and from patients (n = 13) with various etiologies of viral hepatitis: type A, B, delta, enterically transmitted non-A, non-B, and non-viral hepatitis liver conditions. Hepatocellular reactivity of HCAg was not host-derived as evidenced by absorptions of anti-HCAg FITC-labeled probes with normal liver homogenates, IgG, fibrin/fibrinogen, or red blood cells.

The reactivity of FITC-labeled antibodies with hepatocellular HCAg was blocked by serum samples from patients and chimpanzees experimentally infected with HCV, but not by preinoculation serum samples from chimpanzees or by control samples from either primates or patients infected with hepatotropic viruses other than HCV.

Fluorescent antibody blocking studies of HCAg in hepatocytes with sera from chimpanzees (n = 11) and patients (n = 6) showed that the HCAg fluorescence, although specific for HCV infection, was unrelated to anti-HCV (C100-3, NS4) reactivity. Serum samples from posttransfusion hepatitis cases used for preparation of FITC-labeled reagents and found reactive with HCAg in hepatocytes were positive for anti-HCAg with titers of 1:100 or higher.

Absorptions of the fluorescein-labeled anti-HCAg immunoglobulins with HCV nonstructural recombinant proteins (F3 (NS3), F5 (NS4), and F6 (NS5)) revealed that HCAg fluorescence in hepatocytes depends on reactivity other than that of nonstructural proteins. Minimal inhibition of HCAg fluorescence was observed only after absorption with E3 (NS3) proteins. These observations suggested that hepatocellular reactivity of anti-HCAg may be related to HCV structural proteins (nucleocapsid or envelope).

A fluorescent antibody blocking assay on liver cryostat sections containing large deposits of HCAg was used to determine the presence of anti-HCAg in serum samples from experimental primates and patients. Profiles of anti-HCAg and anti-HCV (C100-3, NS4) in serum in relation to HCAg in hepatocytes were established in sera obtained from chimpanzees (n = 7) with acute HCV infection.

HCAg was detected in liver biopsy specimens obtained during the chronic phase of the disease from experimentally infected chimpanzees and from patients with various clinico-pathologic forms of chronic non-A, non-B hepatitis seropositive for anti-HCV. The antigen was identified in 5 of 10 chimpanzees with elevated ALT values and/or histopathologic changes in the liver characteristic for chronic HCV infection. HCAg was found in 11 of 12 patients with chronic persistent hepatitis (1/1), chronic active hepatitis (8/9), and active liver cirrhosis (2/2). All the animals with HCAg in hepatocytes and 10 of 11 patients with HCAg-positive liver biopsy specimens were positive for anti-HCAg and anti-HCV (C100-3) antibodies. The only serologically negative patient was infected with HIV and had very large deposits of HCAg in several liver cells.

The immunomorphologic data indicate specific association of hepatocellular HCAg with HCV-induced non-A, non-B hepatitis and the antigen(s) of hepatitis C virus. Morphologic identification of HCAg as a specific morphologic marker of HCV infection is important for clinical and experimental studies of the pathogenesis of hepatitis C and natural history of the HCV infection.

References

1. Choo Q-L, Kuo G, Weiner A, Overby LR, Bradley DW, Houghton M (1989) Isolation of a cDNA clone derived from a blood-borne non-A, non-B hepatitis virus genome. Science 244: 359–362
2. Kuo G, Choo Q-L, Alter H, Gitnick GL, Redeker AG, Purcell RH, Miyamura T, Dienstag JL, Alter MJ, Stevens CE, Tegtmeier GE, Bonino M, Colombo M, Lee WS, Berger K, Shuster JR, Overby LR, Bradley DW, Houghton M (1989) An assay for circulating antibodies to a major etiologic virus of human non-A, non-B hepatitis. Science 244: 362–364

Author's address: Dr. K. Krawczynski, Hepatitis Branch, Centers for Disease Control, Atlanta, GA 30333, USA.

VII   Hepatitis C virus and liver disease

Arch Virol (1992) [Suppl] 4: 201–204
© Springer-Verlag 1992

Hepatitis C virus (HCV) and autoimmune liver diseases

K.-H. Meyer zum Büschenfelde, G. Gerken, and **M. Manns**

I. Medizinische Klinik und Poliklinik der Johannes Gutenberg Universität Mainz,
Federal Republic of Germany

Summary. Anti-HCV tests were positive in 18–45% of sera from patients with autoimmune chronic active hepatitis. High gammaglobulin levels may result in false positive results, however, some sera show true positivity. PCR testing of such sera is necessary in order to determine whether HCV is directly involved in specific forms of the disease.

Introduction

Several subgroups of autoimmune type chronic active hepatitis can be distinguished by different circulating autoantibodies (Table 1) [5]. Recently, hepatitis C virus antibodies (anti-HCV) have been detected in patients with autoimmune liver diseases [3]. In this study we evaluated hepatitis C virus antibodies in our patients with various defined subgroups of autoimmune type chronic active hepatitis, primary biliary cirrhosis, and other hepatic and non-hepatic diseases.

Material and methods

Sera were tested from patients with different hepatic and non-hepatic liver diseases as listed in Table 2. Autoantibodies were determined by established and previously published techniques [5]. Furthermore, LKM antibodies were detected by Western blot against human liver microsomes and against recombinant human cytochrome P450 IID6 Liver Kidney Microsomal (LKM)-1 antigen [6]. Hepatitis C virus antibodies were determined by commercially available ORTHO-ELISA, using recombinant HCV polypeptide C-100-3 on the solid phase. As a confirmation assay, anti-HCV antibodies were also detected by recently available recombinant immunoblotting assay (RIBA) from Ortho. In this assay recombinant viral proteins 5-1-1 and C-100-3 are used as antigens.

Table 1. Heterogeneity of HBsAg negative chronic active hepatitis (CAH)

	ANA	LKM	SLA	SMA	AMA	Immunosuppressive treatment
CAH nonA, nonB	−	−	−	−	−	−
classical autoimmune type (lupoid) CAH	+	−	−	+	−	+
LKM antibody positive CAH	−	+	−	−	−	+
SLA antibody positive CAH	−	−	+	+/−	+/−	+
SMA antibody positive CAH	−	−	−	+	−	+
primary biliary cirrhosis	−	−	−	−	+	−

Table 2. Proportion of positive anti-HCV results in autoimmune liver diseases

	ELISA		RIBA-Assay
Autoimmune CAH			
ANA/SMA	12/67	18%	0%
LKM	20/44	45%	20%
SLA	7/123	34%	5%
Primary biliary cirrhosis	7/123	6%	0%
Primary sclerosing cholangitis	1/16	6%	0%
Systemic lupus erythematosus	0/14	0%	0%
Extrahepatic cholestasis	0/15	0%	0%
Inflammatory bowel disease	0/6	0%	0%
Healthy controls	2/100	2%	0%

Results

Sera were positive for anti-HCV from 42% (73/173) of patients with autoimmune CAH (Table 2). Between the various subgroups of autoimmune type chronic active hepatitis highest incidence was found in subgroup II associated with LKM antibodies where 45% (20/44) were positive. Furthermore, we tested serum samples obtained before and under immunosuppressive treatment from 5 patients with autoimmune CAH who were anti-HCV positive before treatment. In 4 of these patients, anti-HCV tests became negative under immunosuppressive therapy. The clinical course of one of these patients is demonstrated in Fig. 1. However, one patient with high titer anti-HCV antibodies did not show a significant reduction in anti-HCV titers. Moreover, OD-values of the anti-HCV ELISA correlated well with serum immunoglobulin levels. Correlation coefficients ranged between 0.615 for

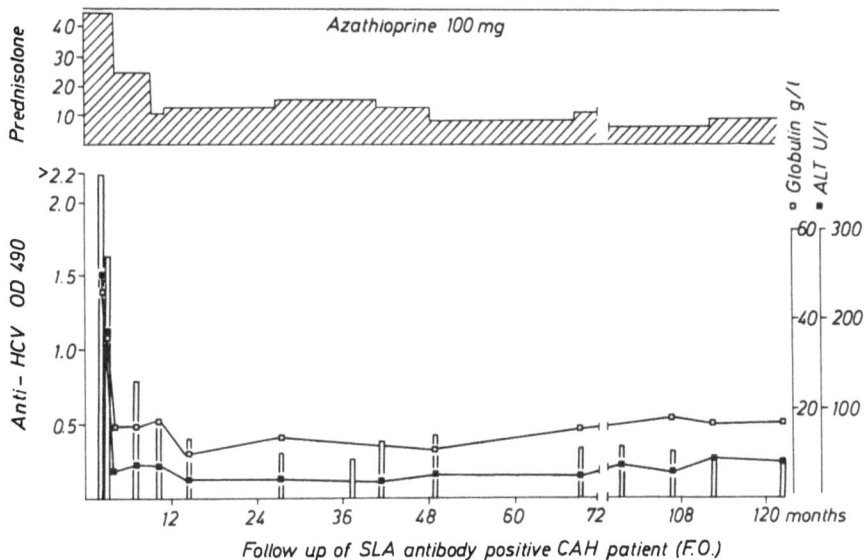

Fig. 1. Follow-up of an SLA antibody positive patient, whose serum became anti-HCV negative under immunosuppressive treatment

PBC sera and 0.951 for serum samples obtained from a Soluble Liver Antigen (SLA) positive patient responding well to immunosuppression.

Conclusion

In patients with various subgroups of autoimmune type chronic active hapatitis we found an incidence of positive anti-HCV tests ranging from 18 to 45%. Also, we found highest incidences in LKM-1 antibody positive CAH. We did not find as high an incidence as has been reported from Italy [3]. Furthermore, we could confirm that high gammaglobulin levels may result in false positive anti-HCV test [4]. However, some patients may have true positive HCV antibodies as confirmed by normal globulin levels and positive RIBA assay. In these patients, only the direct evaluation of HCV specific RNA applying PCR technology will clarify whether simultaneous infection with hepatitis C virus may contribute to this liver disease and whether ongoing HCV infection may induce specific forms of autoimmune type chronic active hepatitis.

Acknowledgements

The authors are indebted to Ms. U. Dang for competent technical assistance and Thomas Philipp for help in statistical evaluation. The authors thank Dr. M. Houghton and Ortho Diagnostics, USA, for the donation of anti-HCV ELISA test kits and RIBA test. Our work was supported by the Deutsche Forschungsgemeinschaft, Sonderforschungsbereich 311, Projekt A1.

References

1. Choo OL, Kuo G, Weiner AJ, Overby LR, Bradley DW, Houghton M (1989) Isolation of cDNA clone derived from a blood borne non-A, non-B viral hepatitis genome. Science 244: 359–361
2. Kuo G, Choo QL, Alter HJ, Gitnick GL, Redeker AG, Purcell RH et al (1989) An assay for circulating antibodies to a major aetiologic virus of human non-A, non-B hepatitis. Science 244: 362–364
3. Lenzi M, Ballardini G, Fusconi M, Cassani F, Selleri L, Volta U, Zauli D, Bianchi FB (1990) Type 2 autoimmune hepatitis and hepatitis C virus infection. Lancet i: 258–259
4. McFarlane IG, Smith HM, Johnson PJ, Bray GP, Vergani D, Williams R (1990) Hepatitis C virus antibodies in chronic active hepatitis: pathogenic factor or false positive result? Lancet i: 754–757
5. Manns M, Gerken G, Kyriatsoulis A, Staritz M, Meyer zum Büschenfelde K-H (1987) Characterization of a new subgroup of autoimmune chronic active hepatitis by auto-antibodies against a soluble liver antigen. Lancet i: 292–294
6. Manns M, Johnson EF, Griffin KJ, Tan EM, Sullivan KF (1989) Major antigen of liver-kidney microsomal autoantibodies in idiopathiac autoimmune hepatitis is cytochrome P450 db1. J Clin Invest 83: 1066–1072

Authors' address: Prof. Dr. K.-H. Meyer zum Büschenfelde, I. Medizinische Klinik und Poliklinik, Johannes Gutenberg Universität Mainz, Langenbeckstrasse 1, D-W-6500 Mainz 1, Federal Republic of Germany.

Arch Virol (1992) [Suppl] 4: 205–209
© Springer-Verlag 1992

Antibodies to hepatitis C virus in primary biliary cirrhosis

E. Bertolini[1], F. Marelli, P. Zermiani, P. M. Battezzati, M. Zuin[1], G. A. Moroni, and
M. Podda

Department of Internal Medicine, Istituto di Scienze Biomediche Ospedale
S. Paolo, University of Milan, and [1]Blood Transfusion Center, Ospedale S. Paolo,
Milan, Italy

Summary. We investigated the prevalence of anti-HCV in 160 consecutive patients with primary biliary cirrhosis. By ELISA, 19 (12%) were positive, as compared to a 68% prevalence in 135 patients with chronic non-A, non-B hepatitis. Serum IgG levels were significantly higher in the anti-HCV positive group. By RIBA, seropositivity was confirmed for 4 patients, whereas 7 were indeterminate. A slight, non-significant reduction of life expectancy was found in anti-HCV positive patients. Until reliable and independent confirmatory tests become available, definitive conclusions on the importance of anti-HCV positivity in primary biliary cirrhosis are improper.

Introduction

A high prevalence of antibodies against Hepatitis C Virus (anti-HCV), as determined by enzyme immunoassay (ELISA) in patients with autoimmune chronic hepatitis, has been reported by several investigators [1–3]. In Primary Biliary Cirrhosis (PBC), a chronic, progressive, cholestatic liver disease probably related to abnormalities of immune regulation [4], highly variable rates of seropositivity, from 0 to 42%, have been reported [1, 2, 5–8]. Indeed, the significance of ELISA anti-HCV positivity in autoimmune liver diseases, as well as in other immunologically-related diseases, is controversial [9, 10]. Lack of specificity of the assay may be responsible for the uncertainty. It has also been suggested that viral infections might trigger the immune mechanisms responsible for the pathogenesis of autoimmune liver disease [3]. On the other hand, abnormalities of immune regulation may render these patients more susceptible to viral infection or impair virus clearance. There is one report of concomitant anti-HCV positivity associated with a more severe outcome of PBC [8].

Our study was aimed at investigating the prevalence of anti-HCV in a large series of patients with PBC. Supplemental tests were used to confirm positive results obtained by ELISA. To obtain more information about the relationship between PBC and hepatitis viruses, we also investigated serum markers for hepatitis B virus (HBV) in these patients.

Patients and methods

Patient population and study design

We studied a series of 160 patients (147 women) with histological and clinical diagnosis of PBC [11], who came to our Unit between January 1980 and December 1989. Serum samples had been collected from all patients at the first visit and stored at $-20\,^{\circ}$C until tested for anti-HCV. Fourteen (9%) patients had been transfused, none had histories of drug or alcohol abuse at any time. At presentation, ages ranged from 27 to 77 years (median 53), 68 (43%) patients had stage IV by histology and 144 (90%) had serum antimitochondrial antibodies (titer > 40). We also tested sera collected from each of the 88 patients in this series who reported for follow-up in 1990.

A population of 135 patients with clinical and histological diagnoses of chronic non-A, non-B hepatitis (NANBH), currently followed by us, and a hospitalized population of 80 patients with no evidence of liver disease were selected as control groups.

Antibody testing

All samples were tested for anti-HCV by conventional ELISA (Ortho Diagnostics, Raritan, New Jersey, USA), following the manufacturer's instructions. Results were read at 492 nm with an ELISA plate reader (Auto Reader II-Ortho) with upper limit of 3.0 optical density (OD) units and a mean cut-off level of 0.459. Positive samples were retested in duplicate, then by the Chiron/Ortho Recombinant Immunoblot Assay (RIBA C100) [12], and finally by conventional ELISA after a preliminary wash with 8 M urea [13].

Serum HBV markers had been tested for by commercial radioimmunoassay (Abbott, Chicago, USA) at the time of presentation.

Statistics

The Mann-Whitney test and Wilcoxon signed-rank test were used for statistical analysis of quantitative variables, and Fisher's exact test of dichotomous variables. Pearson's correlation coefficients were calculated by the usual procedure. Observed survival time was the interval between presentation and death or liver transplantation and was computed by the Kaplan-Meier method. Patient survivals were compared by the log-rank test [14].

Results

Anti-HCV was detected by ELISA in 19 patients with PBC (12%, mean OD $\pm$ SD 1.4 ± 0.9), compared to a 68% prevalence (mean OD 2.7 ± 0.7) for the NANBH control group, and to a 0% prevalence in the control hospitalized population. None of the patients showed evidence of HBV infection, as denoted by the presence of serum hepatitis B surface antigen (HBsAg); 46

(29%) had serum antibodies to hepatitis B core antigen (anti-HBc). The patients' characteristics are reported in Table 1 according to anti-HCV status. Serum IgG levels were significantly higher in the anti-HCV positive group and there was a significant correlation ($r = 0.502$, $P < 0.001$) between OD values and serum IgG concentrations. Survival analysis showed no significant differences according to anti-HCV positivity (Fig. 1).

When sera collected in 1990 from the 88 patients who reported for follow-up (median follow-up period, 31 months; range, 2 to 125) were tested, 9 (10%, mean OD 1.7 ± 1.1) were positive for anti-HCV by ELISA. Ten patients in this group had been positive at the time of the first visit: Two had become negative (OD values 0.648 and 1.760 at first visit, 0.230 and 0.197 in 1990) and 1 patient, who had been negative, became positive (OD 0.136 at first visit and 3.0 in 1990). The latter had no identifiable source of infection.

ELISA OD values in relation to RIBA data are analyzed in Table 2. In the NANBH control group, 19 of 37 ELISA positive samples were RIBA reactive, 11 were indeterminate, and 7 were non-reactive.

Table 1. Characteristics of patients according to anti-HCV status

	Anti-HCV +ve (n = 19)	−ve (n = 141)
Age (yrs, m ± SD)	54 ± 9	54 ± 11
Cirrhosis (%)	53	41
Bilirubin (mg/dl, m ± SD)	2.9 ± 2.9	2.4 ± 4.3
IgM (mg/dl, m ± SD)	753 ± 552	566 ± 563
IgG (mg/dl, m ± SD)	2767 ± 888[a]	1774 ± 626
Transfusion (%)	16	8
HBsAg +ve (%)	0	0
Anti-HBc +ve (%)	21	30

[a] $P < 0.001$ vs. anti-HCV −ve

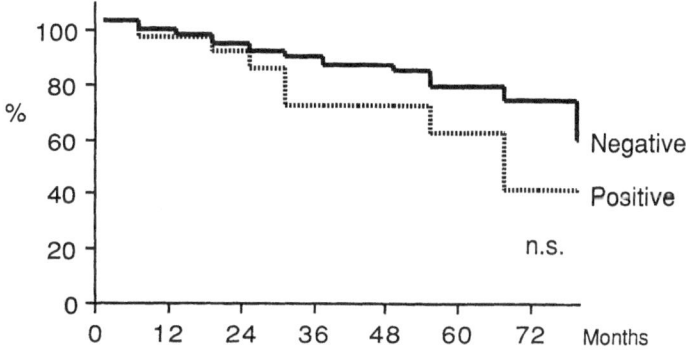

Fig. 1. Survival of PBC patients according to anti-HCV positivity (n.s. not significant)

Table 2. RIBA test in relation to ELISA OD values

ELISA OD (mean of two readings)	RIBA		
	Reactive	Indeterminate	Non-reactive
< 1.0	1	3	10
1.0–2.9	0	8	4
> 2.9	6	0	0

Data for the 32 patients whose sera were ELISA positive in 1990 or at the time of the first visit

Six ELISA positive patients with PBC were retested after the urea wash. Of these, the OD of one decreased to below the cut-off level. In the NANBH control group, 5 of the 81 ELISA positive patients who were retested with the urea wash became negative. All the patients whose OD decreased to below the cut-off level had absorbance values lower than 3.0 in conventional ELISA.

Discussion

The prevalence of anti-HCV as determined by enzyme immunoassay in our 160 patients with PBC was lower than that reported in other studies carried out in smaller series of Italian patients [7, 8]. A similar variability of prevalence rates is also found in other studies from different European countries [1, 2, 5, 6]. It is unlikely that such discrepancy is explained by differences in the populations studied. It is more likely that some shortcomings in the assay play a major role. In fact, the association between positive results by ELISA and serum IgG levels, observed by us as well as by others [1, 5], suggests that non-specific binding to the ELISA solid-phase antigen may give false-positive results. Patients with autoimmune diseases are likely to have high levels of immunoglobulins, with high rates of false-positive results [10].

After the confirmation of ELISA positivity by RIBA, the anti-HCV prevalence (2.5%) was still substantially higher than that reported for the general Italian population [15]. Since RIBA uses the same antigenic preparation as ELISA, cross-reactivity with serum immunoglobulins might also invalidate RIBA. Alternatively, it is possible that patients with a chronically evolving disease like PBC are at higher risk for HCV infection, due to frequent medical treatment and hospitalization, especially in areas with relatively wide diffusion of hepatitis viruses, like Italy. However, the prevalence of HBV antibodies in our patients with PBC is similar to that found in the general Italian population [16], and, at least for HBV, impaired virus clearance in PBC should be excluded, since none of our patients was positive for HBsAg.

Finally, our data do not definitively support poor prognosis as a result of anti-HCV positivity, since life-table analysis indicated only a slight, non-significant reduction in life expectancy in anti-HCV positive patients. Furthermore, patients with higher IgG levels usually have more advanced disease and are more likely to be anti-HCV positive by ELISA.

Until reliable and independent confirmatory tests become available, we cannot draw any definitive conclusions on the importance of anti-HCV positivity in PBC, although our findings do suggest that HCV infection does not play a major role in the pathogenesis and clinical course of PBC.

References

1. McFarlane IG, Smith HM, Johnson PJ, Bray GP, Vergani D, Williams R (1990) Hepatitis C virus antibodies in chronic active hepatitis: pathogenic factor or false-positive result? Lancet i: 754–757
2. Esteban JI, Esteban R, Viladomiu L, et al (1989) Hepatitis C virus antibodies among risk groups in Spain. Lancet ii: 294–296
3. Lenzi M, Ballardini G, Fusconi M, et al (1990) Type 2 autoimmune hepatitis and hepatitis C virus infection. Lancet i: 258–259
4. Kaplan MM (1987) Primary biliary cirrhosis. N Engl J Med 316: 521–528
5. Housset C, Hirschauer C, Courouce AM, Calvo A, Degos F, Benhamou JP (1990) High prevalence of false positive anti-HCV tests in primary biliary cirrhosis. J Hepatol 11: 30
6. Schrumpf E, Elgjo K, Fausa O, Haukenes G, Rollag H (1990) The significance of anti-HCV antibodies measured in chronic liver disease. J Hepatol 11: 111
7. Fusconi M, Lenzi M, Ballardini G, Miniero R, Cassani F, Zauli D, Bianchi FB (1990) Anti-HCV testing in autoimmune hepatitis and primary biliary cirrhosis. Lancet ii: 823
8. Chiaramonte M, Floreani A, Giacomini A, et al (1990) Anti-HCV in primary biliary cirrhosis. Gut 31: A626
9. Theilmann L, Blazek M, Goeser T, Gmelin K, Kommerell B, Fiehn W (1990) False-positive anti-HCV tests in rheumatoid arthritis. Lancet ii: 1346
10. Boudart D, Lucas JC, Muller JY, LeCarrer D, Planchon B, Harousseau JL (1990) False-positive hepatitis C virus antibody tests in paraproteinaemia. Lancet ii: 63
11. Schaffner F, Popper H (1982) Clinical-pathologic relations in primary biliary cirrhosis. In: Popper H, Schaffner F (eds) Progress in Liver Diseases. Grune and Stratton, New York, pp 529–554
12. Weiner AJ, Truett MA, Rosenblatt J, et al (1990) HCV testing in low-risk population. Lancet ii: 695
13. Gray JJ, Wreghitt TG, Friend PJ, Wight DGD, Sundaresan V, Calne RY (1990) Differentiation between specific and non-specific hepatitis C antibodies in chronic liver disease. Lancet i: 609–610
14. Fleiss JL (1973) Statistical methods for rates and proportions. Wiley, New York
15. Sirchia G, Bellobuono A, Giovannetti A, Marconi M (1989) Antibodies to hepatitis C virus in italian blood donors. Lancet ii: 797
16. Giusti G, Galanti B, Gaeta GB (1980) Epidemiology of viral hepatitis in Italy. Boll Ist Sieroter Milan 59: 571–580

Authors' address: Prof. Mauro Podda, Istituto di Scienze Biomediche Ospedale S. Paolo, via A. Di Rudinì, 8, I-20142 Milano, Italy.

Arch Virol (1992) [Suppl] 4: 210–211
© Springer-Verlag 1992

Prevalence of antibodies against hepatitis C virus in primary biliary cirrhosis and autoimmune chronic hepatitis

A. Suárez[1], S. Riestra[1], M. Rodríguez[1], C. A. Navascués[1], F. Tévar[2], R. Pérez[1], L. Rodrigo[1], and J. L. Sánchez Lombroña[1]

[1] Gastroenterology Section, Covadonga Hospital
[2] Communitary Transfusion Center, Oviedo, Spain

Summary. We have determined the prevalence of antibodies against hepatitis C virus (anti-HCV) in 45 patients with primary biliary cirrhosis (PBC) and 6 with autoimmune chronic active hepatitis (AI-CAH). Anti-HCV was positive in two cases of PBC, both with a history of previous blood transfusion, and in one patient with AI-CAH, only during an active phase of the disease.

*

The aim of this study was to determine the prevalence of antibodies against hepatitis C virus (anti-HCV) in patients with chronic liver disease due to an immunological mechanism: Primary biliary cirrhosis (PBC) and auto-immune chronic active hepatitis (AI-CAH). Both diseases present an auto-immune pathogenesis, but many details of the etiology are still not completely understood, and could be associated with another mechanism. Our aim was to assess the potential role of hepatitis C virus in the development of the liver damage in these patients.

A total of 51 patients, 45 with PBC and 6 with AI-CAH were tested for the presence of anti-HCV (Ortho). Thirteen patients, all with PBC, were previously transfused. Sera obtained from 90 chronic active non-A, non-B (NANBH) hepatitis patients were tested for comparison.

Of the patients with PBC 2/45 (4%) and 1/6 (16%) with AI-CAH were positive for anti-HCV. Both patients that were anti-HCV positive with PBC were multitransfused many years ago. None of the non-transfused patients with PBC were anti-HCV positive. The third anti-HCV positive case, a patient with AI-CAH, was analyzed during an active phase of the disease, and a serum sample obtained later, in an inactive phase, was negative.

Of the patients with chronic active NANBH hepatitis 85/90 (94%) were positive for anti-HCV.

The prevalence of anti-HCV in PBC seems to be low, in agreement with the 9% (2/22) reported by Gray et al. [1]. Our findings (4% (2/45)) confirm this point in Spanish patients, and although the prevalence is higher than in blood donors (132/17000 (0.78%)) [3], the antecedent of a parenteral exposure can explain this finding, a point that must be always investigated.

Initial reports showed a high prevalence of anti-HCV in AI-CAH, but retrospective studies [2] suggested that serum from AI-CAH patients may contain a component that gives false positive results during the active phase of the disease. We agree with this, as we found a case that was anti-HCV positive transitorily in exacerbation, without treatment.

As we expected, anti-HCV prevalence was significantly higher in chronic NANBH than in PBC (85/90 (94%) vs. 2/45 (4%); $p < 0.001$) and AI-CAH (85/90 (94%) vs. 1/6 (16%), $p < 0.001$). There was no difference in prevalence of anti-HCV between PBC and AI-CAH.

Finally, the low prevalence of anti-HCV in patients with PBC or AI-CAH and the possibility of explaining all cases by a previous parenteral exposure or a transitory false positive strongly suggests that the hepatitis C virus probably is not implicated in the pathogenesis of the liver damage in these patients.

References

1. Gray JJ, Wreghitt TG, Friend PJ, Wight DGD, Sundaresan V, Calne RY (1990) Differentiation between specific and non-specific hepatitis C antibodies in chronic liver disease. Lancet i: 609–610
2. Mc Farlane JG, Smith HM, Johnson PJ, Bray GP, Vergani D, Williams R (1990) Hepatitis C virus antibodies in chronic active hepatitis: pathogenetic factor or false-positive result? Lancet i: 754–757
3. Riestra S, Suárez A, Rodrigo L (1990) Transmission of Hepatitis C virus. Ann Intern Med 113: 411–412

Authors' address: Dr. A. Suárez, Gastroenterology Section, Covadonga Hospital, Oviedo, Spain.

Arch Virol (1992) [Suppl] 4: 212–214
© Springer-Verlag 1992

HCV infection and chronic active hepatitis in alcoholics

S. Brillanti, C. Masci, S. Siringo, G. Di Febo, M. Miglioli, and L. Barbara

Istituto di Clinica Medica e Gastroenterologia, University of Bologna, Bologna, Italy

Summary. Histological signs of chronic active hepatitis were found in 11/41 (27%) patients with chronic alcoholic liver disease. All these 11 patients tested positive for antibodies to HCV and no other causes of chronic hepatitis were found.

*

The presence of serum antibodies to hepatitis C virus (HCV) in patients with transfusion-associated and community-acquired non-A, non-B hepatitis seems to indicate that HCV is the predominant agent of these kinds of liver diseases [1, 2]. The presence of serum antibodies to HCV has, however, also been reported in different groups of alcoholic patients with chronic liver disease [2, 3]. The real meaning of these findings is not clear.

Histological patterns of alcoholic liver disease are: 1) fatty liver or steatosis, 2) alcoholic hepatitis, and 3) cirrhosis. Alcohol abuse is not known to induce chronic active hepatitis (CAH) and the pathogenesis of this lesion in alcoholics is not well understood [4, 5].

The overall histological pattern of chronic hepatitis C is commonly that of CAH. If HCV infection plays a role in the development of chronic liver disease in alcoholics, histological signs of CAH have to be expected in these subjects. The aim of this study was to assess the correlation between the presence of CAH and the positivity for serum antibodies to HCV in alcoholic patients with chronic liver disease. We studied 41 consecutive cirrhotic patients (31 were males, ages ranged from 31 to 74) presenting a history of alcohol intake above 80 grams per day for over 5 years. All patients were prospectively followed for 6 months before entering the study, and alcohol intake was very low or absent in each patient. No blood transfusion or intravenous drug abuse was present in the history of these patients. Serum HBsAg and antinuclear factor was negative, and ferritin was normal or slightly elevated in each subject.

Serum samples and liver biopsy specimens were obtained from all subjects. Serological testing included determination of HBV markers, by using commercial radioimmunoassays (Abbott Laboratories), detection of antibodies to HCV, by using Ortho HCV ELISA Test (Ortho Diagnostic Systems), and routine biochemical determinations, including AST, ALT, alkaline phosphatase (APh), and gamma-Glutamyl-Transpeptidase (GGT). The evaluation of liver histology was done by an expert pathologist without knowledge of the clinical and serological data.

Testing sera for anti-HBc and anti-HBs gave positive results in 16 patients (39% of our study), and serum antibodies to HCV were found in 15 patients (37% of our study). Histological changes of CAH were present in 11 subjects (27% of all cirrhotic patients).

There was neither a significant correlation between anti-HBc and anti-HCV status, nor was there a significant correlation between anti-HBc positivity and the presence of CAH (assessed by chi-square test).

A very strong correlation was found between the presence of histological changes of CAH and the positivity for serum antibodies to HCV. In fact, 11 subjects presented histological signs of CAH: all were positive for anti-HCV, whereas none of the 26 anti-HCV negative patients had CAH (Table 1).

Positive results for anti-HCV were confirmed, by using recombinant immunoblot assay procedure (RIBA, Ortho Diagnostic Systems), in all subjects with CAH.

Results of biochemical tests showed that a significant increase in serum ALT level and a significant decrease in the AST/ALT ratio and in GGT level was present in cirrhotic patients with CAH, in comparison with cirrhotic patients without CAH. This study indicates that a histological picture of chronic active hepatitis, a lesion that is not known to be induced by alcohol abuse, is present in about 30% of alcoholic patients with chronic liver disease. The presence of CAH is associated with the positivity for antibody to HCV and the absence of other possible causes of CAH.

Table 1. Histological changes and anti-HCV status in alcoholic cirrhotic patients

	Anti-HCV	
	Positive	Negative
Cirrhosis with CAH	11	0
Cirrhosis without CAH	4	26

CAH Chronic active hepatitis

In conclusion, sporadic exposure to HCV infection seems to be the cause of chronic active hepatitis in alcoholic patients.

References

1. Alter HJ, Purcell RH, Shih JW, Melpolder JC, Houghton M, Choo Q-L, Kuo G (1989) Detection of antibody to hepatitis C virus in prospectively followed transfusion recipients with acute and chronic non-A, non-B hepatitis. N Engl J Med 321: 1494–1500
2. Bruix J, Barrera JM, Calvet X, Ercilla G, Costa J, Sanchez-Tapias JM, Ventura M, Vall M, Bruguera M, Bru C, Castillo R, Rodes J (1989) Prevalence of antibodies to hepatitis C virus in Spanish patients with hepatocellular carcinoma and hepatic cirrhosis. Lancet ii: 1004–1006
3. Esteban JI, Esteban R, Viladomiu L, Lopez-Talavera JC, Gonzales A, Hernandez JM, Roget M, Vargas V, Genesca J, Buti M, Guardia J (1989) Hepatitis C virus antibodies among risk groups in Spain. Lancet ii: 294–297
4. Goldberg SJ, Mendenhall CL, Connel AM, Chedid A (1977) Non-alcoholic chronic hepatitis in the alcoholic. Gastroenterology 72: 598–604
5. Nei J, Matsuda Y, Takada A (1983) Chronic hepatitis induced by alcohol. Dig Dis Sci 28: 207–215

Arch Virol (1992) [Suppl] 4: 215–216
© Springer-Verlag 1992

Hepatitis C virus infection in patients with idiopathic hemochromatosis (IH) and porphyria cutanea tarda (PCT)

A. Piperno[1], **R. D'Alba**[1], **L. Roffi**[1], **M. Pozzi**[1], **A. Farina**[1], **L. Vecchi**[2], and **G. Fiorelli**[1]

[1] Institute of Biomedical Sciences, Clinical Medicine, Monza, University of Milan
[2] Department of Clinical Pathology, S. Gerardo Hospital, Monza (Milan), Italy

Summary. The prevalence of HCV antibodies in IH and PCT patients was examined. It was found that both groups are characterized by increased incidence of HCV infection. These results suggest a possible connection between HCV and iron overload.

*

In addition to iron overload, a number of patients with IH and PCT show biochemical and histological signs of chronic active liver diseases (CALD). Elevated alcohol intake is also common in both diseases and a high incidence of HBV infection has been already described in iron-overloaded patients [1]. In this study we explored the prevalence of anti-HCV in patients with IH and PCT and evaluated the relationship between HCV infection and presence of CALD in these cases.

Thirty patients with IH and twelve patients with PCT were studied. Their mean age was $50.4 + 11.3$ ($28 + 68$) and $57.8 + 7.1$ ($51–70$) years, respectively. CALD with or without cirrhosis was present in 8 patients with IH (26%) and 8 with PCT (72%). One patient with PCT refused liver biopsy.

Iron parameters, urinary porphyrins and liver function tests were performed using standard methods. Serum ferritin, HBV antigens and antibodies were determined by radioimmunoassays. Anti-HCV was determined by an ELISA test (Ortho). Liver histology was defined by two independent observers (L.R. and A.F.).

Seven patients with IH (23%) and seven with PCT (58%) were positive for anti-HCV (ELISA mean optical density/cut-off ratio = 5), indicating a high prevalence of anti-HCV in the two groups of patients in the absence of exposure to blood. These prevalences are similar or even higher (in PCT patients) to those observed in patients with chronic active hepatitis or cirrhosis of unknown or alcoholic origin at low risk of blood-borne viral infection [2].

The prevalence found in PCT may be overestimated, however, due to the relatively small number of PCT patients studied.

No significant differences were observed between anti-HCV positive and negative patients in regard to age, HBsAg, antiHBc alone and alcohol abuse. CALD with moderate to marked activity (Knodell histologic activity index higher than 10) was much more frequent in anti-HCV positive than in anti-HCV negative patients, both in IH (57% vs 13%, p = 0.033) and PCT (83% vs 40%). A statistical significance was also attained considering IH + PCT groups taken together (p = 0.005).

Since the finding of anti-HCV seems to be associated with chronic infection, HCV may be important in the pathogenesis of liver damage in about one fourth of patients with IH and even more in patients with PCT. In both diseases, HCV can act synergistically with iron in accelerating the development of hepatic tissue damage. It has been previously reported that a relationship exists between increased amounts of available iron and HBV infection [3]. The high prevalence of anti-HCV observed in both IH and PCT patients, may suggest that the association between HCV infection and iron overload may not be coincidental. It remains also to be established whether HCV infection may play a role in unmasking the existence of the underlying defect in PCT.

References

1. Conte D, Piperno A, Mandelli C, Fargion S et al (1986) Clinical, biochemical and histological features of primary hemochromatosis: a report of 67 cases. Liver 6: 310–315
2. Esteban JI, Viladomiu L, Gonzalez A et al (1989) Hepatitis C virus antibodies among risk groups in Spain. Lancet ii: 294–296
3. Felton C, Lustbader ED, Merten C, Blumberg BS (1981) Serum iron levels and response to hepatitis B virus. Proc Natl Acad Sci USA 76: 2438–2441

Authors' address: Dr. A. Piperno, Istituto di Scienze Biomediche S. Gerardo, Clinica Medica, Via Donizetti 106, I-20052, Monza (Milano), Italy.

VIII   Diagnosis of hepatitis C virus

Arch Virol (1992) [Suppl] 4: 219–221
© Springer-Verlag 1992

Diagnostic reagents for hepatitis C virus

**J. A. Glazebrook, B. C. Rodgers, T. Corbishley, J. A. Garson, D. Parker,
J. A. J. Barbara, R. S. Tedder**, and **P. E. Highfield**

Wellcome Diagnostics R&D, Beckenham, Kent Medical Microbiology, UCMSM,
London NLBTC, Colindale, UK

Summary. The development of diagnostic methods for hepatitis C virus is presented. Special attention is paid to the selection of antigenic markers, the type of assay selected and the interpretation of results. A few of the pitfalls and ambiguities of various assays are discussed and possible future methods are described.

*

Parenterally-transmitted non-A, non-B hepatitis (PT-NANBH) is an important public health problem. With effective screening for HBV, PT-NANB accounts for more than 90% of post-transfusional hepatitis with perhaps 10% of transfusions resulting in disease. Additional routes of transmission include intravenous drug abuse, contaminated blood products and other parenteral transmissions. The virus cannot be cultured and there are only low levels of virus in infectious material (e.g. blood or liver).

For many years there have been false alarms concerning the identification of the infectious agent. Recently PT-NANB-specific DNA was isolated from cDNA libraries prepared from infectious chimpanzee plasma [1]. The initial clone (5-1-1) was used to identify overlapping clones and the sequence of the non-structural gene region of the virus was published. The virus has been designated hepatitis C virus (HCV) and is described as having a single-stranded, positive sense RNA genome which expresses its gene products as a single polyprotein.

The putative NS4 region of HCV has been expressed as a fusion protein with human superoxide dismutase in yeast and this recombinant protein (C-100-3) has been used in immunoassays to screen for the presence of HCV-specific antibodies [2]. Previously surrogate markers (anti-HBc and

ALT) have been used to screen blood donations for potential NANB-infected units in some countries. A confirmatory assay (RIBA) using printed strips of test and control antigens is now available to supplement the EIA. A second EIA, using the same antigen but on beads, plus a neutralisation-based confirmatory assay are now available from a different supplier. These assays represent a step forward in PT-NANBH research but in some ways have raised as many questions as they have answered.

Very little is known about the immune response to HCV antigens; indeed little is known about the HCV antigens themselves. It would be remarkable if C-100-3 were the most effective antigen to use in screening blood donations. In addition, what are the implications for an individual found to seropositive for anti-C100-3 antibodies? Are RIBA and the neutralisation test the best ways to confirm HCV? We have attempted to answer some of these questions using HCV-specific reagents produced from our own independent clones.

The published HCV sequence has been used by a number of groups who have presented partial sequences from different isolates [3, 4] including the structural protein region from Japanese patients [5, 6]. Independently of the published sequences, we have isolated cDNA clones from known human carriers of PT-NANBH. In particular, we have identified two overlapping clones, JG2 and JG3, which come from the putative NS5 region as well as a single clone, BR11, which contains structural sequences. These clones were recombined with a baculovirus based expression vector to produce the encoded antigens in insect cells. The non-structural (NS) recombinant was designated BHC-7 and the structural (S) recombinant BHC-9.

The two antigens were first used independently in an anti-human immunoglobulin format EIA to determine the antibody status of various sera from individuals at high-risk of HCV-infection e.g. haemophiliacs. We found that BHC-9 detected antibody in a higher proportion of the samples than did BHC-7 (25/32 vs 17/32 haemophiliacs); however using both antigens together 26/32 were positive. As might be expected, the structural antigen was more effective than the non-structural but there are some sera which appear to have antibodies only to non-structural regions. These observations were borne out by other risk groups and all future work has used both antigens together.

Initially the combined antigens were compared with the commercially-available, C-100-3-based EIA. In general the two antigens together detect antibody in a higher proportion of any patient group than C-100-3 alone even though there are some samples which are C-100-3 positive but negative with BHC7 + 9. These discrepancies must be resolved. The first point to note is that we are not comparing like with like and, given our current state of knowledge, it is possible that sera which react with one HCV antigen might not react with all other HCV antigens. Those samples which are BHC7 + 9 positive and C-100-3 negative have been assayed with the two antigens separately. We find that the majority of these samples react with only the structural antigen BHC-9. Whilst the majority of reactive sera contain antibodies to all three antigens, significant numbers of sera have antibodies to only one or two of the antigens. As yet the prognostic value of these different patterns of antibody response is unknown.

Another related approach to analysing these discrepant samples is to use a confirmatory assay such as an immunoblot or a neutralisation test. We have our own in-house western blot using various purified HCV recombinant proteins to confirm BHC7+9 reactive sera. In general, many of the C-100-3-positive, BHC7+9-negative samples are RIBA negative or indeterminant whereas we find that only a few BHC7+9 reactive samples are falsely positive due to reaction with non-HCV-protein contaminants.

PCR amplification can be used to determine the presence of viral RNA in the sample. Care must be exercised when using PCR, not only technically to avoid contamination but also in interpreting the results. A positive signal under properly-controlled conditions will indicate current infection; but a negative result may mean that an infection has resolved or that the level of virus is below the limit of the method or that the sequence of the virus is sufficiently distinct that the amplification primers do not function efficiently. We do not know enough about the biology of HCV yet to address these points. Nevertheless we currently find that about 45% of the samples which are reactive with both C-100-3 and BHC7+9 are PCR-positive. Of those which react with only BHC7+9, about 30% are PCR-positive; whereas only 6% of those reactive only with C-100-3 are PCR-positive. Expressed another way, of 25 PCR-positive samples, 24 react with BHC7+9 and only 18 react with C-100-3.

In conclusion, we are at an exciting stage of the study of PT-NANBH. The most important agent of this disease has been identified and reagents are becoming available to study the response to infection. As more and more different antigens of HCV are used, it is clear that the antibody profile can differ from individual to individual and we may be able to identify those who will progress to chronic disease. Additionally we should be able to identify prognostic markers which will provide an early indicator of changes in the disease state.

References

1. Choo et al (1989) Science 244: 359–362
2. Kuo et al (1989) Science 244: 362–364
3. Kubo et al (1989) Nucleic Acids Res 17: 10367–10372
4. Maeno et al (1990) Nucleic Acids Res 18: 2685–2689
5. Okamoto et al (1990) Japan J Exp Med 60: 167–177
6. Takeuchi et al (1990) Nucleic Acids Res 18: 4626

Authors' address: Dr. P. E. Highfield, Wellcome Diagnostics R&D, Beckenham, Kent, U.K.

Arch Virol (1992) [Suppl] 4: 222–226
© Springer-Verlag 1992

Follow-up of patients with hepatitis non-A, non-B: incidence and persistence of anti-HCV depend on route of transmission

K. Gmelin*, F. Kurzen, B. Kallinowski, T. Goeser, J. Arnold, B. Kommerell, and
L. Theilmann

Department of Internal Medicine, University of Heidelberg, Heidelberg,
Federal Republic of Germany

Summary. Of 32 patients with non-A, non-B hepatitis, 10 (31%) were still anti-HCV-positive 12.8 years after the acute phase of the disease. Seven of the patients (21.9%) still had elevated ALT levels, and among these, 5 out of 5 patients who had been subject to parenteral risk were anti-HCV-positive. In contrast, none of the patients who had not been subject to parenteral risks were positive.

Introduction

In Germany, as many as 23% of clinically apparent cases of acute hepatitis are classified as non-A, non-B hepatitis [7]. Recently, an assay to detect antibodies to the hepatitis C virus (anti-HCV) has been developed [2, 6]. The antibodies appear late after the acute episode of hepatitis, and little is known about their persistence thereafter [1, 6]. We therefore followed-up patients with acute non-A, non-B hepatitis up to fifteen years after the initial admission to the hospital, to study the role of hepatitis C virus (HCV) in chronic hepatitis and to evaluate the persistence of anti-HCV in these cases.

Materials and methods

Patients studied

Heidelberg is a medium-sized German city with about 150,000 inhabitants. Only inpatients admitted to the Department of Gastroenterology of the University of Heidelberg from 1974 to 1981 were included into the study.

Acute viral hepatitis

The diagnosis of acute viral hepatitis was made on the basis of anamnestic, clinical, and biochemical findings. Possible risks of infection were taken from the patients' records. Parenteral risks of included blood transfusions, illicit drug-abuse and being on a health-care staff. Non-parenteral routes included close contact to jaundiced individuals, and foreign travels. A sporadic form of hepatitis was assumed if any possible exposure was denied by the patient.

Follow-up studies

Patients with non-A, non-B hepatitis were followed-up in 1983 and in 1990. The follow-up study included clinical and anamnestic reevaluation, determination of both liver function tests and virological markers. Chronicity was suggested if liver function tests were still abnormal; liver histology could be determined in only a few patients.

Serum samples used

Sera of the patients were taken for routine laboratory investigations both during acute hepatitis and at the time of the follow-up and were stored at $-20\,°C$.

Laboratory tests

HBsAg, anti-HBs, anti-HBc, anti-HBc IgM, anti-HAV, anti-HAV IgM were tested by radio- or enzyme immunoassays (Ausria, Ausab, Corab, Havab, Havab M, and Corzyme M; Abbott, Wiesbaden FRG). Anti-CMV or Anti-EBV of the immunoglobulin M class were determined by ELISA or immunofluorescence technique. Serum transaminases, pro-thrombin time, autoantibodies and serum immunoglobulin fractions were determined by routine techniques. Antibodies to hepatitis C virus were determined by use of an assay manufactured by Ortho Diagnostics (Neckargemünd, Germany).

Results

Between 1974 and 1981, 112 patients had been hospitalized because of non-A, non-B hepatitis. Sera of 29 patients taken during the acute phase of hepatitis, as well as taken in 1983 and in 1990, were still available to be retested for presence of antibodies to hepatitis C virus (see Table 1). Twenty-one sera of the total of 29 patients taken during the acute hepatitis phase were positive for antibodies to hepatitis C virus.

Of the 112 patients, 49 (= 43.8%) and 32 (= 28.6%) could be followed-up in 1983 and in 1990, respectively. The mean follow-up time in 1990 ranged from 9 to 15 years (mean = 12.8 years). Sera of all patients with non-A, non-B hepatitis remained negative for HBsAg, anti-HBc IgM, anti-HAV IgM, anti-EBV IgM and anti-CMV IgM. Sera of 49 patients followed-up in 1983 were available to be retested for anti-HCV; 57.1% of these sera were anti-HCV-positive. Liver function tests were elevated in 19 of 49 patients (= 38.8%) and in 7 of 32 patients (= 21.9%) in 1983 and in 1990, respectively (see Table 2). In 1983, 9 out of 10 patients with a chronic course of non-A, non-B hepatitis were anti-HCV-positive in case of parenteral transmission and in 6/9 in non-

Table 1. Anti-HCV in follow-up sera from patients with non-A, non-B hepatitis

Route of transmission	aVH[a] N	Follow-up in 1983 n	anti-HCV-pos. (%)	Follow-up in 1990 n	anti-HCV-pos. (%)
Parenteral	46	22	18 (81.8)	19	8 (42.1)
Non-parenteral	56	27	10 (37.0)	13	2 (15.4)
Total	112	49	28 (57.1)	32	10 (31.3)

[a] aVH = acute viral hepatitis

Table 2. Chronic liver disease (cld) in the follow-up of patients with acute non-A, non-B hepatitis

Route of transmission	aVH[a] N	Follow-up in 1983 n	cld n (%)	Follow-up in 1990 n	cld n (%)
Parenteral	46	22	10 (45.6)	19	5 (26.3)
Non-parenteral	56	27	9 (33.3)	13	2 (15.4)
Total	112	49	19 (38.8)	32	7 (21.9)

[a] aVH = acute viral hepatitis

parenteral transmission. In 1990, all five patients who had parenteral transmission with still elevated ALT values were positive for anti-HCV whereas neither two patients with non-parenteral risks were positive (see Table 3).

Discussion

Studies on the frequency of non-A, non-B hepatitis are nearly impossible to compare as there are marked discrepancies in respect to geographic variation, hospitalization, age and criteria for chronicity of disease [3]. In acute viral hepatitis, diagnosis of hepatitis C is difficult, because the appearance of antibodies to hepatitis C virus only appear after an extended delay [6]. The assay of Ortho Diagnostics uses a synthetic polypeptide that binds non-

Table 3. Anti-HCV in follow-up sera from patients with chronic liver disease

Route of transmission	Follow-up in 1983			Follow-up in 1990		
	Total	Chronic liver disease		Total	Chronic liver disease	
	N	n	(anti-HCV-positive) (n)	N	n	(anti-HCV-positive) (n)
Parenteral	22	10	(9)	19	5	(5)
Non-parenteral	27	9	(6)	13	2	(0)
Total	49	19	(15 = 78.9%)	32	7	(5 = 71.4%)

neutralizing antibodies. This anti-HCV assay does not enable differentiation between present or anamnestic HCV infection. In some cases of acute resolving non-A, non-B hepatitis this antibody does not appear at all. The incidence of anti-HCV is high in patients with chronic liver disease following transfusion-related hepatitis [1, 4, 5]. Both post-transfusional and sporadic chronic hepatitis C may be followed by a long-term persistence of anti-HCV [8, 9].

Our results show that anti-HCV may be lost with normalization of liver function tests. This was preferentially observed in those patients with non-parenterally transmitted non-A, non-B hepatitis. The low rate of anti-HCV-positive patients with sporadic non-A, non-B hepatitis suggests the existence of an agent other than hepatitis C virus, or alternatively, of another pattern of humoral response to HCV.

References

1. Alter HJ, Purcell RH, Shih JW, Melpolder JC, Houghton M, Choo QL, Kuo G (1989) Detection of antibodies to hepatitis C virus in prospectively followed transfusion recipients with acute and chronic non-A, non-B hepatitis. N Engl J Med 321: 1494–1500
2. Choo QL, Kuo G, Weiner AJ, Overby LR, Bradley DW, Houghton M (1989) Isolation of a cDNA clone derived from a blood-borne non-A, non-B viral hepatitis genome. Science 244: 359–362
3. Dienstag JL (1983) Non-A, non-B hepatitis. I. Recognition, epidemiology, and clinical features. Gastroenterology 85: 439–462
4. Esteban JI, Esteban R, Viladomiu L, Lopez-Talavera JC, Gonzalez A, Hernandez JM, Roget M, Vargas V, Genesca J, Buti M, Guardia J, Houghton M, Choo Q-L, Kuo G (1989) Hepatitis C virus antibodies among risk groups in Spain. Lancet ii: 294–297
5. Hopf U, Möller B, Küther D, Stemerowicz R, Lobeck H, Lüdtke-Handjerry A, Walter E, Blum HE, Roggendorf M, Deinhardt F (1990) Long-term follow-up of post-transfusion and sporadic chronic hepatitis non-A, non-B and frequency of circulating antibodies to hepatitis C virus (HCV). J Hepatol 10: 69–76

6. Kuo G, Choo Q-L, Alter HJ, Gitnick GL, Redeker AG, Purcell RH, Miyamura T, Dienstag JL, Alter MJ, Stevens CE, Tegtmeier GE, Bonino F, Colombo M, Lee W-S, Kuo C, Berger K. Shuster JR, Overby LR, Bradley DW, Houghton M (1989) An assay for circulating antibodies to a major etiologic virus of human non-A, non-B hepatitis. Science 244: 362–364
7. Müller R, Willers H, Knocke KW, Sipos S, Höpken W (1979) Epidemiologie und Prognose der Hepatitis Non-A, non-B. Dtsch Med Wschr 104: 1471–1474
8. Roggendorf M, Deinhardt F, Rasshofer R, Eberle J, Hopf U, Möller B, Zachoval R, Pape G, Schramm W, Rommel F (1989) Antibodies to hepatitis C virus. Lancet ii: 324–325
9. Van der Poel CL, Reesink HW, Lelie PN, Leentvaar-Kuypers A, Cho Q-L, Kuo G, Houghton M (1989) Anti-hepatitis C virus antibodies and non-A, non-B post-transfusion hepatitis in the Netherlands. Lancet ii: 297–299

Authors' address: Dr. K. Gmelin, Luitpold Kliniken, Bismarckstrasse 24/38, D-W-8730 Bad Kissingen, Federal Republic of Germany.

Arch Virol (1992) [Suppl] 4: 227–231
© Springer-Verlag 1992

Confirmation of anti-HCV EIA reactivities by RIBA and neutralization assay among blood donors and patients with chronic liver disease and hepatocellular carcinoma

E. Tanzi[1], C. Galli[1], M. Delaito[2], T. Bertin[2], G. Pizzocolo[3], A. Rodella[3], L. Buscarini[4], G. Sbolli[4], U. Rossi[5], L. Romanò[1], M. Chiaramonte[2], and A. R. Zanetti[6]

[1] Institute of Virology, University of Milano
[2] Department of Gastroenterology, University of Padova
Hospitals of [3] Brescia, [4] Piacenza and [5] Legnano
[6] Department of Hygiene, University of Camerino, Italy

Summary. The aim of our study was to confirm by Recombinant Immuno-blot Assay (RIBA) and by neutralization assay the repeat positive reactions found by two commercially available EIAs (Ortho and Abbott) when testing samples from volunteer blood donors, patients with chronic liver disease and with hepatocellular carcinoma. Our data show a high confirmatory rate among patients with chronic viral NANBH and HCC, while among donors and patients with CLD other than NANBH the percentage of presumptive EIA positive reactions confirmed by RIBA and/or neutralization assay is much lower. In our experience, the neutralization assay appears to be somewhat more sensitive than RIBA, especially when samples show low EIA optical densities.

*

Currently available anti-HCV enzyme immunoassays are designed to detect the reactivity against a recombinant antigen (c100-3), expressed in yeast (*S. cerevisiae*), encoded by part of the non-structural portion of HCV genome. Although the clinical significance of a positive anti-HCV result has not yet been completely defined, these assays have proven to be useful in prevention of post-transfusion non-A, non-B (NANB) hepatitis and to recognize HCV etiology in most cases of NANB chronic liver diseases [1, 3].

The specificity issue, that has been raised by several studies, achieves a particular relevance in the screening of blood donors, due to the supposed low prevalence of anti-HCV antibodies in this population and to the ensuing low positive predictive value of screening tests [4, 5].

The aim of our study was to confirm anti-HCV reactivity among donors and patients with chronic liver disease (CLD) and hepatocellular carcinoma

(HCC) by two currently available methods: the Recombinant Immunoblot Assay (Chiron-Ortho RIBA) and the EIA test based on the blocking procedure (Abbott HCV Neutralization).

To this end we assayed by RIBA and neutralization test samples collected from 187 volunteer blood donors found anti-HCV repeat reactive (RR) by Ortho HCV EIA, after a screening of 13,314 (1.4%) donors attending several Blood Transfusion Centres in northern Italy. Sera collected from 131 patients with chronic liver disease of various etiologies (all biopsy proven) and classified according to the conventional criteria [2] and from 72 patients with hepatocellular carcinoma, were also available for the study.

Both Abbott and Ortho HCV EIAs were employed in parallel on CLD and HCC samples. The 187 blood donors who were found to be reactive by Ortho EIA were subsequently assayed by Abbott HCV EIA. During the few months period that elapsed between the first screening by Ortho EIA and the subsequent testing by Abbott EIA, sera were stored at $-20\,^{\circ}$C. All Ortho EIA RR samples were then tested by RIBA, while Abbott EIA RR samples were confirmed by the neutralization assay. Both assays are designed to confirm the presence of anti-c100 antibodies by excluding possible re-activities against *S. cerevisiae* or the fusion protein (SOD) employed in the production of the c100 peptide. In the RIBA test, SOD, c100-3 and 5-1-1 (43 of the 363 amino acids of c100 expressed in *E. coli*) peptides are applied to a nitrocellulose strip, along with two procedural positive controls (low and high levels of human IgG). A sample is confirmed positive for anti-HCV if it reacts against both c100 and 5-1-1 with at least the same intensity as the low IgG control. A reaction against only one of the two HCV peptides is considered as "indeterminate", while samples showing no reaction (except possibly against SOD) are considered as negative.

The Abbott confirmatory assay is based on the blocking principle. Briefly, two aliquots of the sample are incubated, one with a buffered solution and the other one with the same solution containing a recombinant peptide that includes 256 of the 363 amino acids of the c100-3 sequence, expressed in *E. coli*, employing CMP-KDO synthetase (CKS) instead of SOD as a fusion protein. After this preincubation, both aliquots are analyzed by the conventional Abbott HCV EIA procedure. The samples, in which a reduction of the optical density (OD) of 50% or more in the blocked sample compared to the unblocked is observed, are considered positive for anti-c100 antibodies.

Statistical analysis was performed by the chi-square test.

Of the 187 blood donors found to be repeatedly reactive by Ortho EIA, 60 (32%) were confirmed positive by RIBA, 39 (21%) were indeterminate and 88 (47%) were RIBA negative.

Only 129 out of the 187 (69%) samples found repeat reactive by Ortho EIA resulted also RR when tested by Abbott EIA. Of the 58 discordant samples, 54 (93%) were negative and 4 (7%) were indeterminate when tested

Table 1. Confirmation by RIBA and neutralization assay of 129 samples found RR by both EIAs

		N.	Neutralization assay	
			Neutralized (%)	Non neutralized (%)
R	POS	60	60 (100)	0 (0)
I B	IND	35	25 (71.4)	10 (28.6)
A	NEG	34	10 (29.4)	24 (70.6)
	Total	129	95 (73.6)	34 (26.4)

Table 2. Confirmation by RIBA and neutralization assay of EIA repeat reactive samples subdivided according to their OD (sample/cut-off)

OD s/co	RIBA+ (%)		Neutralization+ (%)	
1–2	2/63	(3.2)	8/23	(34.8)
>2–4	10/46	(21.7)	24/32	(75)
>4	48/78	(61.5)	63/74	(85.1)
Total	60/187	(32.1)[a]	95/129	(73.6)[a]

[a] $p < 0.001$

by RIBA. As is shown in Table 1, all 60 donor specimens which were reactive by RIBA were confirmed as positive by the neutralization assay. Moreover, 10 of 34 (29.4%) RIBA negative and 25 of 35 (71.4%) RIBA indeterminate samples were found positive when tested by the neutralization test.

As shown in Table 2, only 32.1% (60/187) of donor specimens repeatedly reactive by Ortho EIA could be confirmed by RIBA, while 73.6% (95/129) of those found RR by Abbott EIA were neutralized by the blocking assay ($p < 0.001$). These data indicate that the confirmation rates by RIBA raised from 3.2% (2/63) in samples which showed low Ortho EIA optical density (sample/cut-off ratio ≤ 2) to 61.5% in samples showing a stronger reactivity (s/co > 4). The correlation between increasing OD and frequency in confirmation of positive reactivities was also observed by the blocking test, although to a lesser extent, since 34.8% (8/23) of the weak reactivities and 85.1% (63/74) of the strong reactivities obtained by Abbott EIA were confirmed as positive (Table 2).

230   E. Tanzi et al.

Table 3. Anti-HCV among patients with CLD as detected by two commercially available EIAs and by RIBA and neutralization test

CLD	N.	Ortho EIA RR (%)	RIBA+ (%)	Abbott EIA RR (%)	Neutralization+ (%)
Cryptogenic and post-transfusion	64	44 (68.7)	9/10 (90)	41 (64.0)	41/41 (100)
In drug addicts	13	13 (100)	1/2 (50)	13 (100)	13/13 (100)
ALD	20	14 (70.0)	4/7 (57.1)	11 (55)	11/11 (100)
PBC	18	13 (72.2)	10/13 (76.9)	10 (55.5)	8/10 (80)
Autoimmune	8	3 (37.5)	NT	3 (37.5)	3/3 (100)
HBsAg+	8	3 (37.5)	1/1 (100)	4 (50)	4/4 (100)
Total	131	90 (68.7)	25/33[a] (75.8)	82 (62.6)	80/82 (97.6)

[a] Four samples from drug addicts (n=1), ALD (n=2), PBC (n=1) gave indeterminate results by RIBA

Among patients with CLD, the anti-HCV prevalence was 68.7% (90/131) when their serum samples were tested by Ortho EIA and 62.6% (82/131) when tested by Abbott EIA (Table 3). RIBA was performed on only 33 cases, out of which 25 (75.8%) were confirmed as positive, while 4 gave indeterminate results. On the other hand, 80 out of 82 (97.6%) samples found RR by Abbott EIA could be confirmed positive by neutralization test. Discrepancies in the results obtained with the two EIAs and/or with the confirmatory assays were more frequent among patients with a diagnosis of primary biliary cirrhosis (PBC) and alcoholic liver disease (ALD), raising the crucial problem of the interpretation of anti-HCV positive reactions in such patients.

Among the 72 HCC cases, anti-HCV was found in 43 (59.7%) patients when tested by Ortho EIA and in 45 (62.5%) patients when tested by Abbott EIA. RIBA was not available for these sera. By the blocking assay, 42 out of 45 (93.3%) samples found RR by Abbott EIA were confirmed as positive. It is noteworthy that all 42 samples confirmed positive by the neutralization test were also RR by Ortho EIA.

These preliminary data indicate that both EIAs are reliable for detecting anti-HCV antibodies in patients with cryptogenic and post-transfusion NANB hepatitis and in hepatocellular carcinoma, as demonstrated by the high confirmatory rate obtained with the blocking assay, while among volunteer blood donors and in patients with CLD other than NANB hepatitis, confirmatory assays are necessary due to the high number of non-specific reactions. For this purpose, the neutralization assay appears to be somewhat more sensitive than RIBA in the confirmation of EIA repeat reactive samples, especially when these show a low optical density.

Further analysis of RIBA indeterminates, some of which have been found capable of transmitting HCV [6], as well as of samples that gave discordant results by the two methods, is mandatory.

References

1. Alter HJ, Purcell RH, Shih JW, Melpolder JC, Houghton M, Choo QL, Kuo G (1989) Detection of an antibody to hepatitis C virus in prospectively followed transfusion recipients with acute and chronic non-A, non-B hepatitis. N Engl J Med 321: 1494–1500
2. Fogarty International Center Proceedings (1976) Nomenclature diagnostic criteria and diagnostic methodology for disease of the liver and biliary tract. Washington D.C.; U.S. Government Printing Office, 22
3. Kou G, Choo QL, Alter HJ, Gitnick GL, Redeker AG, Purcell RH, Miyamura T, Dienstag JL, Alter HJ, Stevens CE, Tegtmeier GE, Bonino F, Colombo M, Lee WS, Kuo C, Berger K, Shuster JR, Overby LR, Bradley DW, Houghton M (1989) An assay for circulating antibodies to a major etiologic virus of human non-A, non-B hepatitis. Science 244: 363–364
4. Weiner AJ, Truett MA, Rosenblatt J, Han J, Quan S, Polito AJ, Kuo G, Choo QL, Houghton M, Agius C, Page E, Nelles MJ (1990) HCV testing in low-risk population. Lancet 336: 695
5. Wong DC, Diwan AR, Rosen L, Gerin JL, Johnson RG, Polito A, Purcell RH (1990) Non-specificity of anti-HCV test for seroepidemiological analysis. Lancet 336: 750–751
6. Zanetti AR, Tanzi E, Zehender G, Magni E, Incarbone C, Zonaro A, Primi D, Cariani E (1990) Hepatitis C virus RNA in symptomless donors implicated in post-transfusion non-A, non-B hepatitis. Lancet 336: 448

Authors' address: Prof. Alessandro R. Zanetti, Department of Hygiene, University of Camerino, Viale E. Betti 3, 62032 Camerino, Italy.

Arch Virol (1992) [Suppl] 4: 232–233
© Springer-Verlag 1992

Recombinant immunoblot assay for hepatitis C virus antibody in chronic hepatitis

G. Taliani, M. C. Badolato, R Lecce, C. De Bac, E. De Marzio, C. Balsano, M. Artini,
and **M. Levrero**

Institute of Tropical Diseases, "La Sapienza" University and Fondazione A. Cesalpino,
I Clinica Medica, "La Sapienza" University, Rome, Italy

Summary. Testing for hepatitis C virus by ELISA requires confirmation by recombinant immunoblot assay (RIBA). The first-generation RIBA uses the same antigen as used in the ELISA and one further antigen. A second-generation RIBA in which two further antigens are present, detects positivity that is not found by either the ELISA or the original RIBA. Consequently, although it is adequate to test Elisa positive sera with the first-generation RIBA, the second-generation assay is recommended for confirming negativity.

*

At present, the only available test for confirming the specificity of the ELISA positivity for antibodies to Hepatitis C Virus (anti-HCV) is an immunoblotting test (RIBA, Chiron Corporation) in which two or four recombinant HCV antigens are immobilized on nitrocellulose strips. In the first generation RIBA (RIBA-I), two non-structural HCV antigens are employed: the same antigen present in the HCV-ELISA (c100) and a subsequence of this antigen (5-1-1), while in the second generation RIBA (RIBA-II), two more HCV antigens are present: a non-structural (c33) and a structural (c22) antigen. By these tests, the serum sample is confirmed as anti-HCV positive if reactivity to any two antigen bands is observed, it is indeterminate if only one band is reactive and it is not confirmed if no reactivity is found [1, 2].

We employed RIBA-II to confirm the specificity of the anti-HCV positivity revealed by ELISA (Ortho Diagnostics) in 15 sera from HBsAg-positive and in 89 sera from HBsAg negative patients with histologically proven chronic liver disease.

Out of the 15 HBsAg positive patients, 4 (26.7%) have not been confirmed (i.e. no visible bands to any antigen were present). On the other hand, 87 out of 89 (97.7%) HBsAg negative, anti-HCV positive sera were confirmed by RIBA (Fisher's exact test, P < 0.001), and the remaining two were indeterminate (i.e. only the c22 band was reactive).

Interestingly, 2 of 4 HBsAg positive not-confirmed samples had ELISA optical density (OD) values greater than 2.500 (cut-off value range 0.425–0.492) and they were still positive at a dilution of 1 in 100, while 11 out of 87 HBsAg negative confirmed samples were weakly positive by ELISA (OD < 1.000). This indicates that, also among chronic patients, the ODs can be unrelated to the specificity of the ELISA reaction [3].

Since the confirmation rate was significantly lower among HBsAg positive (73.3%) compared to HBsAg negative (97.7%), the anti-HCV positive result should be always confirmed in HBV infected patients.

Moreover, by RIBA-II we have examined 18 sera from anti-HCV repeatedly negative patients with histologically proven chronic liver disease. Of these, 7 (38.8%) reacted to both c33 and c22 antigens which are not present in the HCV-ELISA, therefore, more than one third of the anti-HCV negative chronic patients turned out to have antibodies to HCV antigens different from c100, and this implies that using only this antigen for screening purposes, the prevalence of HCV related chronic liver diseases could have been underestimated.

It remains to be clarified whether the serological pattern characterized by the absence of anti-c100 antibodies, but presence of anti-c33 and anti-c22 corresponds to a particular biological and immunological status, although the mean ALT level + Standard Deviation (SD) of these patients was not significantly different than that of anti-c100 positive patients (respectively 148 + 72 and 156 + 68, Student's t test, p = n.s.).

In conclusion, our data show that all HBsAg positive sera that are anti-HCV positive by ELISA should be confirmed by RIBA, while the ELISA anti-HCV negative sera from chronic patients should be retested by a second-generation assay that includes also c33 and c22 antigens.

References

1. Skidmore S. Recombinant immunoblot assay for hepatitis C antibody (1990) Lancet 335: 1346
2. Van Der Poel CL, Cuypers HTM, Reesink HW et al (1991) Confirmation of hepatitis C virus infection by new four antigen recombinant immunoblot assay. Lancet 337: 317–319
3. Wong DC, Diwan AR, Rosen L et al (1990) Non-specificity of anti-HCV test for seroepidemiological analysis. Lancet i: 750–751

Authors' address: Dr. G. Taliani, Institute of Tropical Diseases, "La Sapienza" University and Fondazione A. Cesalpino, I Clinica Medica, "La Sapienza" University, Rome, Italy.

Arch Virol (1992) [Suppl] 4: 234–237
© Springer-Verlag 1992

PCR detection of HCV RNA among French non-A, non-B hepatitis patients

J. Li, L. Vitvitski, S. Tong, and C. Trépo

Unité de Recherche sur les Hépatites, INSERM U.271 Lyon, France

Summary. Hepatitis C virus (HCV) cDNA was amplified from serum of 26/40 French chronic non-A, non-B hepatitis patients by the nested polymerase chain reaction. Compared with anti-C100, viral cDNA represents a more reliable marker of active HCV replication.

*

Hepatitis C virus (HCV) has been identified as the etiological agent responsible for the majority of post-transfusion and sporadic non-A, non-B (NANB) hepatitis cases around the world [1]. Until now, the only marker of HCV infection was an antibody directed against a part of the viral nonstructural protein (anti-C100; 2). This antibody develops only late after infection [3] and does not reflect active HCV replication. False positivity of the ELISA assay used to detect anti-C100 has been reported, especially in patients with auto-immune hepatitis [4–6]. An alternative method would be to directly assay for the presence of viral genomic RNA. Since HCV RNA is a component of infectious virus particles, its presence in the serum reflects active virus replication, and infectivity of the serum. Since the nearly complete cDNA sequence of the prototype HCV strain has been disclosed [7], detection of HCV infection by polymerase chain reaction has become feasible. We report here the detection of HCV RNA in serum of 40 French NANB hepatitis patients by the so-called "nested PCR". The results were compared with those of anti-C100 serology.

To detect HCV RNA, RNA was extracted from 200 μl of fresh or stored serum by guanidinium/phenol/chloroform solution (20% guanidinium, 50% phenol, 12.5% chloroform, and 17.5% H_2O), and precipitated by ethanol. RNA was dissolved in 20 μl DEPE treated water, with the addition of

Sense 1 (S1): 5'-GGCTATACCGGCGACTTCGA-3'(2456-2475)

Sense 2 (S2): 5'-GCAATACGTGTGTCAC-3'(2488-2503)

Anti-sense 1 (AS1):5'-AGCTCATACCAAGCACAGCC-3'(2697-2716)

Anti-sense 2 (AS2):5'-TCATAGCACTCACAGAGGAC-3'(2673-2692)

Fig. 1. Schematic representation of part of the NS3 region of HCV genome. The locations and sequences of the primers used for nested PCR are indicated. S1/AS1 are the first-round PCR primer pair and S2/AS2 the second-round PCR primer pair

RNasin (1 U/μl final concentration). Ten μl was mixed with 1 μl 100 pM each of the sense and anti-sense primers (Fig. 1, S1 and AS1), and denatured at 65 °C for 10 min. cDNA synthesis was performed at 37 °C for 1 hr using cloned MuLV reverse transcriptase (GIBCO, BRL). The cDNA reaction was diluted with 50 μl H_2O, boiled for 10 min to inactivate the enzyme. The cDNA was amplified in 100 μl volume containing 200 μM dNTP, 1 u of Taq DNA polymerase, and 1 μg of RNase A. Thirty five cycles were carried out. Each cycle consisted of denaturation at 94 °C for 1 min, annealing at 37 °C for 2 min, and chain elongation at 72 °C for 3 min. Ten μl of PCR product was reamplified by a second round of PCR using internal primer pairs (S2 and AS2, see Fig. 1). The 2 pairs of PCR primers were located in the nonstructural region NS3 of HCV genome, according to the prototype HCV sequence (Houghton et al., 1989). The PCR product was electrophoresed in a 2% agarose gel and visualized by ethidium bromide staining.

After the first round of PCR, only one out of the 40 NANB samples was found to have the expected amplification band in a 2% agarose gel. After the 2nd round of PCR amplification, however, the expected 204 bp amplification band was observed in 26 out of 40 NANB sera tested (Fig. 2A and Table 1). After Southern transfer, all these 26 samples hybridized with a cloned HCV cDNA probe provided by Dr A. J. Weiner from Chiron Corporation (Fig. 2B), suggesting the specificity of amplification. The PCR results were compared with those of the anti-C100 ELISA test (Table 1). Of the PCR positive cases, 69% were positive for anti-C100, while 72% of anti-C100 positive cases were positive for HCV RNA by PCR. The concordance of HCV RNA and anti-C100 was 63%.

Detection of HCV sequences by PCR was originally reported by Weiner. et al. [8]. However, the amount of HCV sequences circulating in the blood is so low that detection by two successive rounds of PCR using 2 sets of primer pairs (nested PCR) was more successful for serum samples [9–11]. The high detection rate of HCV cDNA (65%) in our samples suggests that the primer pairs used in this study are relatively conserved between the original Chiron

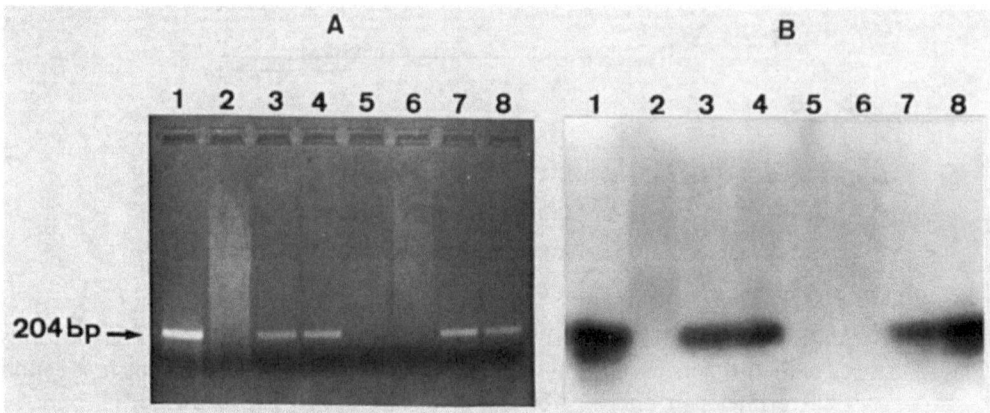

Fig. 2. PCR amplification of HCV sequence from French NANB hepatitis patients (**A**) and hybridization with an HCV cDNA probe (**B**). The serum samples are numbered as lanes 1–8. The location of the expected amplification band is indicated by an arrow

Table 1. Correlation between HCV RNA and anti-C100

	RNA+	RNA−	Total
Anti-C100	18	7	25
Anti-C100	8	7	15
Total	26	14	40

HCV strain and the predominant HCV strains circulating in France. This is consistent with the idea that NS3 is a relatively conserved region of the HCV genome. Moreover, we have recently determined PCR amplified HCV sequences from several French patients. Sequence data showed a high degree of homology between the major French HCV strain and the prototype Chiron strain [12]. However, sequence homology between major French HCV isolates and a Japanese isolate is low. Recently, a 1.5 kb sequence of a French HCV isolate has been determined (Li et al., submitted). This will allow better clinical PCR detection of HCV infection in European countries by using primers directly based on European HCV strains.

Comparison of the results between HCV RNA detection and anti-C100 serology revealed about 37% discordance between those two HCV markers. It is clear that the anti-C100 marker is unable to discriminate by itself between past or currently active infection. Since we could detect HCV RNA using only 100 μl serum equivalent, the method reported here should find wide application in clinical research. One major application of the PCR technique will be the study of the different therapeutic approaches aimed at the suppression of HCV replication [10]. In one of the cases that we tested, HCV RNA was detectable before interferon therapy but became undetectable in the two samples obtained during and after treatment. Further study based on large number of samples will allow us to evaluate the effectiveness of this method for the follow up of interferon therapy in chronic hepatitis C.

Acknowledgements

The authors wish to thank Dr A. J. Weiner (Chiron Corporation, CA) for kindly providing the HCV cDNA clone and the primer "sense 2" (2488–2503) used in the present study. J.S.L is a recipient of a fellowship from the Marcel Mérieux Foundation.

References

1. Choo Q-L, Kuo G, Weiner AJ, Overby LR, Bradley DW, Houghton M (1989) Isolation of a cDNA clone derived from a blood-borne non-A, non-B viral hepatitis genome. Science 244: 359–362
2. Kuo G, Choo Q-L, Alter HJ, Gitnick GL, Redeker AG, Purcell RH, Miyamura T, Dienstag JL, Alter MJ, Stevens CE, Tegmeier GE, Bonino F, Colombo M, Lee W-S, Kuo C, Berger K, Shuster JR, Overby LR, Bradley DW, Houghton M (1989) An assay for circulating antibodies to a major etiologic virus of human non-A, non-B hepatitis. Science 244: 362–364
3. Alter HJ, Purcell RH, Shih JW, Melpolder JC, Houghton M, Choo Q-L, Kuo G (1989) Detection of antibody to hepatitis C virus in prospectively followed transfusion recipients with acute and chronic non-A, non-B hepatitis. Lancet ii: 1494–1500
4. Boudart D, Lucas JC, Muller JY, Carrer DL, Planchon B, Harousseau JL (1990) False-positive hepatitis C virus antibody tests in paraproteinaemia. Lancet ii: 63
5. McFarlane IG, Smith HM, Johnson PJ, Bray GP, Vergani D, Williams R (1990) Hepatitis C virus antibodies in chronic active hepatitis: pathogenetic factor or false-positive result? Lancet i: 754–755
6. Theilman L, Blazek M, Goeser T, Gmelin K, Komerell B, Fiehn W (1990) False-positive anti-HCV tests in rheumatoid arthritis. Lancet ii: 1346
7. Houghton M, Choo Q-L, Kuo G (1989) European patent application No. 0318216A1
8. Weiner AJ, Kuo G, Bradley DW, Bonino F, Saracco G, Lee C, Rosenblatt J, Choo QL, Houghton M (1990) Detection of hepatitis C viral sequences in non-A, non-B hepatitis. Lancet i: 1–3
9. Garson JA, Tedder RS, Briggs M, Tuke P, Glazebrook JA, Trute A, Parker D, Barbara JAJ, Conteras M, Aloysius S (1990) Detection of hepatitis C viral sequences in blood donations by "nested" polymerase chain reaction and prediction of infectivity. Lancet i: 1419–1422
10. Kanai K, Iwata K, Nakao K, Kako M, Okamoto H (1990) Suppression of hepatitis C virus RNA by interferon-α. Lancet ii: 245
11. Kaneko S, Unoura M, Kobayashi K, Kuno K, Murakami S, Hattori N (1990) Detection of serum hepatitis C virus RNA. Lancet i: 976
12. Li JS, Vitvitski L, Tong SP, Trépo C (1990) PCR amplification of HCV sequences from French serum samples and preliminary sequence analysis (1990) Hepatitis B Virus Meeting, La Jolla, California, USA, Abstracts, p 43

Authors' address: Dr. J. Li, Unité de Recherche sur les Hépatites, INSERM U271, 151 Cours A. Thomas, F-69003 Lyon, France.

IX Hepatitis C virus and blood donation

Arch Virol (1992) [Suppl] 4: 241–243
© Springer-Verlag 1992

HCV and blood transfusion

H. W. Reesink[1], **C. L. van der Poel**[1], **H. T. M. Cuypers**[2], and **P. N. Lelie**[2]

[1] Red Cross Blood Bank Amsterdam
[2] Central Laboratory of the Netherlands, Amsterdam, The Netherlands

Summary. Posttransfusion hepatitis remains a threat to transfusion therapy. Testing for increased ALT levels has been used in an attempt to reduce this risk. Presence of the infectious agent, hepatitis C virus (HCV), appears to be a much more sensitive criterion. Stored serum samples from transfusion blood as well as recipients of transfusion were tested by ELISA, RIBA and PCR for the presence of HCV. The results show that RIBA and PCR are about equally sensitive and are able to detect HCV positivity in many sera that might have been otherwise transfused. Routine screening for the presence of virus will dramatically reduce the danger of hepatitis infection to transfusion patients.

*

Posttransfusion hepatitis (PTH) is one of the major complications of blood transfusion therapy. Recipients of blood products have about a 10% risk of developing PTH in the United States, in Northern Europe 2–4% and in Southern Europe 15–20%.

All studies show that 80–90% of all PTH cases are attributed to non-A/non-B (NANB) hepatitis [1]. Several independent studies demonstrated that a proportion of the donors carrying the NANB agent had increased levels of alanine amino transferase (ALT). When donors with elevated ALT were excluded about 30% of PTH-NANB would be prevented. Some older studies indicated that anti-hepatitis-B core (anti-HBc) positive donors may have an increased risk to transmit PTH-NANB, but more recent studies do not confirm this [1, 2].

Researchers at the Chiron Corporation (USA) recently isolated a cDNA clone from a parenterally transmitted NANB-hepatitis viral genome. This virus was named hepatitis C virus (HCV). A polypeptide antigen (C100-3) was expressed in yeast, with which first a RIA and later an ELISA antibody test (anti-HCV) could be developed. The specificity, sensitivity and predictive value of the new anti-HCV ELISA was established by studying sera from donors and recipients, implicated and not-implicated in PTH-NANB [3, 4].

From 1984 through 1986 a prospective study was conducted in Amsterdam to establish the incidence of PTH-NANB in patients who underwent cardiac surgery [2].

In 1989 stored serum samples of 5150 blood product transfusions and 383 recipients of the prospective study were tested with anti-HCV ELISA.

6 of 9 (67%) recipients with PTH-NANB and 9 of 374 (2.4%) without PTH-NANB seroconverted and became anti-HCV positive (Chi square, $p < 0.001$).

6 of 151 (3.9%) bloodproducts transfused to recipients with PTH-NANB and 31 of 4999 (0.6%) bloodproducts transfused to recipients without PTH-NANB were anti-HCV-reactive (Chi square, $p < 0.001$).

Of 35 anti-HCV positive blood products 9 (26%) were associated with patients developing PTH-NANB and/or anti-HCV seroconversion.

Donor co-factors associated with infectivity of anti-HCV positive blood products were: raised ALT (6 of 9 infective vs 1 of 26 not-infective); a mean ELISA optical density/cut-off ratio ≥ 2 (7 of 9 vs 9 of 26); and persistent donor anti-HCV seropositivity (7 of 14 vs 0 of 14) [5].

A recently developed recombinant immunoblot assay (RIBA) applying the 5-1-1 antigen and C100-3 antigen (Ortho) may also be a tool to discriminate infectious from non-infectious anti-HCV positive blood donors [6].

Of the anti-HCV positive blood products associated with recipient PTH-NANB, 4 of 6 were RIBA positive as compared to 2 of 31 anti-HCV positive bloodproducts not associated with recipient PTH-NANB [7].

In 1990, fresh plasma could be obtained from the recipients and part of the anti-HCV positive donors. A polymerase chain reaction (PCR) to detect viral RNA sequences (developed in our institutes) was applied to these fresh sera. PCR-positivity correlated strongly with RIBA positivity in donors as well as in recipients.

The use of anti-HCV blood donor screening to prevent PTH-NANB, was compared with ALT screening. A corrected efficacy of 63% and 65%, a specificity of 93% and 64% and a positive predictive value of 16.2% and 3.6% were found, respectively; 0.7% or 3.8% of blood donations would be discarded, respectively.

The introduction of routine screening of blood donors for anti-HCV antibodies will dramatically reduce the risk of developing PTH-NANB for recipients.

A problem for the bloodtransfusion organisations will be that at present no practical confirmatory assays are available to discriminate infectious from non-infectious donors, thus causing the rejection of a high proportion (in the Netherlands 75%) of non-infectious anti-HCV reactive blood donors.

References

1. Reesink HW, van der Poel CL (1989) Blood transfusion and hepatitis. Still a threat? Blut 58: 1–6
2. Reesink HW, Leentvaar-Kuypers A, van der Poel CL, et al (1988) Non-A, non-B posttransfusion hepatitis in open heart surgery patients in the Netherlands: preliminary results of a prospective study. In: Zuckerman AJ (ed) Viral hepatitis and liver disease. Liss, New York, pp 558–560
3. Van der Poel CL, Reesink HW, Lelie PN, et al (1989) Anti-hepatitis C antibodies and non-A, non-B post-transfusion hepatitis in the Netherlands. Lancet ii: 294–297
4. Kuo G, Choo Ql, Alter HJ, et al (1989) An assay for circulating antibodies to a major etiologic virus of human non-A, non-B hepatitis. Science 244: 362–364
5. Van der Poel CL, Reesink HW, Lelie PN, et al (1990) Infectivity of blood seropositive for hepatitis C virus antibodies. Lancet 355: 558–560
6. Ebeling F, Naukkarinen R, Leikola J (1990) Recombinant immunoblot assay for hepatitis C antibody as predictor of infectivity. Lancet 335: 982–993
7. Van der Poel CL, Reesink HW, Lelie PN, et al (1990) Anti-HCV and transaminase testing of blood donors. Lancet 336: 187–188

Author's address: Dr. H. W. Reesink, Red Cross Blood Bank Amsterdam, NL-1006 AC Amsterdam, The Netherlands.

Arch Virol (1992) [Suppl] 4: 244–246
© Springer-Verlag 1992

Evaluation of anti-HCV positive blood donors identified during routine screening

J. I. Esteban, J. C. Lopez-Talavera, J. Genescà, A. Gonzalez, V. Vargas, M. Buti, J. Guardia, and **R. Esteban**

Liver Unit, Department of Medicine, Hospital General Universitari Vall d'Hebrón, Barcelona, Spain

Summary. Of 30,231 donors tested, 368 (1.2%) were anti-HCV positive. Of these, 254 have been evaluated, with the following results: only 25% have a history of parenteral risk, seroprevalence increases with age and approximately 80% of those that are anti-HCV positive in our population are probably infected with HCV. In addition, an unexpectedly large number of these persons have chronic and/or severe liver disease and will require combined diagnostic approaches for accurate evaluation.

*

The identification of the hepatitis C virus (HCV) as the major causative agent of non A, non B hepatitis, and the development of a recombinant ELISA to detect antibodies directed to this agent will provide an invaluable tool to better define the natural history of this chronic disease.

The significance of a positive antibody test in asymptomatic persons detected during screening, regarding both potential infectivity and presence of liver disease, remains largely unknown. We have therefore investigated the epidemiological, clinical and histological features of a large cohort of anti-HCV positive patients identified at blood donation.

Of the 30,231 donors tested between July 1989 and April 1990, 368 (1.2%) were found repeatedly reactive for anti-HCV (Ortho). Their mean age was 45.8 ± 0.82 (range 19 to 65), and only 29% were first time donors. Although the overall prevalence of anti-HCV among male and female donors was similar, male donors aged $\leqslant 40$ had a significantly higher prevalence of seropositivity than female donors of similar age (1% vs 0.45%, respectively; $p < 0.0001$). Similarly, the anti-HCV prevalence among donors aged > 50

was almost three times as high as that found in donors $\leqslant 50$ irrespective of gender (2.5% vs 0.9%, respectively; $p < 0.0001$).

Of 254 seropositive patients evaluated so far, only 24% had a history of evident percutaneous blood exposure (transfusion or i.v. drug abuse). When compared to sex and age-matched seronegative donors, a family history of liver disease, a history of blood transfusion, and previous i.v. drug abuse and/or tattoing were all variables independently associated with seropositivity. A history of drug abuse and/or tattoing was the only factor associated with seropositivity among donors aged < 30, but only among males. Also, a history of surgery and/or transfusion correlated with seropositivity in all donors older than 30, irrespective of sex.

At evaluation, all patients were asymptomatic, and abnormalities of the physical exam were found in 9% of them. Thirty-eight percent of the seropositive donors had an elevated ALT level compared to only 4% of seronegative controls. This proportion of patients with ALT elevation increased to 58% during subsequent follow-up. Moreover, during interview 15% of anti-HCV positive donors had a history of previous ALT abnormal levels in the preceding years compared to only 1% of controls ($p < 0.0001$).

Patients were classified into 4 groups according to their mean initial ELISA ratio (ER) (sample OD divided by cut-off value): 112 (44%) with ER > 5 (group I); 64 (25%) with ER between 2.1 and 5 (group II); 55 (22%) with ER between 1 and 2 (group III); and 23 (9%) with ER between 0.6 and 0.9 (group IV). Initial samples from 113 seropositive donors were also tested with a recombinant immunoblot assay (RIBA, Chiron). Sixty-one percent were considered positive (reactive for both the 5.1.1. and C-100 bands), 31% gave indeterminate results (unreactive for one band or faintly reactive for one or both bands) and 8% were negative (unreactive for both bands). Indeterminate results were classified in 2 groups: those with a detectable 5.1.1. band (type A) and those reactive only for the C-100 band (type B). The proportion of patients with each RIBA pattern was significantly correlated with their ER group: group I: 90% positives, 6% indeterminate A, 4% indeterminate B; group II: 54% positives, 25% indeterminate A, 21% indeterminate B; groups III + IV: 24% positives, 42% indeterminate A, 10% indeterminate B, 24% negatives.

The proportion of patients with elevated ALT significantly correlated with the ER group ($p = 0.005$). Sixty-one percent of patients in group I had abnormal ALT; in group II, the ALT was elevated in 55% of the cases; and only 40% of donors had a raised ALT in group III + IV. At the same time, an elevated ALT profile was observed in 71% of donors reactive for the 5.1.1. band (RIBA positive + type A indeterminates), whereas only 10% of donors unreactive for this band (RIBA negative + type B indeterminates) had a raised ALT ($p < 0.0001$).

Liver biopsy was randomly performed to 83 seropositive patients; this group was not different from the total group of donors evaluated. Some degree of histological abnormality was found in 81% of them: 13% had minimal changes, 25% chronic persistent hepatitis, 36% chronic active hepatitis and 6% active hepatic cirrhosis. The percentage of donors with histological abnormalities increased with the ER: 57% of those with ER $\leqslant 2$, 75% of those with ER between 2.1 and 5, and 98% of those with ER > 5. All patients with an elevated ALT had liver damage, irrespective of their ER, while all patients with a normal liver had normal ALT. The presence of a 5.1.1. band in the RIBA was also predective of liver damage, specially among donors with an ER > 2. This band was detectable in 10 donors of this group of 12 with liver damage, it was absent in 5 of the 6 with a normal histology, and 10 of the 11 with a detectable 5.1.1. band had abnormal histology.

Investigation of infectivity was done through "look-back" analysis of a subset of 43 donors. 36 (84%) of them were considered infectious: 22 had been implicated in cases of posttransfusion hepatitis, and at least two recipients of previous donations from the remaining 14 were found anti-HCV positive. Several variables were investigated to identify markers of potential infectivity in these implicated donors. An elevated ALT level, a 5.1.1. band present RIBA pattern, and the presence of histological abnormalities were all significantly correlated with infectivity. However, none of them was found independently associated with infectivity.

In summary, three are the most remarkable findings of this study. First, the confirmation that only 25% of asymptomatic anti-HCV positive persons have a history of parenteral exposure to blood, and that seroprevalence increases with age. Second, that about 80% of anti-HCV positive persons are very likely persistently infected with HCV. And third, that an unexpectedly high proportion of these people have chronic, and often severe liver disease. The combination of ALT determination, ELISA ratio, RIBA pattern and liver biopsy will help to better evaluate persons who test positive for anti-HCV during routine screening.

Authors' address: Dr. J. I. Esteban, Liver Unit, Department of Medicine, Hospital General Universitari Vall d'Hebrón, Barcelona, Spain.

Arch Virol (1992) [Suppl] 4: 247–248
© Springer-Verlag 1992

Presence of HCV RNA in serum of asymptomatic blood donors involved in post-transfusion hepatitis (PTH)

E. Villa[1], M. Melegari[1], I. Ferretti[1], C. Vecchi[2], P. P. Scaglioni[1], M. De Palma[2], F. Manenti[1]

Chair of Gastroenterology[1] and Blood Bank[2], University of Modena, Italy

Summary. To study the causes of residual posttransfusion hepatitis, serum from implicated donors was tested by PCR for the presence of HCV RNA. Of 20 anti HCV negative donors, 4 were HCV RNA positive and thus, infective. The results suggest that higher-level investigations are necessary for prospective donors who present blood enzyme abnormalities or other questionable characteristics.

*

Since HCV RNA is a direct marker of HCV infection, we investigated a group of blood donors involved in PTH for the presence in serum of HCV RNA [1].

In a group of 27 blood donors involved in eight cases of HCV-positive PTH, routine blood tests and HCV antibody (ELISA and RIBA, Ortho Diagnostic Systems) were studied. For the detection of viral sequences, serum RNA was extracted according to the method of Chomczynski and Sacchi [2], reverse-transcribed and amplified using primers belonging to the non-structural region of HCV.

Results

The amplified product of the PCR reaction was 582 base pairs, as predicted from the HCV sequence. Of the 27 donors examined, 7 (25.9%) were found to be anti-HCV positive by ELISA, HCV RNA was present in 5 (71.4%) of these anti-HCV positive subjects, all were RIBA positive. Among the 20 anti-HCV negative blood donors (all RIBA negative), 4 (20.0%) were HCV RNA positive.

ALT levels were found to be below 45 UI/l in 18 donors while the other 7 had ALTs over the limit accepted for transfusion. The anti-HCV negative, HCV RNA positive blood donors had repeated abnormal tests in the past but now had normal ALTs.

Discussion

Our study offers a direct explanation for the substantial proportion of residual cases of anti-HCV positive PTH: 37% of the anti-HCV negative blood donors involved were HCV RNA positive and therefore infective. Another relevant characteristic of these subjects was the normality of ALT values in 3 out of 4: none of the current screening tests, therefore, would have revealed the risk associated with the transfusion of this blood. This suggests the opportunity of creating a register of blood donors who have presented blood enzyme abnormalities and who have been reevaluated by applying second-level investigations such as HCV RNA.

Acknowledgements

This study was supported by Associazione Italiana per la Ricerca sul Cancro (AIRC).

References

1. Esteban JI, Gonzalez A, Hernandez JM et al (1990) Evaluation of antibodies to hepatitis C virus in a study of transfusion-associated hepatitis. N Engl J Med 323: 1107–1112
2. Chomczynski P, Sacchi N (1987) Single step method of RNA isolation by acid guanidinium thiocyanate–phenol–chloroform extraction. Anal Biochem 162: 156–159

Authors' address: Dr. Erica Villa, Chair of Gastroenterology, Via del Pozzo 71, I-41100 Modena, Italy.

Arch Virol (1992) [Suppl] 4: 249–252
© Springer-Verlag 1992

Epidemiology of anti-HCV antibodies in France

O. Agulles, C. Janot, and
the Viral Hepatitis Study Group of the French Blood Transfusion Society

CRTS – Vandoeuvre, France

Summary. The aim of this large survey which covered 173,038 unselected blood donors, was to determine the seroprevalence of anti-HCV antibodies and surrogate markers (ALT and anti-HBc) in France. The results revealed a frequency of 0.63% of anti-HCV positive donors. The correlation with surrogate markers was very poor but since we know nothing about the infectivity of anti-HCV negative donations, screening of surrogate markers must still be performed to prevent post-transfusional hepatitis.

*

Serological testing for anti-HCV antibodies has been mandatory for every blood donation in France since March 1990. A cooperative study involving twelve blood transfusion centres was carried out to determine the seroprevalence of anti-HCV in blood donors and the correlation between anti-HCV and the surrogate markers (SM) alanine aminotransferase (ALT) and anti-HBc for which French blood donors have been systematically tested since 1988.

This large survey covered 173,038 unselected donors from various areas of France evenly distributed throughout the country. The Ortho HCV antibody ELISA test system was used according to the manufacturer's instructions. Each serum was tested once and every reactive specimen was tested in duplicate. 1,099 samples were repeatedly reactive for anti-HCV and the frequency (0.63%) did not differ significantly in the various parts of the country (Fig. 1), not even in Toulouse where anti-HBc has already been tested for over a long time period or in large metropolitan areas such as Paris or Marseille.

Of 173,038 samples tested, 165,393 had no SM but 7,645 had one or both. In the anti-HCV negative population 96% had no SM but 4% had one. In

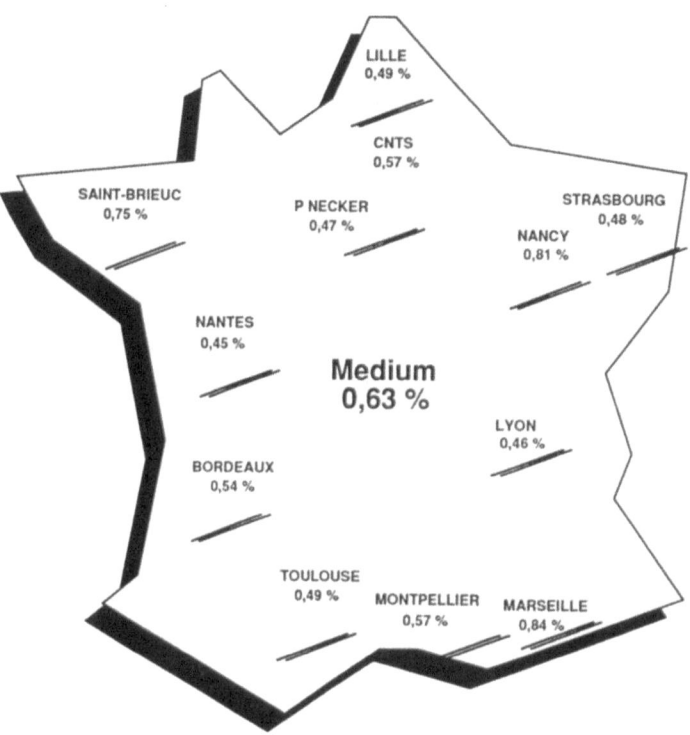

Fig. 1. Seroprevalence of anti-HCV antibodies in France

the anti-HCV positive population, 84% had no SM but 16% had one. Anti-HCV positive donors represent only 0.55% of the SM negative population (917/165,393). Anti-HCV and SM positive samples represent 2.38% of the SM positive population (182/7,645) and we can say that the SM frequency is four times greater in the anti-HCV positive population than in the negative population. But, as the SM positive population represents only 5% of the total (7,645/173,038) most of the anti-HCV positive donors are in the SM negative population. The correlation between anti-HCV and raised ALT was better than that between anti-HCV and anti-HBc (5.45% versus 2.43%) if we consider that raised ALT is 2 times less frequent than anti-HBc in the general population. As donors with SM constituted only 5% of the total, only 16.5% of anti-HCV seropositive donors would have been eliminated from blood donation by testing for anti-HBc/ALT.

These results are summarized in Table 1 and Fig. 2. They were obtained in a country with a low incidence of anti-HBc (about 3% of blood donors in France). By screening anti-HCV and SM, 5% of blood donations are destroyed: 0.1% are anti-HCV and SM positive, 0.5% are only anti-HCV positive and 4.4% are SM positive without anti-HCV antibodies.

The ELISA tests for virus detection have now reached satisfactory sensitivity levels. However, falsely positive reactions remain a matter of concern. The Recombinant Immuno Blot Assay (RIBA) or the neutralization

Table 1. Distribution of surrogate markers (SM) in anti-HCV positive and negative populations

		SM neg.	SM pos.	Total
aHCV neg.	N	164 476	7 463	171 939
	%	95.6%	4.35%	100%
aHCV pos.	N	917	172	1 099
	%	83.4%	16.6%	100%
Total		165 393	7 645	173 038

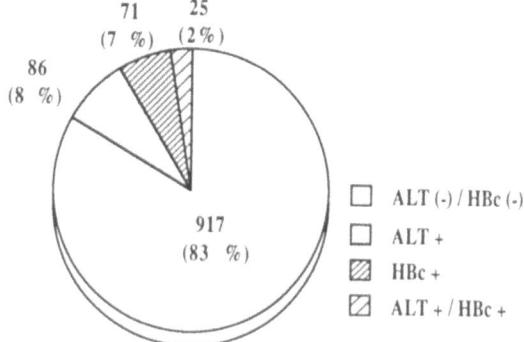

Fig. 2. Distribution of surrogate markers in the anti-HCV positive population (1,099 samples)

test might help to check the results of ELISA tests. In two groups of ELISA anti-HCV positive donors, only 45% [1] and 44% [2] of samples were found RIBA seropositive. Such results indicate that epidemiological studies using solely the ELISA test may have led to overestimation of the real frequency of anti-HCV seropositive donors.

Large prospective studies are needed to evaluate the efficiency of anti-HCV testing in blood donors and to find out how many anti-HCV negative donations are infectious.

References

1. Weiner AJ, Truett MA, Rosenblatt J, Han J, Quan S, Polito AJ, Kŭo G, Choo Q-L, Houghton M (1991) HCV: Immunologic and hybridization-based diagnostics. In: Hollinger FB, Lemon SM, Margolis H (eds) Viral hepatitis and liver disease. Williams & Wilkins, Baltimore, pp 360–363

2. Courouce AM, Janot C et le Groupe de Travail "Hépatites Virales" de la Societé
Nationale de Transfusion Sanguine (1990) Test de confirmation de l'anticorps anti-VHC.
XVème Congrès de la Société Nationale de Transfusion Sanguine. Besancon

Authors' address: Dr. O. Agulles, CRTS, Avenue de Bourgogne, F-54500 Vandoeuvre,
France.

Arch Virol (1992) [Suppl] 4: 253–255
© Springer-Verlag 1992

Assay of antibodies to hepatitis C virus protein C100-3 in blood donors from Northern Germany

G. Caspari[1,*], H.-J. Beyer[1], W. H. Gerlich[2,*], H. Schmitt[1]

[1] German Red Cross Blood Transfusion Service of Lower Saxony, Springe
[2] Department of Medical Microbiology, Georg-August-University, Göttingen

Summary. The prevalence of anti-C100-3 increases with age from 0.41% to 1.26%. It is more frequent in donors with elevated ALT (4.5%). Most ALT elevations, however, are not related to anti-C100-3. Low EIA signals ($<3\times$ cutoff) are often non-specific. The cutoff value should be 2.5 times higher. High EIA signals correlate with ALT elevations.

*

The German Red Cross Blood Transfusion Service of Lower Saxony (Northern Germany) introduced routine screening for hepatitis C virus (HCV) infections on May 2, 1990. During the first three months, 116,700 different donors were tested by an enzyme immune assay for antibodies to recombinant HCV protein C100-3 from yeast (Ortho Diagnostics), and 842 donors ($=0.72\%$) were repeatably reactive. This rate is relatively low and poses some problems for the effectiveness of the routine screening procedure, as shown below.

The repeatable reactivity rate increased with *age* from 0.41% in the youngest age group to 1.26% for those over 58 years old (see Table 1). The difference between the repeatable reactivity rates for men and women of 58 years or older was not significant.

The repeatable reactivity rate increased also with the *ALT level* of the *current donation*. It was 0.66% for donors with normal ALT compared to 4.56% for donors with an ALT of more than twice the upper limit of the

---

* Present address: Institute of Medical Virology, University of Giessen, Federal Republic of Germany.

Table 1. Correlation of anti-HCV to sex and age in blood donors from Lower Saxony (Percentage of anti-HCV positives and absolute numbers)

Age (Years)	Sex male	female	Sum
All age groups	0.71% 515/72,212	0.73% 327/44,488	0.72% 842/116,700
18–27	0.42% 70/16556	0.41% 46/11,160	0.41% 116/27,716
28–37	0.48% 65/13,526	0.48% 41/8,376	0.48% 106/21,902
38–47	0.68% 107/15,707	0.72% 71/9,795	0.69% 178/25,502
48–57	0.98% 184/18,771	0.98% 106/10,813	0.98% 290/29,584
≥58	1.16% 89/7,652	1.45% 63/4,344	1.26% 152/11,996

normal range. Women with slightly elevated ALT were much more often reactive than men.

We have stored in our computer ALT results of all blood donations since 1975. In fact, many reactive donors had *ALT* elevations *in previous donations*. However, 95–99% of the ALT elevations in the past were not associated with anti-C100-3, and thus were probably not due to HCV (Table 2).

We then further evaluated the Ortho anti-C100-3 test. Optimally, it should have a minimal number of results around its cutoff value between positive and negative. However, in 15,545 blood donors we found a frequency minimum of optical densities between 2.5 and 3 times the cutoff. Further analysis of 529 positive and grey-zone results confirmed this finding. The reproducibility was very good for the strongly reactive samples. Out of

Table 2. Anti-C100-3 and ALT elevations in previous blood donations

Highest ALT (IU/l) for previous donations	Anti-C100-3 in current donation pos.	neg.	% pos.
≤19[a] or ≤23[b]	447	69,248	0.64
≤37[a] or ≤45[b]	260	29,725	0.87
<100	58	5,087	1.14
≥100	18	355	4.83

[a] female donors   [b] male donors

122 sera with an initial O.D. of ⩾ 2.200, 97 had the same O.D. in *both* retests. Four initially strongly reactive results were not reproducible. This false positivity, even if very strong, may be due to occasional difficulties in handling the Ortho ELISA plates with the Behring ELISA processor. The reproducibility was much lower for low positives and grey-zone results. These findings correspond to those of Weiner et al. [1].

With the new cutoff of 2.5 times the former one, there would have been 179 initially reactives instead of 457 and 156 repeatably reactives instead of 375.

Serum-ALT is significantly more often elevated in donors strongly anti-C100-3 positive than in low anti-C100-3 positive donors. If an optical density of more than three times the original cutoff is the marker for strong positivity, then 33.1% of these donors have an ALT of ⩾ 19 IU/l compared to 11.9% of donors with a low anti-C100-3 positive result ($Chi^2 = 29.9$). The difference remains significant for all ALT-limits between 17 IU/l and 40 IU/l.

A total of 27,508 donors donated twice between May 2 and October 8, 1990. Of these, 27,276 or 99.15% were anti-HCV negative on both occasions; 41 converted from negative to repeatably positive; 46 were repeatably positive at the time of the first and negative at the time of the second donation; 145 were positive for both donations. This means that possibly 22% of donors repeatably reactive for one donation are found negative for the next donation about 3 months later.

From the repeatably Ortho anti-C100-3 reactive samples 379 were available for retesting in the corresponding Abbott test. Of these, 105 samples or 28% were initially negative. Twenty-eight samples were in the Abbott grey zone of cutoff minus 20 percent and only 246 of the 379 samples (65%) were initially reactive in the Abbott test. The initial Abbott results correspond to some extent to the strength of the initial Ortho results. We did no retesting with the Abbott test and we had no possibility to crosscheck Abbott positive samples with the Ortho test.

References

1. Weiner AJ, Truett MA, Rosenblatt J, Han J, Quan S, Polito AJ, Kuo G, Choo Q-L, Houghton M, Agius C, Page E, Nelles MJ (1990) HCV testing in low-risk population. Lancet 336: 695

Authors' present address: Dr. G. Caspari, Institute of Medical Virology, University of Giessen, D-W-6300 Giessen, Federal Republic of Germany.

X Chronic hepatitis in childhood

Arch Virol (1992) [Suppl] 4: 259–262
© Springer-Verlag 1992

Chronic viral hepatitis in children

G. Maggiore and **C. De Giacoma**

Clinica Pediatrica dell'Università, IRCCS Policlinico San Matteo, Pavia, Italy

Summary. Hepatitis B virus is the most common causative agent of chronic viral hepatitis in children. The disease may take an aggressive course, but remains mostly asymptomatic. HDV infection occurs in about 13% of those children who are chronic carriers and are HBsAg positive. HCV infection is generally related to parenteral risk and generally remains asymptomatic. In addition to describing the course of the various diseases, treatment and control measures are discussed.

*

Chronic viral hepatitis in children is mainly related to Hepatitis B virus (HBV) infection and in some cases to infection with Hepatitis D virus (IIDV) or with the newly described Hepatitis C virus (HCV).

Chronic HBV infection occurs in three phases [1]. In the replicative phase, the child exhibits all serum markers of active viral replication (HBeAg, HBV DNA). In liver tissue, HBcAg is detectable in the nuclei of the hepatocytes and free HBV DNA is detectable by molecular hybridization techniques. Aminotransferases are increased and liver histology shows inflammation and necrosis with the picture typical of a chronic aggressive hepatitis in almost 60% of patients. The spontaneous seroconversion phase occurs after a mean period of 4 to 5 years which can however be as long as 8 years or more. It is characterized by the appearance of anti-HBe antibodies and by the progressive disappearance of serum and tissue markers of HBV replication. Aminotransferases tend to normalize often after an abrupt increase which is generally asymptomatic. The mean annual rate of HBe seroconversion is about 16%, with a wide range from 10 to 30% [2].

Ongoing seroconversion could be predicted if: 1. the child is more than 3 years of age; 2. the mother is HBsAg negative; 3. there is history of an acute disease; 4. the child has an active disease with aggressive hepatitis on liver biopsy and/or with focal pattern of distribution of HBcAg in liver tissue [3]; 5. High titers of anti-HBc IgM are present in the serum [4]. In the non-replicative phase, seroconversion to anti-HBe is stable and persistent, serum HBV DNA is absent, aminotransferases are normal and histologic activity is absent. Spontaneous clearance of serum HBsAg is exceptional in these patients.

Chronic hepatitis B in children is rarely symptomatic or associated with clinical evidence of severe chronic liver disease. In most children, the condition is detected fortuitously by the finding of increased serum amino-transferase activities and/or of a moderate, asymptomatic liver enlargement. In the vast majority of children HBeAg is present at diagnosis, however about 15% of patients may present with increased aminotransferases activity and anti-HBe in the serum. In these patients serum markers of HDV infection and serum HBV DNA must be evaluated. Children with increased aminotransferases, anti-HBe, but without HDV infection represent a hetero-geneous group with a variety of histological patterns including cirrhosis [5].

The course of the liver disease related to the HBV infection depends mostly on the severity of liver damage. A liver biopsy in a child with chronic HBV infection can demonstrate, along the duration of the disease, different patterns of liver injury. Cirrhosis is one of the most severe complications of HBV infection in children; it may occur at any age and at any phase of the disease. Rapid evolution toward cirrhosis may be demonstrated particularly during infancy and may be associated with very little or absent clinical and biochemical evidence of liver damage [6]. The prevalence of cirrhosis in children with chronic HBV infection is wide, ranging from 3–32% in the largest published series [7–9]. This difference may depend on different factors, such as the number of patients studied, geographic or ethnic origin and the different degree of sensitivity of the methods employed for dia-gnosing cirrhosis. In children, blind percutaneous liver biopsy is, in fact, a low-sensitivity test with up to 50% false negative results [9]. Risk factors for severe liver lesions and cirrhosis include: 1. Male sex; 2. Infection in early age; 3. Evidence of HDV infection and 4. Early seroconversion to anti-HBe [7, 8].

HBV infection plays a major role in the development of hepatocellular carcinoma [10]. Integration of HBV DNA sequences into the liver cell genome has been demonstrated in children with chronic hepatitis, however this event does not seem to be related to the duration or the severity of the disease [11]. Although hepatocarcinoma is a rare event in childhood, it has been described with increasing frequency in chronic carriers of HBsAg as young as 6 years of age. Early diagnosis is based on frequent evaluation by ultrasonography and serum alpha-fetoprotein level. When hepatocarcinoma is clinically symptomatic the prognosis is poor.

The prevalence of HDV infection in children with HBsAg positive chronic hepatitis is about 13% and association with severe active disease or cirrhosis is frequent [12]. The risk of HDV infection for childhood HBsAg chronic carriers increases with age. Other risk factors are geographic and familial environment and exposure to blood derivatives.

Prevalence studies of HCV infection in children using the recently available serological test for anti-HCV are in progress. Sporadic infection seems infrequent, however patients frequently exposed to blood transfusion or derivatives are at risk of HCV infection. The disease is rarely symptomatic and persistent increase of serum aminotransferases with variable inflammatory activity in the liver is the main feature.

Treatment

The best way to control the disease related to HBV infection is active prevention with HBV vaccination. In adults, alpha-interferon therapy results in a higher HBe seroconversion rate and in a significant loss of HBsAg in patients treated versus controls. Two controlled trials in children failed to confirm these encouraging results, showing a similar rate of HBe seroconversion in patients treated and controls without disappearance of HBsAg from serum [13, 14].

A pilot study of alpha-interferon treatment in children with HDV infection is in progress in Italy. No data are available on interferon therapy in children with anti-HCV positive chronic hepatitis.

References

1. Alagille D, Hahchouel M, Maggiore G, Bernard O (1989) Viral hepatitis in children. In: Lebenthal E (ed) Textbook of gastroenterology and nutrition in infancy, 2nd edn. Raven Press, New York, pp 1005–1015
2. Bortolotti F, Cadrobbi P, Crivellaro C et al (1983) Changes in hepatitis B e antigen antibody system in children with chronic hepatitis B virus infection. J Pediatr 103: 718–722
3. Bortolotti F, Alberti A, Cadrobbi P, Rugge M, Armigliato M, Realdi G (1985) Prognostic value of hepatitis B core antigen (HBcAg) expression in the liver of children with chronic hepatitis type B. Liver 5: 40–47
4. Bortolotti F, Bertaggia A, Rude L et al (1987) IgM antibody to hepatitis B core antigen in children with chronic type B hepatitis. Eur J Pediatr 146: 394–397
5. Bortolotti F, Calzia R, Cadrobbi P, Crivellaro C, Alberti A, Marazzi MG (1990) Long term evolution of chronic hepatitis B in children with antibody to hepatitis e antigen. J Pediatr 116: 552–555
6. Maggiore G, De Giacomo C, Marzani MD, Sessa F, Scotta MS (1983) Chronic viral hepatitis in infancy. J Pediatr 103: 749–752
7. Bortolotti F, Calzia R, Cadrobbi P et al (1986) Liver cirrhosis associated with chronic hepatitis B virus infection in childhood. J Pediatr 108: 224–227
8. The Italian Pediatric Study Group on Chronic Hepatitis (1986) Chronic hepatitis B in children: a multicenter study in Italy. (Abstract) Pediatr Res 20: 699
9. Vajro P, Hadchouel P, Hadchouel M, Bernard O, Alagille D (1990) Incidence of cirrhosis in children with chronic hepatitis. J Pediatr 117: 392–396

10. Wu TC, Tong MJ, Hwang B, Lee SD, Hu MM (1987) Primary hepatocellular carcinoma and Hepatitis B infection during childhood. Hepatology 7: 46–48
11. Scotto J, Hadchouel M, Hery C et al (1983) Hepatitis B virus DNA in children's liver diseases: detection by blot hybridisation in liver and serum. Gut 24: 618–624
12. Maggiore G, Hadchouel M, Sessa F et al (1985) A retrospective study of the role of delta agent infection in children with HBsAg-positive chronic hepatitis. Hepatology 5: 7–9
13. Lok ASF, Lai CL, Wu PC, Leung EKY (1988) Long term follow-up in a randomized controlled trial of recombinant alpha-2-interferon in chinese patients with chronic hepatitis B infection. Lancet ii: 298–302
14. La Banda F, Ruiz Moreno M, Carreno V et al (1988) Recombinant alpha-2-interferon treatment in children with chronic hepatitis B. Lancet i: 250

Authors' address: Dr. G. Maggiore, Clinica Pediatrica, Policlinico San Matteo, I-27100 Pavia, Italy.

Arch Virol (1992) [Suppl] 4: 263–264
© Springer-Verlag 1992

Long-term outcome of chronic type B hepatitis in childhood

M. Mengoli[1], M. E. Balli[2], S. Tolomelli[1], A. Ghirarduzzi[1], and F. Balli[2]

[1] Unità di Epatologia, Divisione di Medicina Interna Ospedale S. Sebastiano,
Correggio (RE)
[2] Clinica Pediatrica Università di Modena, Italy

Summary. The course of hepatitis B was followed in 35 children. Various prognostic factors are evaluated. The long-term outcome of the disease is poor, often progressing to cirrhosis.

*

Although we have only a partial knowledge of the natural history of chronic type B hepatitis in childhood, some authors believe that the outcome of this liver disease is generally good and cirrhosis is a rare early complication of HBV infection [1].

Clinical, biochemical and histological features of chronic type B hepatitis were evaluated in 35 children (M 28 and F 7; mean age 6.3 y; 23 CAH, 12 CPH) hospitalized in the period 1970–82 and followed up for a mean period of 12 years (range 8–20 years). Twenty-one patients (14 CAH and 7 CPH) had previously recognized acute hepatitis (anicteric in 10 cases). The possible source of infection was believed to be household contact in 16 patients, but it remained unknown in 19. Hepatomegaly was detected in all, associated with splenomegaly in 9 cases. Eight children were symptomatic while all patients had raised aminotransferase levels. All cases were HBsAg and anti-HBc +: 25 HBeAg and 5 anti-HBe +, 5 HBeAg and anti-HBe −. In all patients one or more subsequent liver biopsies were performed during the study. No therapy was instituted in CPH; at the beginning of follow-up all children with histological features of CAH received immunosuppressive treatment (prednisone plus azathioprine), which was discontinued after 4 to 36 months in all cases.

Aminotransferase levels, both in CPH and CAH patients, decreased to almost normal levels independent of the histological outcome. Fourteen children (11 CAH, 3 CPH) seroconverted from HBeAg to anti-HBe; the difference of mean seroconversion time in CPH, in children with CAH who improved and who became worse during follow-up, was not significant. None of the children cleared HBsAg from serum during the study. In all but one cases of CPH, a histological improvement occurred during follow-up; whereas 15 (65.2%) cases of CAH showed aggravation of the lesions, with progression to cirrhosis in 6 cases (26.1%), 8 cases (34.8%) presented an improvement of the histological picture.

In conclusion, it seems that:

— There is no correlation between entity of hepatic damage and clinical or histological course.
— Seroconversion from HBeAg to anti-HBe does not necessarily imply a favorable prognosis. In our series this event seemed to be a duration marker of HBV infection, which was not significantly related to the outcome of chronic hepatitis.
— CPH have, but not invariably, good prognosis.
— Contrary to previous reports [1], our study on long-term outcome of chronic active hepatitis type B in childhood suggested that prognosis of this disease is poor, with frequent disease progression to cirrhosis.

Reference

1. Bortolotti F, Calzia R, Cadrobbi P, et al (1986) Liver cirrhosis associated with chronic hepatitis B virus infection in childhood. J Ped 108: 224

Authors' address: Dr. M. Mengoli, Clinica Pediatrica Università di Modena, Ospedale S. Sebastiano, Correggio (RE), Italy.

Arch Virol (1992) [Suppl] 4: 265–267
© Springer-Verlag 1992

Hepatitis in children with thalassemia major

G. Nigro[1], G. Taliani[2], U. Bartmann[1], R. Vitolo[1], T. Perrone[1],
S. Mattia[1], P. Pisano[1], A. Petruccelli[1], S. Maiozzi[1], and M. Midulla[1]

[1] Pediatric Institute of "La Sapienza" University and Institute of Experimental Medicine
of the National Council of Research, Rome
[2] Institute of Tropical Diseases, "La Sapienza" University, Rome, Italy

Summary. Since thalassemia major patients are transfusion dependent, they
are at a particularly high risk of contracting post-transfusion hepatitis. In
this study, 36 transfusion-dependent children were followed up for evidence
of viral hepatitis. Of 23 with increased ALT levels, 17 were anti-CMV and 12
were anti-HCV positive, 9 were positive for both CMV and HCV. Of 13
children with normal transaminase levels, 5 were CMV positive and 3 were
HCV positive. These results show that CMV may be a very common cause of
non-A, non-B hepatitis in transfusion dependent thalassemic children.

Introduction

HCV has been recently shown as the major cause of post-transfusion non-A,
non-B hepatitis [1]. Since patients with β-thalassemia major are life-long
transfusion dependent, we tested sera from these patients for antibodies to
HCV and other potentially involved viruses.

Material and methods

For 18 months, 36 transfusion-dependent children (23 males) with β-thalassemia major,
aged 3 months to 15 years (mean 7.2 years) were followed up for clinical and serological
evidence of viral hepatitis. Serial serum samples were examined for cytomegalovirus (CMV),
Epstein-Barr virus (EBV), hepatitis A (HAV), B (HBV), and C (HCV) viruses, by detection of
class-specific (IgG, IgA, IgM) CMV antibodies [2], by detection of anti-VCA and anti-
EBNA IgM, by viral isolation from urine and/or saliva and by detection of class-specific
(IgG, IgA, IgM) antibodies [2], by detection of anti-HAV IgM, HBsAg, HBeAg, anti-HBe,
IgG and IgM anti-HBc, HBV-DNA, and anti-C100-3 antibodies, respectively.

Results

Based on the ALT values, the thalassemic patients were divided into two groups: a) 23 subjects (63.8%) with persistently increased ALT values (>twice upper limit); b) 13 subjects (26.2%) with persistently normal transaminases. Among the patients of group a), 17 subjects (73.9%) showed antibodies to CMV and 12 subjects (52.2%) had antibodies to HCV; of these, 9 subjects (39.1%) were seropositive to both CMV and HCV. Among the patients of group b), 5 subjects (38.5%) showed antibodies to CMV and 3 subjects (23%) had antibodies to HCV; of these, only one child (7.7%) was seropositive to both CMV and HCV. A significantly ($p < 0.01$) higher rate of CMV and HCV infections was, therefore, revealed in the patients with persistent hypertransaminasemia in comparison to the patients with normal ALT values.

During the follow-up, 7 out of 23 patients (30.4%) with persistent hypertransaminasemia showed 8 episodes of high and rapid ALT increase. These episodes were correlated with infections caused by CMV (3), HCV (1), HAV (1), EBV (1); two cases remained of unknown etiology. A 6-year old boy had two episodes associated with HAV and CMV, respectively.

Discussion

Our study showed a high prevalence of HCV and CMV infections in the thalassemic patients with persistent hypertransaminasemia. CMV has been also associated, more frequently than other viruses, with episodes of high and rapid increase of transaminases. Based on pre-existing specific IgG and appearance of IgM and/or IgA antibodies, these episodes were likely related to reactivated endogenous CMV or to reinfections caused by transfusion-acquired different strains [3]. However, relapse of chronic HCV hepatitis might have also occurred, since both patients with unknown episodes of acute hypertransaminasemia were already HCV-positive. In conclusion, in addition to HCV, CMV should be also considered as a possible cause of non-A, non-B hepatitis in transfusion-dependent thalassemic patients.

Acknowledgements

This work was in part supported by the National Council of Research, Progetto Finalizzato "Prevenzione e controllo dei fattori di malattia (FATMA), contract no. 1991".

References

1. Alter HJ, Purcell RH, Shih JW, Melpolder JC, Houghton M, Choo Q-L, Kuo G (1989) Detection of antibody to hepatitis C virus in prospectively followed transfusion recipients with acute and chronic non-A, non-B hepatitis. N Engl J Med 321: 1494–1500

2. Nigro G, Mattia S, Midulla M (1989) Simultaneous detection of specific serum IgM and IgA antibodies for rapid serodiagnosis of congenital or acquired cytomegalovirus infection. Serodiagn Immunother Infect Dis 3: 355–361
3. Nigro G, Lionetti P, Digilio G, Multari G, Vania A, Midulla M (1990) Viral infections in transfusion-dependent patients with β-thalassemia major: the predominant role of cytomegalovirus. Transfusion 30: 808–813

Authors' address: Dr. G. Nigro, Pediatric Institute of "La Sapienza" University and Institute of Experimental Medicine of the National Council of Research, Rome, Italy.

Arch Virol (1992) [Suppl] 4: 268–272
© Springer-Verlag 1992

Hepatitis in pre-school children: prevalent role of cytomegalovirus

G. Nigro[2], **S. Mattia**[1], **R. Vitolo**[2], **U. Bartmann**[2], and **M. Midulla**[1]

[1] Pediatric Institute of "La Sapienza" University, and Rome
[2] Institute of Experimental Medicine of the National Council of Research, Rome, Italy

Summary. Virological and serological investigations were performed on 8 children with clinical and/or laboratory signs of hepatitis. Cytomegalovirus (CMV) appeared as the most frequently involved etiologic agent, since it was associated with 5 severe or chronically-evolving cases. Out of the other 3 patients with non-CMV associated hepatitis, all completely recovering, two had clinically typical Epstein-Barr virus infections, while the remaining patient had an asymptomatic HBV infection.

Introduction

Symptomatic hepatitis is not a frequent event in childhood; in fact, both acute and chronic hepatitis may often occur subclinically. Hepatitis A virus (HAV), hepatitis B virus (HBV) and non-A non-B viruses, including the recently recognized hepatitis C virus (HCV), are reported to be the most frequently implicated etiologic agents [1, 13]. Cytomegalovirus (CMV) and Epstein-Barr virus (EBV) are generally considered as possible causes of hepatic disease concomitantly with other, often predominant, clinical manifestations [4, 7, 8].

However, little is known about hepatitis occurring in children aged 1 to 6 years. The aim of this study was to investigate on epidemiological, virological and clinical features of hepatitis in pre-school children.

Material and methods

From January 1984 to December 1988, virological and serological investigations were performed on 2817 (1521 males) patients, aged 1 day to 15 years, suspected to be affected by viral infections. When clinical and/or laboratory findings suggested possible hepatic involvement, tests for main viral agents of hepatitis were also performed. Markers for HBV and IgM antibodies against HAV were detected by enzyme immunoassays (EIAs) and/or radioimmunoassays (RIAs) from Abbott (North Chicago, IL USA) and Sorin (Saluggia, Italy). Antibodies to HCV were detected by a second generation EIA (Ortho Diagnostic Systems, NJ USA). Testing for EBV infection used the indirect immunofluorescence (IIF) method to detect antibodies to viral capsid and EBV-associated nuclear antigens. CMV infection was diagnosed by viral isolation from saliva and/or urine and by detection of complement-fixing (CF) and class-specific CMV antibodies (IgG, IgA, IgM) using a previously described EIA [9]. CMV-specific antibodies were also detected by commercial u-capture EIA (ETI-Cytok-M reverse, Sorin, Italy).

The diagnosis of acute hepatitis was based on clinical history and biochemical features (ten-fold or higher increase of transaminases) and on the exclusion of other potential causes of acute liver damage. Chronic hepatitis was suggested when transaminases were more than two-fold higher than the upper normal levels for at least 6 months.

Results

Among all tested patients, 8 previously healthy children (6 males), aged 18 months to 6 years (mean: 3.2 years), were found to have clinical and/or laboratory signs of hepatitis. Of these, 5 patients had concomitant CMV infection, two had EBV infection and the remaining patient had HBV hepatitis.

HBV hepatitis was detected by chance in an anicteric 18 month-old boy brought to the pediatric ward because he was found to have eye-shadow on his lips and tongue. Since he also had moderate hepatosplenomegaly, transaminases and other enzymes were examined and all appeared increased: in particular, ALT was 1,284 U/L. Moreover, positivity of HBsAg, HBeAg and HBV-DNA were detected. Normal values of transaminases were revealed after 3 months, and appearance of anti-HBs after 6 months.

EBV-associated hepatitis occurred in two boys, aged 2 and 4 years, respectively. They showed relevant hepatosplenomegaly and clinical manifestations such as fever or adenomegaly, but no jaundice. The transaminases were moderately increased (about 150–300 U/L) and persisted for less than one month.

CMV infection was associated with two cases of acute hepatitis, of which one was symptomatic, and three asymptomatic cases of persistently moderate hypertransaminasemia. The only symptomatic case of hepatitis occurred in a 4 year-old girl, who deceased after 50 days of disease [15]. This patient showed progressive worsening, including neurologic and respiratory features, concomitantly with increasing levels of blood bilirubin (17.5 mg%), mostly conjugated (15.5 mg%), and transaminases (AST 6,560 U/L; ALT 1,180 U/L). Immunological studies revealed decrease of OKT4 cells (43%; normal values 55–65%) with low T4/T8 ratio (0.84). Virological examinations revealed CMV-specific IgG, IgA and IgM, CF titre of 1:64, and virus isolation from urine. Submassive necrosis, diffuse fibrosis and CMV inclusions in hepatocytes and biliary cells were revealed by necroscopy.

Asymptomatic acute hepatitis occurred in a 4-year old boy with high AST (894 U/L) and ALT (1,134 U/L), which normalized only after 5 months, and very low OKT4/T8 cells ratio (0.43); CMV infection was revealed by virus isolation from urine and by low CMV-specific IgA levels.

Chronic persistence of moderate hypertransaminasemia (ALT = 100–200 U/L) was found in three anicteric children, aged 1, 2 and 5 years, respectively. Laboratory examinations were performed in two cases because of protracted diarrhea in one and meteorism and hepatosplenomegaly in the other. After disappearance of diarrhea and meteorism, anicteric hypertransaminasemia persisted in all three patients. No immunologic abnormalities were revealed. Two patients recovered after about 2 years, while the remaining patient still has, after 6 years, hypertransaminasemia. Light and electron microscopic examinations from the liver biopsy recently obtained from this latter patient are shown in Figs. 1 and 2.

Persistent and active CMV infection was detected in all three patients by stable CF titre of 1:64, and high specific IgG levels; in addition, in two of these patients, CMV was isolated from urine. Epidemiological studies showed an intrafamilial CMV circulation, demonstrated by CF-seropositivity in all cases of CMV-associated hepatitis.

Discussion

Our study showed that hepatitis in pre-school children is predominantly anicteric and prognostically related to the implicated virus. In fact, both HBV and EBV-associated hepatitis were anicteric and self-limited, while CMV-associated hepatitis was severe and/or chronically-evolving, although generally anicteric.

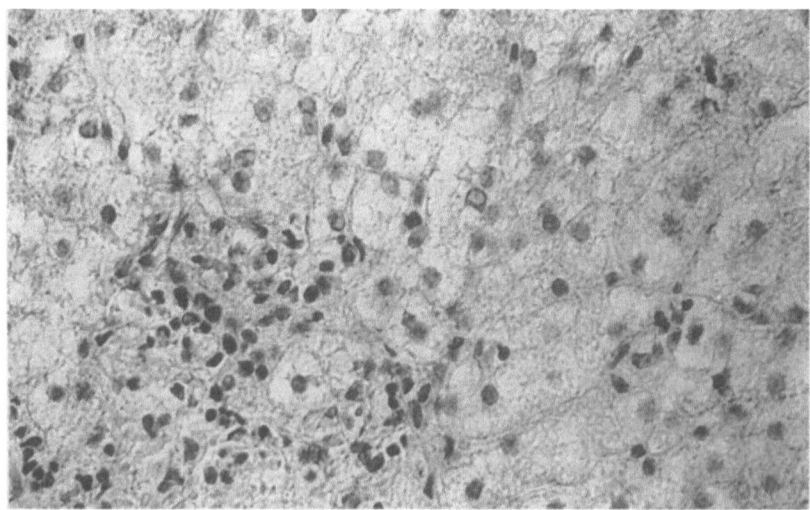

Fig. 1. Liver histological examination (×2300) from a girl with hypertransaminasemia persisting for 6 years, showing steatosis, initial fibrosis and lymphomononuclear cells

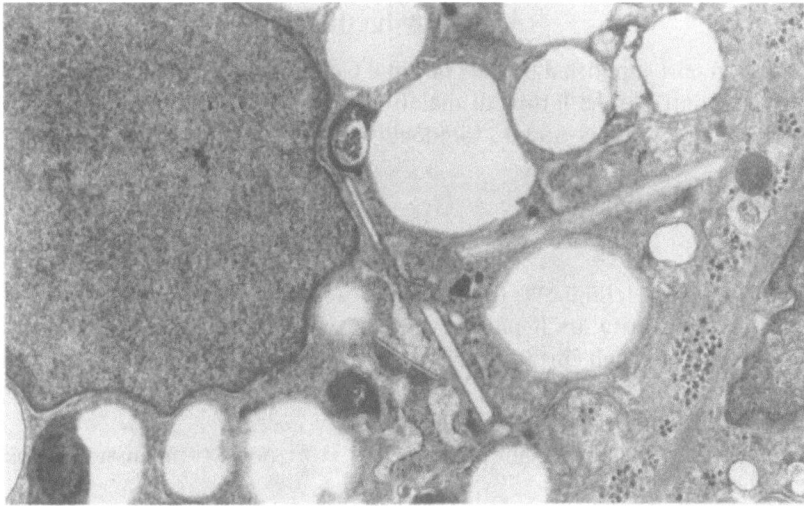

Fig. 2. Electron micrograph (× 25000), from the same patient, showing lipid vacuoles of different size, crystalloid forms suggesting a cholesterol origin, and aggregates of collagen fibres in the extracellular space between the hepatocytes

EBV infection was suspected because of clinical manifestations like adenomegaly, hepatosplenomegaly and fever, but hypertransaminasemia was a surprising finding because of the absence of jaundice [6, 7, 12]. On the contrary, HBV infection was diagnosed only by chance and had an undefinable duration. In fact, from the time the diagnosis was made, hypertransaminasemia persisted for three months but it is not known when it began. This case, presumably related to the transmission from a HBsAg carrier girl attending the same nursery of the patient [8], may account for an undiscovered rate of horizontally-transmitted hepatitis B [3].

The occurrence of CMV-associated hepatitis was higher than expected, as it has been also shown in thalassemic patients [10]. Since CMV is the most frequent cause of neonatal hepatitis, that may occasionally be severe and protracted [2, 4, 5, 12, 14], a congenital or perinatally-acquired CMV infection may be suggested for the three patients with persistent hyper-transaminasemia. On the other hand, in both patients with acute CMV hepatitis, the absence of CMV-specific IgM in the asymptomatic case and the autoptic finding of diffuse fibrosis in the fatal case suggest that CMV infection began some months or even years before the appearance of hepatitis. Therefore, a congenital or perinatally-acquired CMV infection could not be ruled out even in these patients.

In conclusion, as shown by the occurrence of a fatal outcome, CMV hepatitis may be extremely severe also in presumably immunocompetent subjects. Therefore, the need to perform tests for CMV infection in all cases of hepatic disease and, moreover, in all neonates with jaundice, hepatospleno-megaly or other possible signs of CMV infection is stressed.

Acknowledgements

This work was in part supported by the National Council of Research, Progetto Finalizzato "Prevenzione e controllo dei fattori di malattia (FATMA)", contract no. 1991. We thank Prof. A. Ceccamea, S. Valia and L. Simonelli for light and electron microscopy examinations.

References

1. Alter HJ, Purcell RH, Shih JW, Melpolder JC, Houghton M, Choo Q-L, Kuo G (1989) Detection of antibody to hepatitis C virus in prospectively followed transfusion recipients with acute and chronic non-A, non-B hepatitis. N Engl J Med 321: 1494–1500
2. Chang M-H, Hsu H-C, Lee C-Y, Wang T-R, Kao C-L (1987) Neonatal hepatitis: a follow-up study. J Pediatr Gastroenterol Nutr 6: 203–207
3. Gray Davis L, Weber DJ, Lemon SM (1989) Horizontal transmission of hepatitis B virus. Lancet i: 889–893
4. Griffiths PD, Ellis DS, Zuckerman AJ (1990) Other common types of viral hepatitis and exotic infections. Br Med Bull 46: 512–532
5. Ho M (1990) Cytomegalovirus. In: Mandell GL, Douglas RGJr, Bennett JE (eds) Principles and practice of infectious diseases, 3rd edn. Churchill Livingstone, New York Edinburgh London Melbourne
6. Hoagland RJ, McCluskey RT (1955) Hepatitis in mononucleosis. Ann Intern Med 43: 1019–1030
7. Horwitz CA, Burke MD, Grimes P, Tombers J (1980) Hepatic function in mononucleosis induced by Epstein-Barr virus and cytomegalovirus. Clin Chem 26: 243–246
8. Nigro G, Taliani G (1989) Nursery-acquired asymptomatic B hepatitis. Lancet i: 1451–1452
9. Nigro G, Mattia S, Midulla M (1989) Simultaneous detection of specific serum IgM and IgA antibodies for rapid serodiagnosis of congenital or acquired cytomegalovirus infection. Serodiagn Immunother Infect Dis 3: 355–361
10. Nigro G, Lionetti P, Digilio G, Multari G, Vania A, Midulla M (1990) Viral infections in transfusion-dependent patients with thalassemia major: the predominant role of cytomegalovirus. Transfusion 30: 808–813
11. Pass RF, Stagno S, Myers GJ, Alford CA (1980) Outcome of symptomatic congenital cytomegalovirus infection: results of long-term longitudinal follow-up. Pediatrics 66: 758–762
12. Sumaya CV, Ench Y (1985) Epstein-Barr virus infectious mononucleosis in children. I. Clinical and general laboratory findings. Pediatrics 75: 1003–1010
13. Tabor E (1988) Etiology, diagnosis, and treatment of viral hepatitis in children. Adv Pediatr Infect Dis 3: 19–46
14. Toghill PJ, Williams R, Stern H (1969) Cytomegalovirus infection and chronic liver disease. Gastroenterology 56: 635–637
15. Tucciarone L, Felici W, Nigro G, Mechelli A, Castelvetere M, Olivo G (1989) Un caso di malattia citomegalica acquisita a decorso fatale. Riv Ital Pediatr 15: 643–646

Authors' address: Dr. G. Nigro, Pediatric Institute of Experimental Medicine of the National Council of Research, Rome, Italy.

Arch Virol (1992) [Suppl] 4: 273–276
© Springer-Verlag 1992

Repeated course of interferon treatment in chronic hepatitis B in childhood

R. Giacchino[1], A. Timitilli[1], E. Cristina[1], G. Giambartolomei[2], D. Leonardi[1], F. Caocci[1], and C. Cirillo[1]

[1] Clinica Malattie Infettive Università di Genova
[2] Divisione Malattie Infettive I.G.G., Istituto G. Gaslini, Genova, Italy

Summary. Ten children affected by HBV chronic hepatitis, not responding to a previous treatment with interferon (IFN), have been treated with a reiterated IFN therapy. The response obtained is not encouraging and only one patient became negative for HBeAg and HBV-DNA.

Introduction

To date, several therapeutic trials have been performed on the use of IFN in antiviral treatment of pediatric HBV hepatitis. Interferons (IFNs) are proteins which show their antiviral activity in different ways:

— inhibition of viral replication by a ribonuclease which destroys viral mRNA in the infected cell;
— inhibition of viral entry into the hepatocyte;
— inhibition of viral mRNA transfer and of assembly of the whole viral particle.

In chronic HBV hepatitis, IFNs seem to have more immunomodulatory than direct antiviral activity [3]. In fact, they operate on the immune response to the HBV infection by:

— activating those cells which can recognize complexes of HLA antigen with virus peptides on the cellular membrane;
— increasing cytotoxic activity of T lymphocytes and Natural Killer cells;
— selecting and presenting infected hepatocytes to cytotoxic T lymphocytes, thus causing histologic damage.

The aim of our study is to evaluate the efficacy of a repeated course of IFN treatment in patients who were non-responders to a previous IFN therapy.

Patients and methods

Ten paediatric patients were treated with human lymphoblastoid IFN (Wellferon); to enter the trial patients had to show:

— positivity of serum HBV-DNA detected by molecular hybridization technique with ^{32}P radiolabelled cloned HBV-DNA probe;
— positivity of HBeAg and HBsAg;
— increased transaminases;
— histological evidence of chronic hepatitis with hepatocyte expression of HBcAg.

Clinical characteristics of patients at their entry into the trial are described in Table 1.

Previously all these patients had already received human lymphoblastoid IFN according to a random schedule establishing 12 weeks of therapy with IFN at the dose of 5 MU/m three times a week, preceded by a 4 week course of steroids or placebo treatment [4]. At the end of that treatment, all patients turned out to be non-responders (HBeAg, HBsAg and HBV-DNA positive) independently from the above mentioned pretreatment (5/10 patients had received steroids and 5/10 placebo). In all cases we observed a decrease in serological HBV-DNA; a few patients became negative during IFN treatment. This behavior is the result of a temporary or an initial extinction of viral replication due to IFN therapy.

Because of the characteristics of their responses to the first course of IFN therapy, 10 non-responder patients, during their third year of follow up, were included in a new therapeutic trial.

They received three consecutive cycles, each lasting two months, of IFN therapy at a dose of 5 MU/m^2 three times a week without any pretreatment; at the end of each cycle the patients had a two months' rest phase.

Table 1. Characteristics of patients under treatment with IFN

Patients	No 10
Sex	7 M 3 F
Age	1 a—9 aa
Duration of infection	8 m—6 aa (2.5 aa)
Liver histology	CPH 4 cases
	CAH 6 cases
ALT (IU/l)	57.9 (45–165)
AST (IU/l)	89.1 (33–103)
HBV-DNA score 1	3 cases
2	5 cases
3	1 case
4	1 case

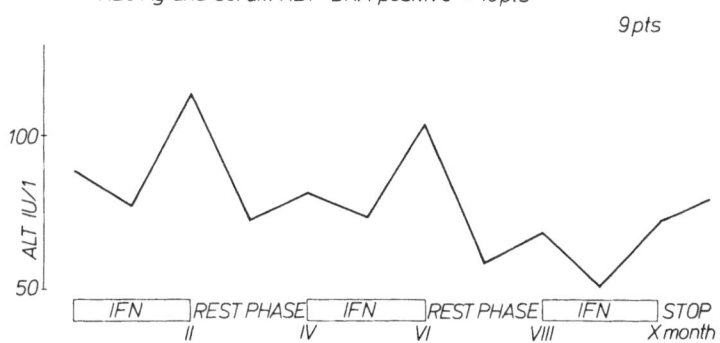

Fig. 1. Transaminase level and virological response to repeated IFN therapy

Results

Virological response (HBeAg and HBV-DNA) and transaminase level (ALT) during treatment and rest phases are shown in Fig. 1.

Only one patient during the last cycle of treatment, showed a complete response (both virological and serological): serum HBV-DNA and HBeAg became negative and transaminases decreased to normal values. The other nine patients never showed any response to treatment, not even temporarily.

Conclusion

In the previous trial 26 children affected with chronic HBV hepatitis had been treated with human lymphoblastoid IFN; 14 (53.8%) had a complete response to therapy during treatment or the first year of follow up.

Such results, similar to those of other European studies [1, 2, 5–7] are more satisfying than the results of the present trial, which are not encouraging. In fact we did not identify any advantage in a second administration of IFN compared with a single cycle of longer therapy.

Patients' response to IFN therapy seems to occur independently of treatment schedule.

In the present study we did not observe the immunomodulant activity which we expected. The multiple cycles of treatment did not stimulate the subjective immune response which one would expect to occur in the withdrawal phase by the enhancement of viral replication after a short course of antiviral treatment.

We have not evaluated yet if a larger dose of IFN (10 MU/m^2) could be more useful (this study is in course).

Characteristics of patients which allow prediction of the response to IFN therapy are still unknown.

Acknowledgements

Il lavoro è stato effettuato nell'ambito del "Progetto di Ricerca di Interesse Nazionale" del Ministero dell'Università e della Ricerca Scientifica e Tecnologica, 1989 (40%): Progetto di Ricerca su "Cirrosi Epatica" (M80).

References

1. Alexander JM, MB, MRCP, William R, F.R.C.P. (1988) Natural history and therapy of chronic hepatitis B virus infection. 85 [Suppl] 2A: 143–146
2. Brook G, Karayiannis P, Thomas HC (1989) Which patients with chronic hepatitis B virus infection will respond to alpha-interferon therapy? A statistical analysis of predictive factors. Hepatology 5: 761–763
3. Davis GL, Hoofnagle JH (1986) Interferon in viral hepatitis: role in the pathogenesis and treatment. Hepatology 6: 1038–1041
4. Giacchino R, Facco F, Giambartolomei G, Navone C, Timitilli A, Cirillo C, Barigione G, Terragna A (1988) Treatment of children with chronic hepatitis B with a combination of steroids and human lymphoblastoid interferon. Chemotherapy 7 [Suppl] 3: 20–25
5. Main J, Thomas HC (1990) Treatment of chronic hepatitis B infection. Pharmac Ther 45: 373–381
6. Moreno MR, Jemenez J, Porres JC, Bartolome J, Moreno A, Carreno V (1990) A controlled trial of recombinant interferon-alpha in caucasian children with chronic hepatitis B. Digestion 45: 26–33
7. Perrillo RP (1989) Interferon therapy for chronic type B hepatitis: The promise comes of age. Gastroenterology 96: 532–536

Authors' address: Prof. Dr. R. Giacchino, I Clinica Malattia Infettive, Università di Genova, Istituto Giannina Gaslini, Largo Gerolamo Gaslini 1, I-16147 Genova, Italy.

Arch Virol (1992) [Suppl] 4: 277–280
© Springer-Verlag 1992

Effect of prednisone priming followed by alfa-interferon in treatment of children with chronic hepatitis B: an interim analysis of a controlled trial

R. Utili, E. Sagnelli, G. Giusti, G. Ruggiero, F. Piccinino, B. Galanti, L. E. Adinolfi, L. Aprea, G. Cesaro, L. Digilio, F. M. Felaco, P. Filippini, G. B. Gaeta, A. Marrone, S. Nardiello, G. Pasquale, T. Pizzella, M. Russo, L. Santarpia, C. Sardaro, and T. D'Amora

Institute of Medical Therapy, Clinic of Infectious Diseases and Tropical/Subtropical Diseases, 1st Medical School University of Naples, Napoli, Italy

Summary. A six-month analysis of a controlled trial on the treatment of chronic hepatitis B in children shows that prednisone priming followed by α-interferon 2A was effective in 6 of 9 treated patients in reducing HBV replication and disease activity.

*

An effective treatment of HBV-induced chronic hepatitis in childhood is presently not available. Previous controlled studies, in children with chronic hepatitis B, using recombinant alfa-interferon (IFN) have shown a limited efficacy in inducing a stable remission of viral replication and of HBeAg/antiHBe seroconversion [1, 2, 3]. These data may be explained by the presence of an immunological tolerance to HBV in children who may exhibit a high degree of viral replication and low disease activity [4]. Recent studies demonstrated that the withdrawal from a short course of prednisone is often associated with an enhanced cellular immune response to HBV and could favor a high response rate to a subsequent IFN treatment [5].

Accordingly, we initiated a prospective controlled study to evaluate the effect of prednisone withdrawal followed by IFN in the treatment of chronic hepatitis B with active viral replication in young patients. Here we report the evaluation of a group of 18 children followed up for a period of 6 months.

The patients (aged 6–15 years; M/F 14/4) were HBsAg, HBeAg, and HBV-DNA positive and have a history of at least 12 months of increased aminotransferase levels and active viral replication. A liver biopsy taken in the preceding 3 months before admission to the study showed chronic active hepatitis in 5 cases and chronic persistent hepatitis in 13 cases. The patients were randomized in two groups of 9 patients each. The first (control group) was left untreated and the second (Pred-IFN group) received a course of one month of prednisone 0.6 mg/kg/day for 3 weeks and 0.3 mg/kg/day in the 4th week, followed by an interval of two weeks and then by treatment with IFN (Roferon-A, Roche) 3 MU/m^2 three times weekly for 12 months. All patients were followed weekly in the first 2 months, and then monthly with clinical, biochemical and virological evaluation. Serum HBV markers were determined by EIA tests (Abbott). Serum HBV-DNA was detected by a slot hybridization technique as previously described [6].

At the initiation of the study the control and Pred-IFN groups were comparable with respect to the age, mean serum levels of AST, ALT and HBV-DNA (Table 1).

In the pred-IFN group 8 of the 9 patients showed a decrease of aminotransferase activities and an increase in serum HBV-DNA levels at the end of prednisone course (Table 2). After the initiation of IFN treatment five patients (n.1, 2, 3, 4, 8) showed a peak of aminotransferase activity (4 to 20 times upper normal levels) within 4–8 weeks and cleared the HBV-DNA shortly thereafter. In these patients enzyme activities returned to 1–2 times the normal limits by the 3rd month and then remained within this range. Four of these patients (nos. 1, 3, 4, 8) also cleared the HBeAg and became antiHBe positive between the 2nd and 6th month of treatment (Table 2). Patient no. 9 had no ALT elevation during treatment, but cleared the HBV-DNA, at the first month, and remained HBeAg positive. In another patient (no. 7) a decline of HBV-DNA and ALT levels has been observed. Two patients (nos. 5 & 6) seem to be non-responders after the 6 months period of

Table 1. Characteristics of patients at admission

Patients features	Control group	Pred + IFN group
Number of patients	9	9
Male/female	6/3	8/1
Median age (years)	12	12
(range)	(6–14)	(6–15)
Mean AST ± s.d.[a]	59 ± 22	54 ± 28
(range)	(32–99)	(36–107)
Mean ALT ± s.d.[a]	106 ± 52	94 ± 62
(range)	(44–205)	(37–245)
Mean HBV-DNA[b]	45	50
(median value)	(35)	(40)
CAH/CPH	2/7	3/6

[a] Normal values up to 40 U/l

[b] pg/ml

Table 2. Serum HBV-DNA and ALT values in the group of treated patients

Patients No.	Months of treatment									
	Prednisone		IFN							
	0	1	0	1	2	3	4	5	6
1 HBV-DNA[a]	50	75	100	25	10	0	0	0	0[c]
ALT[b]	245	59	131	1003	800	50	24	77	77
2 HBV-DNA	25	100	75	25	0	0	0	0	0
ALT	71	48	52	425	240	47	46	32	41
3 HBV-DNA	25	50	75	0	0[c]	0	0	0	0
ALT	87	32	113	689	38	26	28	26	24
4 HBV-DNA	10	25	25	0	0	0	0[c]	0	0
ALT	111	41	115	258	149	112	88	72	45
5 HBV-DNA	50	100	100	100	75	75	75	75	75
ALT	62	30	29	56	38	46	39	48	53
6 HBV-DNA	75	100	100	100	100	75	100	75	100
ALT	57	37	48	44	43	102	66	69	62
7 HBV-DNA	75	100	100	75	50	50	50	50	50
ALT	245	56	122	431	271	171	161	127	119
8 HBV-DNA	10	25	25	0	0	0[c]	0	0	0
ALT	37	26	52	161	122	149	152	130	69
9 HBV-DNA	25	25	0	0	0	0	0	0	0
ALT	77	44	26	33	32	33	41	53	43

[a] pg/ml; [b] normal value < 40 U/l; [c] seroconversion HBeAg/antiHBe

treatment as they showed no significant changes of either ALT or HBV-DNA serum levels.

In the control group none of the 9 patients showed significant changes of serum HBV-DNA, HBeAg/antiHBe or of aminotransferase levels throughout the six months of observation.

The treatment with prednisone and interferon has been so far well tolerated except for a flu-like syndrome which occurred at the beginning of IFN treatment.

In conclusion, our preliminary data seem to indicate that a combination of a short course of prednisone followed by a long course of interferon may be an effective and safe regimen in the treatment of chronic hepatitis B in children.

Acknowledgements

This work was supported in part by Ministero Pubblica Istruzione 40% "Progetto Nazionale Cirrosi Epatica ed Epatiti Virali".

References

1. Giusti G, Ruggiero G, Piccinino F, Galanti B, Sagnelli E, Utili R, Aprea L, Carretta A, Cesaro G, Felaco FM, Filippini P, Gaeta GB, Nardiello S, Pasquale G, Pizzella T, Rosario P, Russo M, Santarpia L, Sardaro C, Digilio L (1989) Long term r IFN α2a treatment in children with chronic type B hepatitis and active viral replication. A prospective randomized study. Ital J Gastroenterol 21: 95
2. La Banda F, Ruiz Moreno M, Carreno V, Bartolome J, Gutiez G, Ramon Y, Cajal S, Moreno A, Mora I, Porres JC (1988) Recombinant alfa-2 interferon treatment in children with chronic hepatitis B. Lancet i: 250
3. Lai CL, Lok ASF, Lin HJ, Wu PC, Yeoh EK, Yeung CY (1987) Placebo-controlled trial of recombinant α-2 interferon in chinese HBsAg carrier children. Lancet ii: 877–880
4. Lok ASF, Lai CL, Wu PC, Leung EKY (1988) Long term follow up in a randomized controlled trial of recombinant gamma-interferon in chinese patients with chronic hepatitis B infection. Lancet ii: 298–302
5. Perrillo RP, Regenstein FG, Peters MG, DeSchryver-Kecskemeti K, Bodicky CJ, Campbell CR, Kuhns MC (1988) Prednisone withdrawal followed by recombinant alpha interferon in the treatment of chronic type B hepatitis. A randomized, controlled trial. Ann Intern Med 109: 95–100
6. Pizzella T, Nardiello S, Toniolo D, Galanti B (1986) Absence of HBV-related DNA sequences in patients with acute NANB hepatitis. J Hepatol 3: 152

Authors' address: Prof. Dr. Riccardo Utili, Istituto di Terapia Medica, I Facoltà di Medicina, Università di Napoli, Via Cotugno 1 (c/o Ospedale Gesù e Maria), I-80135 Napoli, Italy.

Arch Virol (1992) [Suppl] 4: 281–283
© Springer-Verlag 1992

Association between HLA class I antigens and response to interferon therapy in children with chronic HBV hepatitis

R. Giacchino[1], A. Nocera[3], A. Timitilli[1], E. Cristina[1], G. Giambartolomei[2], F. Caocci[1], C. Cirillo[1], and S. Barocci[3]

[1] I Clinica Malattie Infettive, Università di Genova
[2] Divisione di Malttie Infettive, Istituto G. Gaslini
[3] Servizio di Immunologia, Ospedale S. Martino, Genoa, Italy

Summary. In this preliminary study, children with chronic HBV hepatitis, as was also previously shown for adults, respond to interferon therapy in an HLA class I antigen dependent manner. If this can be confirmed on a large scale, HLA typing may serve as a useful indication of interferon-therapy responders.

Introduction

The evolution of hepatitis B virus (HBV) infection is determined by host immune responses that are responsible for hepatocellular injury and viral clearance. As immune responses are regulated by Major Histocompatibility Complex (MHC) genes, many studies have been carried out in recent years to evaluate HLA influence on the course of HBV infection. In chronic HBV positive patients (persistent antigenemia and chronic hepatitis), a significant increase in the frequency of the HLA-B35 allele has been already observed [2] and more recently a significant increase in HLA-DR3, together with a decrease in DQwl, have been reported [1, 3].

In the present study we investigated the presence of possible correlations between frequency of HLA Class I alleles and response to α-interferon (α-IFN) therapy in children affected by HBV chronic hepatitis.

Materials and methods

Patients

Twenty-five patients with high levels of transaminases, serum positivity for HBeAg and HBV-DNA and hepatic histological findings of chronic hepatitis were randomized to receive steroid or placebo pretreatment for a month. After 2 weeks without treatment, all patients then received human lymphoblastoid α-IFN at the dose of 5 MU/m² for 12 weeks. Thirteen patients responded to this therapy: serum HBV-DNA and HBeAg became negative and serum transaminase levels normalized, whereas twelve patients did not respond to this treatment, with persistence of serum HBV-DNA and HBeAg.

HLA Typing

The peripheral blood lymphocytes from all children were isolated from heparinized venous blood by Ficoll-Hypaque density gradient centrifugation. HLA Class I (A,B) alleles were determined using the standard microlymphocytotoxicity assay. The control group was based on a local reference population of 124 healthy subjects. Twelve specificities were tested for the HLA-A locus and twenty for the HLA-B locus.

Statistical analysis

HLA-A,B allele frequencies were evaluated in the following groups:

a) whole population of HBV chronic hepatitis pediatric patients (CHPP);
b) HBV-CHPP responding to α-IFN
c) HBV-CHPP not responding to α-IFN
d) healthy controls

The frequencies within the different groups were compared using the chi-square (X^2) test with Yate's correction. P values were corrected for the number of alleles tested.

Results and discussion

The results are shown in Table I. As previously demonstrated in adult patients [1] HLA-A,B typing showed a significant increase in HLA-B35 also in pediatric patients with HBV positive chronic hepatitis. Interestingly, after dividing the whole patient population into responders and non-responders to α-IFN therapy, the increase in HLA-B35 was found to be significant only in the responder group. The above observations, if confirmed on a larger scale, might offer a useful tool for the identification, within a patient population affected by HBV positive chronic hepatitis, of those subjects capable of responding to α-IFN treatment.

Table 1. Antigen differences between different HBV chronic hepatitis pediatric patient groups and healthy controls

Group	Cases	HLA-B35 frequency (%)	Corrected P values versus group D
A. HBV-CHPP	25	58.6	Pc < 0.01
B. Responders	13	71.4	Pc < 0.01
C. Non-Responders	12	36.3	N.S.
D. Healthy Controls	124	17.5	—

N.S. Not significant

Acknowledgements

This work was supported by a grant "Progetti di Ricerca di Interesse-Nazionale" from Ministero Pubblica Istruzione e Ricerca Scientifica e Tecnologica, 1989 (40%): Progetto Ricerca su "Cirrosi Epatica (M80) and by grant n° CN89.0023.PF.703 (Progetto Finalizza to Biotecnologie e Biostrumentazione) from National Research Council (C.N.R.) Rome.

References

1. Forzani B, Actis GC, Verme G (1984) HLA-DR antigens in HBsAg positive chronic active liver disease with and without associated delta infection. Hepatology 4: 1107–1110
2. Hillis WD, Hillis A, Bias WB, Walker WGB (1977) Association of hepatitis B surface antigenemia with HLA locus B specificities. N Engl J Med 296: 1310–1314
3. van Hattum J, Schreuder GMT, Schalm SW (1987) HLA antigens in patients with various courses after hepatitis B virus infection. Hepatology 7: 11–14

Authors' address: Prof. R. Giacchino, I Clinica Malattie Infettive, Università di Genova, Istituto Giannina Gaslini, Largo Gerolamo Gaslini 1, I-16147 Genova, Italy.

XI Therapy of chronic viral hepatitis

Arch Virol (1992) [Suppl] 4: 287–290
© Springer-Verlag 1992

Treatment of chronic viral hepatitis anno 1990

S. W. Schalm

Department of Internal Medicine and Hepatogastroenterology, University Hospital
Dijkzigt, Rotterdam, The Netherlands

Summary. The therapeutic efficacy of α-interferon therapy in chronic hepatitis is discussed in light of various international studies. Although beneficial to some extent, treatment is of limited value and is accompanied by more-or-less serious side effects. Development of complementary drugs and new therapeutic methods must therefore be continued.

Chronic hepatitis B

Pooled data from the initial studies of interferon therapy with various dosages and durations of treatment suggested a small beneficial effect of interferon therapy with about 25% responses (HBe seroconversion) in treated patients versus about 10% in controls; the percentages of responders among treated patients and controls was highly variable due to variation in selection of patients and the small numbers of patients in each study.

In the recently completed USA multicenter trial the number of patients in each treatment arm exceeded 40; the results are among the most reliable to assess the efficacy of alpha-interferon therapy, given for 16 weeks in a dose of 5 megaU per day (Table 1). HBe seroconversion was observed in 37% of treated patients versus 7% in controls; patients receiving 1 megaU of interferon per day exhibited 17% HBe seroconversion. The study also confirmed earlier observations that, in addition to HBe-seroconversion, loss of HBsAg occurs more frequently (12%) in the 5 megaU interferon treatment group than spontaneously (2%).

Combination of interferon with nucleoside analogues such as acyclovir or adenine arabinoside appears logical in an attempt to enhance the HBe seroconversion rate; pretreatment for 4 weeks with prednisone has also been suggested to improve results of interferon therapy. The outcome of recently completed randomized controlled studies with about 40 patients in each

Table 1. USA multicentre interferon therapy trial for chronic hepatitis B[a]

Percentage with	IFN 5 MU/d	IFN 1 MU/d	Untreated controls
	n=39	n=41	n=41
Loss of HBVDNA/HBeAg	37	17	7
Loss of HBsAg	12	2	0
Reactivation of HBV	0	0	0

[a] Data on file at Schering-Plough

treatment group, however, showed no benefit from either pretreatment with prednisone or from combination with acyclovir (data on file at Schering-Plough and Wellcome, respectively).

Re-analysis of the Rotterdam phase-II study on combination therapy of interferon with acyclovir led to the hypothesis that the impressive results observed were related to repeated courses of antiviral therapy rather than to the combination of interferon and acyclovir. Two independent pilot studies in London (G. J. M. Alexander) and Rotterdam tested a treatment schedule of 4 weeks of interferon pretreatment followed, after 4 weeks of no therapy, by the standard course of 12–16 weeks of interferon therapy. Patients with a partial response (more than 50% decrease in HBeAg titer and HBV-DNA) at week 24 continued interferon therapy for another 4–8 weeks. Results in both studies comprising 10 and 20 patients, respectively showed HBe seroconversion in more than 50% of patients. It appears worthwhile to carry out a randomized trial comparing standard interferon therapy and prolonged interferon therapy with or without interferon pretreatment.

Chronic hepatitis C

A USA multicenter trial including 166 patients confirmed the therapeutic effect of alpha-interferon observed in initial pilot studies and small controlled trials. Normalization of serum ALT was observed in 46% of patients treated with 3 megaU three times a week versus 8% in controls; patients receiving 1 megaU of interferon three times a week showed ALT normalization in 28%. Relapse of active hepatitis (ALT elevation) occurred frequently (>50%) in both interferon groups after therapy. Serum ALT remained persistently normal in about 25% of patients receiving 9 megaU interferon a week for 24 weeks. Enhanced response rates can be expected when interferon therapy is started with higher doses of interferon, such as 5–10 megaU three times per week.

Initially, serum aminotransferases should be monitored every 2 weeks. If there is no improvement in serum ALT levels within 2 months, therapy should be stopped. If serum ALT falls into the normal range, the dose of interferon can be decreased, and therapy should be continued for 6–12 months. An increase in the dose of alpha-interferon may be appropriate in patients who show a partial response only or who have a breakthrough in serum ALT during therapy.

Chronic hepatitis D

An Italian multicenter trial including 48 patients has recently been completed. Homosexual men and persons with HIV antibodies or a history of drug abuse were excluded. During the first 4 months the dose of interferon was 5 megaU/m^2 three times a week, followed by 3 megaU/m^2 a week during the next 8 months.

Initiation of therapy was followed within a few weeks by a decline in serum ALT in 58% of treated patients; the serum ALT, however, increased to pretreatment levels in most patients during therapy with the lower maintenance dose. In only one patient was ALT normalization observed after therapy.

Enhanced response rates have been reported from another Italian controlled trial in which 10 megaU of interferon was given three times a week for 12 months; after therapy a normal serum ALT persisted in about 33% of patients. These and other uncontrolled data suggest that high doses of interferon for prolonged periods may suppress hepatitis D viral replication and the activity of the liver disease, but confirmation of these data is required.

Side-effects of interferon therapy

The side-effects of alpha-interferon may interfere with its therapeutic efficacy. Early side-effects commonly occur during the first few days of treatment, but do not usually present a clinical problem. Late side-effects appear after 2 weeks of treatment and are less common. The medical or psychiatric complications, however, can be of major clinical significance and often require adjustment of the interferon dosage or other medical interventions.

Early side-effects of interferon include fever, chills, fatigue, myalgia, sleep disturbance, loss of concentration and anorexia. These complications can be partly prevented by administration of 500 mg paracetamol every 6 hours for the initial 3 days or by 75 mg indomethacin slow-release tablets twice daily.

Late side-effects can be classified into systemic, hematologic, psychiatric, infectious and autoimmune. Systemic side-effects include weight loss, hair loss and decreased libido in addition to those mentioned above. Hematologic side-effects are neutropenia, thrombocytopenia and anemia. The spectrum of psychiatric side-effects is wide and includes irritability, anxiety, depression and psychosis. Return to drug addiction or alcohol abuse may occur. These side-effects usually have an insidious onset in the second or third month of treatment and occur in 10–20% of patients.

Perhaps secondary to the decrease in neutrophils, patients receiving alpha-interferon are more susceptible to bacterial infections, particularly in the case of pre-existing cirrhosis. Development of auto-antibodies during interferon therapy is common, but overt autoimmune disease is rare. Autoimmune thyroiditis has been reported, as has exacerbation of auto-immune chronic active hepatitis in case of interferon therapy for incorrectly diagnosed chronic hepatitis C.

Management of late side-effects requires extra patient-physician inter-action. The goal of management is to keep the patient on interferon until the maximum benefit has been derived. Often reassurance and encouragement will suffice; occasionally hospital admission for several days may help to regain tolerance. In case of neutropenia (less than $500/mm^3$) or severe psychiatric effects, the dose of interferon should be reduced. Side-effects are rarely so intolerable or medically so severe that interferon will have to be withdrawn completely.

Conclusions

Current results with alpha-interferon indicate a new area in the therapeutic approach to patients with chronic viral hepatitis, but also underline the limited effectiveness of therapy with alpha-interferon alone. Since dosage and duration of interferon therapy are limited by side-effects, a further search for complementary drugs appears essential.

Author's address: Dr. S. W. Schalm, Department of Internal Medicine & Hepato-gastroenterology, University Hospital Dijkzigt, NL-3015 GD Rotterdam, The Netherlands.

Arch Virol (1992) [Suppl] 4: 291–293
© Springer-Verlag 1992

Induction of autoantibodies during alpha interferon treatment in chronic hepatitis B

G. Fattovich[1], C. Betterle[2], L. Brollo[1], G. Giustina[1], B. Pedini[2], and A. Alberti[1]

[1] Istituto di Medicina Clinica, Clinica Medica 2a
[2] Istituto Semeiotica Medica, Università di Padova, Italy

Summary. Thirty-two patients with chronic hepatitis B treated with alpha interferon were tested for 12 different antibodies. Only a minority (18%) of cases developed antinuclear antibodies and none developed clinical signs of autoimmune disease. These data suggest that, at the dose regimen used, interferon therapy of chronic hepatitis B is not associated with triggering of autoimmunity.

*

The availability and large scale production of recombinant and lymphoblastoid interferon (IFN) has resulted in numerous, exploratory clinical trials in chronic HBV infection [3]. The development of autoantibodies and autoimmune disease have been reported to occur in interferon treated patients [1, 2]. On the basis of these observations we have tested for the presence of 12 different antibodies, the sera of 32 patients (23 men, 9 women, mean age 31 years) with biopsy proven hepatitis B surface antigen (HBsAg) positive chronic hepatitis treated with alpha IFN. All patients were hepatitis delta virus (HDV) negative. Seventeen HBeAg and HBV-DNA positive cases received 4.5 MU of recombinant IFN thrice weekly for 4 months and 15 anti-HBe and HBV-DNA positive cases were treated with 5 MU/m^2 of lymphoblastoid IFN thrice weekly for 6 months. Patients were followed at regular intervals during treatment and for 12 months after therapy withdrawal, being investigated for routine serum biochemical tests, for HBV markers, and sera were stored at $-20\,°C$ for autoantibody testing. A response to IFN therapy was defined as sustained termination of HBV replication with ALT normalization.

HBsAg, HBeAg, anti-HBe and anti-HD were detected by commercially available radioimmunoassay or enzyme-linked immunosorbent assays.

Serum HBV-DNA was measured by a dot-blot hybridization technique. Antinuclear (ANA), antimitochondrial (AMA), smooth muscle (SMA), liver-kidney microsomal (LKM) antibodies were detected by indirect immuno-fluorescence technique (IFT) using rat liver and kidney tissue. Antibodies to double-stranded DNA (dsDNA) were measured by IFT on *Crithidia luciliae*. Antibodies to extractable nuclear antigens (ENA) (anti-Sm, anti-RNP) and antibodies to thyroid microsomal antigen (TMA) and to thyroglobulin (TGA) were detected using different commercially available hemag-glutination test reagents. Antibodies to pancreatic islet cells (ICA), to parietal cells (PCA) and to adrenal cortex (AA) were measured by IFT using normal human tissue as antigen source.

Five (15%) patients, all women and HBeAg positive, had ANA and one (3%) anti-HBe and HBV-DNA positive male patient had SMA before treatment. Two of the 5 patients showed an increase in ANA titer during IFN therapy, without clinical manifestations of autoimmune disorders. There was no correlation between ANA positivity before treatment and response to IFN therapy.

ANA appeared during IFN treatment in 5 (18%) of 28 previously negative patients (Table 1). All 5 cases were male. With discontinuation of treatment, titers of ANA fell to undetectable levels in all patients. None of the patients developed clinical signs of autoimmune disease. No patient developed antibodies to endocrine organs or autoantibodies specifically associated with autoimmune liver disease such as LKM and AMA. In conclusion these results, obtained from 32 patients with chronic hepatitis B treated with alpha interferon, indicate that at the dose regimen used only a minority of patients developed autoantibodies and the risk for autoimmunity is very low. However, other therapeutic regimens may behave differently and monitoring for autoantibody occurrence and for signs of autoimmunity

Table 1. Development of antinuclear antibody (ANA) during alpha interferon (IFN) therapy

Patient no.	HBeAg	anti-HBe	Onset after IFN therapy (months)	Response to IFN[a]
1	+	−	4	−
2	+	−	4	−
3	−	+	5	+
4	−	+	3	+
5	−	+	2	−

[a] Loss of both HBeAg and HBV-DNA in HBeAg positive cases or of HBV-DNA in anti-HBe positive patients with alanine aminotransferase normalization

should be planned in ongoing clinical trials of IFN therapy in chronic hepatitis.

References

1. Burman P, Karlsson FA, Oberg K, Alm G (1985) Autoimmune thyroid disease in interferon-treated patients. Lancet ii: 100–101
2. Mayet WJ, Hess G, Gerken G, Rossol S, Voth R, Manns M, Meyer zum Buschenfelde KH (1989) Treatment of chronic type B hepatitis with recombinant-interferon induces autoantibodies not specific for autoimmune chronic hepatitis. Hepatology 10: 24–28
3. Perillo RP (1989) Interferon therapy for chronic type B hepatitis: The promise comes of age. Gastroenterology 96: 532–536

Authors' address: Dr. Giovanna Fattovich, Istituto Medicina Clinica, Clinica Medica 2a, Università di Padova, Via Giustiniani 2, I-35100 Padova, Italy.

Arch Virol (1992) [Suppl] 4: 294–298
© Springer-Verlag 1992

One course versus two courses of recombinant alpha interferon in chronic C hepatitis

G. Taliani, C. Furlan, F. Grimaldi, C. Clementi, R. Lecce, M. Manganaro, F. Duca, and Carlo De Bac

Institute of Tropical and Infectious Diseases, "La Sapienza" University, Rome, Italy

Summary. Fifty-five patients with antibodies to HCV and chronic liver disease have been enrolled in the study. Thirty-four patients were treated with recombinant alpha interferon (IFN, 3 MU daily for 10 days followed by 3 MU twice/week for 3 months), and were compared to 21 untreated controls.

Alanine aminotransferase (ALT) normalization was observed in a significant proportion of treated patients (52.9%), but 66.6% of them experienced a relapse after discontinuation of the therapy. The evaluation of the early ALT behavior after the 10 days priming with daily IFN administration was useful in predicting the response.

The administration of a second IFN course with the same schedule and duration as the first course did not increase the efficacy of the treatment. Increased dosage and/or prolonged administration are probably required.

Introduction

Treatment of chronic hepatitis C with recombinant interferon alfa (alpha-IFN) is generally well tolerated and gives promising results, as it can reduce hepatocellular inflammation and induce normalization of aminotransferases in 30–60% of treated patients [1–5].

Unfortunately, treatment failure and relapse after the end of the therapy are not uncommon [1–5]. We therefore wanted to determine whether administration of a second interferon course after a short withdrawal could increase the percentage of responsiveness and result in a more prolonged remission of the disease.

Materials and methods

Fifty-five patients with antibodies to hepatitis C virus (HCV) and histologically proven chronic liver disease have been assigned to receive 3 million units of recombinant interferon alfa-2b (Intron A, Schering Plough Corporation) daily for 10 days followed by 3 million units twice weekly for 3 months, or no treatment. Thirty-four patients have been treated (19 males, mean age 48.8 yrs) compared to 21 untreated controls (13 males, mean age 37.2 yrs). Among the treated patients, 29 had chronic active hepatitis CAH) and 5 patients had active cirrhosis (AC), while 19 out of 21 untreated controls had CAH and 2 had AC. Anti-HCV positivity was confirmed in all subjects by the first generation recombinant immunoblot assay (RIBA, Chiron Corporation). After the therapy was stopped, all subjects were followed up for four months. The response to IFN was considered complete when alanine aminotransferase (ALT) normalization occurred by the end of treatment, and partial when an ALT reduction of 50% or more from the baseline value without normalization occurred. Relapse was defined as the increase of ALT levels above the normal limit after a complete response.

Twenty treated patients (10 responders who afterwards relapsed and 10 non-responders) underwent a second interferon (IFN) course with the same dose and regimen of the first course, at the end of which an additional 4-month follow-up was completed.

Fisher's exact test and paired or unpaired Student's t-test were employed when appropriate for statistical analysis.

Results

All patients completed the study. At the beginning of the therapy, the median ALT levels + SD of treated patients and controls were $179 + 78$ and $168 + 69$ respectively (p = ns; Fig. 1).

During the administration of IFN, the ALT serum levels decreased in 29 treated patients, no changes were observed in 3 patients, while, in the

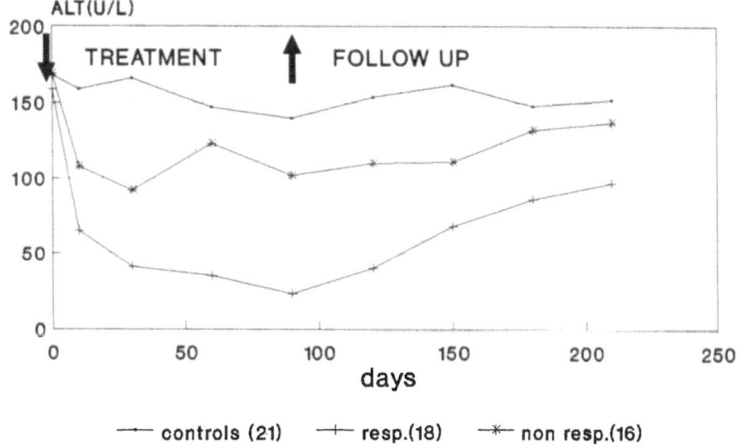

Fig. 1. Mean serum ALT levels during IFN treatment and follow up

remaining 2 patients, a twofold increase of ALT levels occurred during the second month of therapy after a partial early response. At the end of the treatment, 18 treated patients (52.9%; 1 with AC and 17 with CAH) exhibited ALT normalization, and 6 patients exhibited a >50% ALT reduction compared to 1 and 2 untreated controls, respectively. Thus, a complete or partial response to IFN was observed in 24 out of 34 treated patients (70.5%), while the same ALT pattern was observed in 3 out of 21 untreated controls (14.3%; $p = 4.8 \times 10^{-5}$).

After the 10 days of IFN priming 12 out of the 18 responders exhibited either normalization or >50% reduction of ALT levels, and the median ALT value of the patients who responded afterwards to the therapy was, already at that time, significantly lower compared to that of the non-responders (respectively 65 U/L and 108 U/L; $p = 0.02$).

After discontinuation of therapy, the median ALT level of treated subjects increased, although at the end of the follow up it was lower than the base line value (118 U/L versus 179 U/L, $p = 0.011$, Fig. 1). However, 12/18 responders (66.6%) experienced a relapse during the follow up (4 subjects at month 1, 4 at month 2, 2 at month 3 and 2 at month 4) and at the end of the follow up, only one third of them (6/18, 33.3%) still had normal ALT values.

Among the 20 patients who underwent a second IFN course (10 responders and 10 non responders to the first IFN course), 7 patients responded while 10 patients did not respond to both IFN administrations and 3 patients responded to the first but not to the second IFN course. Out of the 7 responders, 5 relapsed during the follow up (2 pts relapsed at month 2, 1 pt at month 3 and 2 pts at month 4). The pattern of ALT level decrease during the second course and increase during the second follow up closely resembled that observed during and after the first IFN administration (Fig. 2). However, ALT normalization was prompter during the first IFN course, since it occurred at the end of the first month of therapy in 70% of responders (7/10), compared to 42.8% (3/7) during the second course.

On the whole, sex, age, histological features and median baseline ALT levels did not affect the response to IFN, while the negativity of antibodies to the 5-1-1 antigen, as demonstrated by the HCV-RIBA confirmatory test, was significantly associated with a better response. In fact, among the treated patients, 5 out of 18 responders were anti-5-1-1 negative compared to none out of 16 non responders ($p = 0.03$).

Discussion

Our results showed that the administration of alpha-interferon therapy was capable of interrupting liver cytonecrosis in 52.9% of treated patients, while in 87% of untreated controls no significant changes were observed (p =

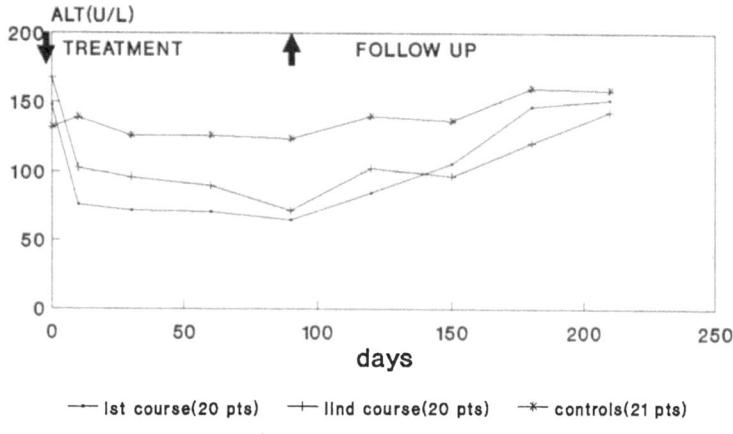

Fig. 2. Mean ALT levels during 1st and 2nd IFN courses

5.6×10^{-6}). However, in the majority of the responders (66.6%) the normalization of ALT levels was followed by a relapse within 4 months after IFN was discontinued.

Interestingly, the early ALT response after the 10 days of IFN priming appeared to be useful in predicting the response to the treatment. In fact, 12/18 responders (66.6%) already had normal ALT levels after the IFN priming.

It has been shown that a good rate of ALT normalization can be obtained also with an IFN dosage as low as 1 million units thrice weekly for 6 months [2, 3], but it is possible that in order to obtain a more sustained biochemical remission, a higher dosage and/or a longer duration of treatment are required [6, 7].

The fact that the presence of cirrhosis did not affect the response to the therapy is probably due to the small number of patients with cirrhosis enrolled in this as in another trial [7], while it is noteworthy that those patients without detectable antibodies to the 5-1-1 antigen by RIBA turned out to be better responders than those with such antibodies [8]. Although this finding is difficult to explain, it could be useful in predicting the response to IFN in a subgroup of anti-HCV positive patients.

Among the 20 patients who underwent a second IFN course, 3 out of 10 subjects who showed ALT normalization during the first treatment did not respond to the second, and the 7 patients who responded to both courses showed a slower ALT normalization during the second IFN administration. In conclusion, our study shows that the administration of a second IFN course with the same schedule as the first course does not improve the response to the treatment. It is conceivable that when a relapse occurs after the IFN withdrawal, a therapy of longer duration and/or with higher IFN dosage would be of greater benefit in sustaining biochemical remission.

References

1. Di Bisceglie AM, Martin P, Kassianides C et al (1989) Recombinant interferon alpha therapy for chronic hepatitis C. A randomized, double-blind, placebo-controlled trial. N Engl J Med 321: 1506–1510
2. Davis GL, Balart LA, Schiff ER et al (1989) Treatment of chronic hepatitis C with recombinant interferon alpha. A multicenter randomized, controlled trial. N Engl J Med 321: 1501–1506
3. Saracco G, Rosina F, Torrani Cerenzia MR et al (1990) A randomized controlled trial of interferon alfa-2b as therapy for chronic non-A, non-B hepatitis. J Hepatol 11 [Suppl]: 43–49
4. Ferenci P, Vogel W, Pristautz H et al (1990) One-year treatment of chronic non-A, non-B hepatitis with interferon alfa 2b. J Hepatol 11 [Suppl]: 50–53
5. Jacyna MR, Brooks MG, Loke RHT et al (1989) Randomized controlled trial of interferon alpha (lymphoblastoid interferon) in chronic non-A, non-B hepatitis. Br Med J 298: 80–82
6. Hoofnagle JH, Mullen KD, Jones DB et al (1986) Treatment of chronic non-A, non-B hepatitis with recombinant human alpha interferon. N Engl J Med 315: 1575–1578
7. Gomez-Rubio M, Porres JC, Castillo I et al (1990) Prolonged treatment (18 months) of chronic hepatitis C with recombinant alpha-interferon in comparison with a control group. J Hepatol 11 [Suppl]: 63–67
8. Taliani G, Lecce R, Clementi C, Furlan C, De Bac C (1991) Antibodies to hepatitis C virus and response to Interferon treatment in non-A, non-B chronic hepatitis. J Hepatol 12: 264–265

Authors' address: Dr. Gloria Taliani, Department of Tropical Diseases, "La Sapienza" University, I-00161, Rome, Italy.

Arch Virol (1992) [Suppl] 4: 299–303
© Springer-Verlag 1992

Interferon therapy of cryptogenic chronic active liver disease and its relationship to anti-HCV

G. Diodati[1], P. Bonetti[1], A. Tagger[2], A. Alberti[1], G. Realdi[3], and A. Ruol[1]

[1] Istituto di Medicina Interna, University di Padova
[2] Istituto di Virologia, University di Milano
[3] Clinica Medica Generale, University di Sassari, Italy

Summary. In a randomized controlled trial of Interferon (IFN) in 60 patients (30 treated and 30 controls) with cryptogenic chronic active liver disease, 70% of treated patients showed complete response, but a high rate of biochemical relapse (62%) was noted. In these cases, a second response to higher doses of IFN has been more difficult and less frequent. A response to IFN was found in 88.5% of anti-HCV positive treated patients and only in 25% of anti-HCV negative. We suggest that serum anti-HCV is a suitable test to predict the response to IFN.

Introduction

On the basis of epidemiological studies, for many years it has been suggested that chronic infection by nonA, nonB viruses may represent a frequent cause of chronic liver disease not only in patients with well documented exposure to parenterally transmitted hepatitis viruses but also in patients with community-acquired (cryptogenic) chronic liver disease, that have no recognizable source of transmission [1].

Furthermore, the positive results obtained from two (one pilot and one controlled) studies [2, 3] with interferon (IFN) in patients with chronic nonA, nonB posttransfusion hepatitis, prompted us to start a randomized controlled clinical trial of recombinant human alpha-2a IFN in a group of patients with cryptogenic chronic active liver disease.

Patients and methods

We enrolled sixty patients (33 males and 27 females), 18–70 years of age, with a morphological diagnosis of chronic active hepatitis (CAH), with or without superimposed cirrhosis. All had abnormal liver function tests (aminotransferases persistently increased at least twice the upper limit, throughout the last year). None had a positive history or clinical or histological evidence of any other known cause of chronic liver disease (including metabolic, genetic, alcoholic, autoimmune and by hepatotropic viruses other than HCV). We excluded patients with advanced or decompensated liver cirrhosis (Child B or C according to Child-Pugh score), pregnant women, patients with positivity for anti-HIV or present drug abuse and severe systemic diseases.

We randomized the patients in a treatment group and in a control group, and the baseline features of all the patients matched perfectly between the two group (Table 1).

In treated patients we administered IFN parenterally thrice weekly for 12 months at tapering doses (6 megaunits (MU) for the first month, 3 MU for three months and 1 MU for further eight months). In case of confirmed and stable biochemical reactivation, IFN was assessed to the previous effective dose. The control group received no treatment and was followed up to a year. All 60 patients underwent clinical and biochemical evaluations every month and at the end of therapy or observation period from all of them a second biopsy was obtained. We used the Fisher Exact Test to evaluate the difference of response to IFN in relation to anti-HCV.

Results

In 21 (70%) of the 30 treated patients, a complete response to therapy was obtained, with a normalization of aminotransferases levels. Three (10%) of these responders showed a partial response to IFN (a significant decrease of ALT to less than 50% of the baseline value). None of the controls showed any significant change in aminotransferase levels, with a statistically significant difference ($p < 0.01$) between treated and control group.

We observed a relevant rate of biochemical reactivation during treatment cycle; 5 (24%) of the 21 complete responders relapsed between the second and the fourth month and a further 8 patients (38%) during the phase at low

Table 1. Baseline features of all enrolled patients

	Treated (30 patients)	Controls (30 patients)
Male/female	23/7	13/17
Age (years)	50 ± 12	55 ± 11
Previous AH	20	20
AST	124 ± 69	125 ± 143
ALT	253 ± 135	202 ± 166
S-IgG (g/dL)	1.9 ± 0.5	1.8 ± 0.5
Anti-HBc ($> 1:100$)	23	10
Histology		
CAH	15 (50%)	18 (60%)
Cirrhosis	15 (50%)	12 (40%)

Age, AST, ALT and S-IgG were expressed by means ± SD; *AH* acute hepatitis; *CAH* chronic active hepatitis dose (1 MU) with an overall relapse rate of 62% (13/21). According to the protocol, we increased the dose of IFN in the reactivating cases and we obtained a complete renormalization of aminotransferases only in 3 (23%) of the 13 cases, while in other 7 (54%), we induced only a significant decrease (more than half of previous values) of aminotransferases, but not a complete normalization (Table 2). No significant change of serum aminotransferases was noted among control patients.

Our study also included evaluation of the histological changes eventually induced by IFN therapy on liver tissue. By now, 28 treated patients and 20 controls underwent a second biopsy. It is important to note that no treated patients have shown any worsening of the histopathological pattern (13 improved, 14 remained unchanged and 1 showed a complete remission); in contrast, we did not see any significant improvement of liver histology among controls (17 were unchanged and 3 worsened).

Recently, a portion of the genome of a nonA, nonB virus (designated as hepatitis C virus) has been cloned [4] and a specific immunoassay for the

Table 2. Primary response, reactivation and secondary response after increasing of IFN therapy

Complete response (normalization of ALT)	N° patients
1°–2° month	20
6°–7° month	1
Total	21/30 (70%)

Partial response (> 50% of baseline ALT	n° patients
1° month	1
3° month	1
12° month	1
Total	3/30 (10%)

Reactivation	n° patients
1°–4° month	5/21 (24%)
5°–12° month	8/21 (38%)
Total	13/21 (62%)

Reinduction after reactivation	n° patients
Complete response	3/13 (23%)
Partial response	7/13 (54%)
No response	3/13 (23%)

antibody to hepatitis C virus (anti-HCV) has been developed [5]. We therefore tested retrospectively the prevalence of anti-HCV in our group of patients, and we tried to study its relationship to the response to IFN therapy. Among treated patients 23/26 (88.5%) anti-HCV positive cases were complete responders to IFN therapy, in comparison with only 1/4 (25%) anti-HCV negative cases (Table 3), with a statistically significant difference between the two groups (p = 0.018).

Discussion

On the basis of epidemiological evidence, we may suggest that most of community-acquired chronic active hepatitis are due to a viral infection of the liver, most likely of the non-A, non-B type.

The recent availability of a simple test to detect anti-HCV may be useful to support this hypothesis. In fact, our preliminary results confirm the high prevalence of anti-HCV in this group of patients and the role of HCV infection in causing chronic active liver disease of unknown etiology.

The effectiveness of interferon therapy in chronic active hepatitis type B and non-A, non-B is well documented by pilot and controlled studies [1, 2, 6]. We have demonstrated the efficacy of IFN treatment in patients with community-acquired chronic active non-A, non-B hepatitis, but we need further investigations to identify the correct schedule (dose and length) of treatment to avoid relapses (during and after treatment) and to obtain stable biochemical and histological remissions.

We also suggest the need for further tests to provide an etiological diagnosis of the disease through detection of viral antigens in serum and in the liver; this will be important to screen patients who may undergo interferon treatment. In our groups of patients, anti-HCV positivity has been predictive for response to therapy and we suggest that in the future, identification of possible prognostic features will become mandatory for the eligibility of the patient to IFN treatment and to obtain the best results.

Table 3. Relationship between serum anti-HCV and response to Interferon therapy

	Anti-HCV pos (26 pts.)	Anti-HCV neg. (4 pts.)
Responders	23	1
	p = 0.018[a]	
Non-responders	3	3

[a] Fisher Exact Test (two sided)

References

1. Alter MJ (1989) NonA, nonB hepatitis: sorting through a diagnosis of exclusion. Ann Intern Med 110: 583–585
2. Hoofnagle JH, Mullen KD, Jones B, et al (1986) Treatment of chronic non-A, non-B hepatitis with recombinant human alpha interferon. A preliminary report. N Engl J Med 315: 1573–1578
3. Jacyna MR, Brooks MG, Loke RHT, et al (1989) Randomized controlled trial of Interferon alpha (lymphoblastoid Interferon) in chronic non-A, non-B hepatitis. Br Med J 298: 80–82
4. Choo QL, Kuo G, Weiner AJ, et al (1989) Isolation of a cDNA derived from a blood borne non-A, non-B viral hepatitis genome. Science 244: 359–362
5. Kuo G, Choo QL, Alter HJ, et al (1989) An assay for circulating antibodies to a major etiologic virus of non-A, non-B hepatitis. Science 244: 362–364
6. Fattovich G, Brollo L, Boscaro S, et al (1989) Long-term effect of low dose recombinant Interferon therapy in patients with chronic hepatitis B. J Hepatol 9: 331–337

Authors' address: Dr. G. Diodati, Istituto di Medicina Clinica, Clinica Medica 2, Policlinico – Università, Via Giustiniani 2, I-35128 Padova, Italy.

Arch Virol (1992) [Suppl] 4: 304–305
© Springer-Verlag 1992

Interferon alpha 2-b therapy of HCV and NonBNonC chronic hepatitis

M. Gargiulo[1], P. Tarquini[1], L. Di Ottavio[1], E. Lattanzi[2], and A. De Nigris-Urbani[2]

[1] Sezione Malattie Infettive and [2] Servizio Trasfusionale e di Immunoematologia
Ospedale Civile "G. Mazzini", Teramo, Italy

Summary. Twenty-four patients with HCV and NonBNonC chronic hepatitis—4 with HIV coinfection—were treated with r-IFN alpha for at least six months. In this period 62.5% of patients show a normalization of ALT but not a sustained remission. Non-responders have histologically more severe and long-lasting chronic hepatitis.

Introduction

Davis [1] and Di Bisceglie [2] have recently established in a multicentric, randomized, controlled trial the efficacy of the treatment of chronic HCV hepatitis with recombinant interferon alpha. The aim of the present study was to determine the efficacy of at least 6 months of alpha r 2-b IFN treatment of our 23 patients with HCV and 4 patients with NBNC chronic hepatitis.

Materials and methods

Twenty-seven patients with chronic hepatitis (21 males, 6 females, age-range 20–77) were enrolled, 18 had biopsy-proven chronic hepatitis (CPH in 9, CAH in 7, CAH/C in 2); 16 patients had a history of i.v. drug addiction, 4 had a post-transfusion hepatitis, 1 was homosexual, 1 had a sibling with HCV positive CAH, in 5 patients risk factors were not determined. None had HBV infection, 7 had HIV co-infection (II-III CDC, 6 in ZDV treatment). Levels of ALT were 1.5–8 × normal value. None had underlying autoimmune conditions. All patients were clinically compensated. Antibody to HCV was detected by Ortho anti-HCV ELISA test and Ortho RIBA confirmatory test. All patients were treated with r-IFN alpha 2-b (Intron A, Schering-Plough) 3 times weekly s.c. for at least 24 weeks, starting with doses of 6 M.U. or 3 M.U. given for 8 weeks; good responders also had 1 M. U. for 8 weeks. We defined a near-complete response ALT abnormal but < 1.5 normal values. Follow-up is in progress.

Results

Antibody to HCV was detected in 23 (85%) of the tested sera. After 6 months 24 patients completed the treatment, 5 stopped it and for 19 therapy is in progress. Twenty one had HCV chronic hepatitis, 3 NBNC chronic hepatitis. Out of 3 dropouts 2 were HCV positive, 1 had NBNC chronic hepatitis. Levels of ALT were normalized in 15, reduced in 4 (2 HIV+), unmodified in 5 (2 HIV+). One patient lost anti-HCV after 8 months treatment. Out of 5 patients who stopped the treatment, two were non-responders and 3 relapsed after 1 month of discontinuing the therapy. No worsening of the disease was observed during the therapy. Mild side effects (fever, myalgias) were usually recorded at the beginning of IFN administration.

Discussion

Six months of r-IFN therapy induced normalization of ALT in 62.5% of our patients. Five non-responders have histologically more severe hepatitis (2 CAH, 2 CAH/C) and a long-lasting disease (>5 yrs. in 4) and 2 had HIV coinfection. Further studies are needed to establish the optimal dose and the duration of treatment in low-responders and to determine if HIV positive status can influence the outcome of the therapy.

References

1. Davis GL, Balart LA, Schiff ER et al (1989) Treatment of chronic hepatitis C with recombinant interferon alpha. A multicentric randomized controlled trial. N Engl J Med 22: 1501–1505
2. Di Bisceglie AM, Martin P, Kassianides C et al (1989) Recombinant alpha interferon therapy for chronic hepatitis C. N Engl J Med 22: 1506–1510

Authors' address: Dr. M. Gargiulo, Sezione Malattie Infettive, Ospedale Civile "G. Mazzini", I-64100 Teramo, Italy.

Arch Virol (1992) [Suppl] 4: 306–307
© Springer-Verlag 1992

Long-term effects of recombinant leukocyte alpha interferon in the treatment of chronic delta hepatitis

G. Tocci, L. Antonelli, E. Boumis, M. Colaiacomo, P. Guarascio, C. Struglia, G. Tossini, and G. Visco

Lazzaro Spallanzani Hospital for Infectious Diseases, Rome, Italy

Summary. A pilot study is described, in which 25 chronic CDH patients were treated with 3 MU recombinant α-interferon per week for 4 months. Improvement was transient and no long-term effects were noted. Side effects were well tolerated and reversible so that longer treatment and higher dosages should be possible.

*

Interferon is the most promising therapy for chronic delta hepatitis (CDH) since it seems to inhibit hepatitis delta virus (HDV) replication [1]. In order to assess the long-term efficacy and safety of interferon in the treatment of CDH we performed this pilot study. Twenty-five patients with CDH, 23 male and 2 female, mean age 31 years (range 18–63), were treated with recombinant leukocyte alpha interferon at 3 MU three times a week for 4 months intramuscularly.

All patients underwent liver biopsy: 15 had chronic active hepatitis, 4 active cirrhosis, 5 chronic active hepatitis with lobular necrosis and 1 chronic persistent hepatitis.

A second liver biopsy was performed in 15 patients 6 months after treatment.

ALT serum levels decreased in 20 out of 25 patients (80%), the mean values decreased by 285 mU/ml to 110 mU/ml during treatment, but 6 months after treatment ALT serum levels increased to initial values.

Among the 15 patients who underwent a second liver biopsy, the histological features improved in 2 (13%), worsened in 6 (40%) and remained unchanged in 7 (47%).

Two out of 5 HBeAg positive patients seroconverted to antiHBe and remained unchanged for at least 6 months after treatment, one became transitorely HBeAg negative and 2 patients, who were also anti-HIV positive, remained unchanged.

No histological improvement was observed in the two HBeAg positive patients who seroconverted to antiHBe, according to the hypothesis that relates the histological activity to HDV replication.

The lack of seroconversion of the HBe system in the 2 anti-HIV positive patients confirms previous observations and emphasizes the importance of the integrity of the immune system in interferon therapy [3].

Side effects were well tolerated and reversible.

In this pilot study no long-term improvement was observed. In order to obtain long-term benefits, further studies with higher doses and/or more-prolonged courses are needed.

References

1. Gregory PB (1986) Interferon in chronic hepatitis B. Gastroenterol 90(1): 237–240
2. Hoofnagle J, Mullen K, Peters M, Avigan M, Park Y, Waggoner J, Gerin J, Hoyer B, Smedile A (1987) Treatment of chronic delta hepatitis with recombinant human alpha interferon. Prog Clin Biol Res 234: 291–298
3. Verme G, Amoroso P, Lettieri G, Pierri P, David E, Sessa F, Rizzi R, Bonino F, Recchia S, Rizzetto M (1986) A histological study of the hepatitis delta virus liver disease. Hepatology 6: 1303–1307

Authors' address: Dr. G. Tocci, Lazzaro Spallanzani Hospital for infectious diseases, USL RM/10, Rome, Italy.

XII Liver transplantation and hepatitis viruses

Arch Virol (1992) [Suppl] 4: 311–316
© Springer-Verlag 1992

Liver transplantation and hepatitis viruses

R. Williams, J. G. O'Grady, S. E. Davies, and **E. Fagan**

Institute of Liver Studies, King's College School of Medicine and Dentistry,
Denmark Hill, London, UK

Summary. Considerations regarding liver transplantation in viral-hepatitis-related acute liver failure and end-stage chronic liver disease are discussed. Parameters of prognosis and indications for transplantation are presented. Differences according to the causative agent are noted, in particular regarding the danger of reinfection. The role of immunoprophylaxis is addressed as is the question of additional antiviral (interferon) treatment.

Introduction

Orthotopic liver transplantation is now widely used in the management of both acute liver failure and end-stage chronic liver disease consequent upon viral infections. The success rate is steadily improving, as evidenced by one-year survival rates of 73–83% in recent years [1, 2]. The longer term outlook is also good as shown by the 5 year actuarial survival rate of 64% in the first 1000 patients treated with cyclosporine in the Pittsburg programme [3]. The survival rates for acute liver failure tend not to be quite as high and most centres report one-year figures in the region of 55–70% [4–8].

Acute liver failure

Viral hepatitis is the commonest cause of acute liver failure, although in the UK it accounts for less than half of the cases seen. The proportion of cases due to the main causative agents—hepatitis A virus (HAV), hepatitis B virus (HBV) $\pm$ delta virus (HDV) and presumed NANB hepatitis—varies geographically. The selection of appropriate patients for transplantation at the earliest possible stage can be difficult, especially as survival rates of 67% and 39% have been achieved in patients with grade III–IV encephalopathy as a consequence of HAV and HBV induced fulminant hepatic failure (FHF), respectively, in this unit [9]. Dependence on encephalopathy as an indicator of prognosis often unnecessarily reduces the time available to find a suitable donor organ. We use five simple clinical and laboratory parameters to assess prognosis early in the course of the disease—aetiology, age, duration of jaundice before the onset of encephalopathy, prothrombin time and serum bilirubin [10]. The precise indications for transplantation are outlined in Table 1. In France there is a great reliance on factor V levels to select candidates for transplantation, with the cut-off point being taken at <20% in a patient with progressive encephalopathy [5]. The reported mortality rate in these patients is 95%. A French study of 115 patients with fulminant HBV also found that HBsAg status and serum alpha-fetoprotein were independent predictors of survival [11]. The survival rate in 21 HBsAg seronegative patients was 47%, as compared to 17% in the 94 HBsAg seropositive patients (p = 0.006).

Recurrence of viral infection is a potential risk after liver transplantation for each of the main causes of acute liver failure—A, B and NANB. HAV has been demonstrated in liver tissue as early as 7 days, and for up to 7 months after liver transplantation using both monoclonal antibody and *in situ* hybridisation techniques [12]. This was associated in one instance with a mild self-resolving hepatitis, during which time HAV was also detectable in faeces.

The majority of patients with fulminant hepatitis B have ceased to have viral replication at the time of admission to hospital. HBeAg was detectable in serum in 12% and 37% respectively of patients in two published series, and in the latter HBV DNA was found in serum in only 9% of cases [13, 14]. These data would suggest that the risk of HBV recurrence after transplantation should be relatively low. However, the available data are limited, and it is interesting that some groups elect to utilise passive immunoprophylaxis in such patients to prevent reinfection. Vogel et al. reported a case of HBV-FHF who remained HBsAg seropositive for 4 months after transplantation before seroconverting to anti-HBs, even though there was no evidence of viral replication or graft dysfunction related to HBV infection [15].

Table 1. Indications for liver transplantation in patients with fulminant viral hepatitis at King's College hospital

Absolute indication
Prothrombin time >100 sec

Otherwise
Any 3 of the following:

Aetiology—NANB hepatitis, halothane hepatitis, idiosyncratic drug reaction,
age—<10 or >40 years,
duration of jaundice before the onset of encephalopathy >7 days,
prothrombin time >50 sec,
serum bilirubin >300 µmol/l

The situation is more difficult with respect to possible NANB recurrence because of the absence of a reliable diagnostic serological test. The development of the new antibody tests for hepatitis C virus (HCV) are unlikely to solve this problem as only 15% of patients in one study were positive in the acute phase of the hepatitis [96]. Furthermore, there is evidence that the positivity rate of the HCV antibody test is unusually low after transplantation [17, 18]. There is however circumstantial evidence that a NANB virus infection may recur after transplantation. The cloned genome of the HCV is similar to flaviviridae [19]. A toga-like virus has been isolated in liver tissue from patients with NANB-FHF [20, 21]. In our experience, severe coagulative necrosis of the graft in the absence of features of primary graft nonfunction, major vascular problems or acute cellular rejection occurs with increased frequency in patients with NANB-FHF 5–10 days post-transplant, but the precise role of virus reinfection remains to be established. Later probable recurrence of NANB hepatitis has also been seen in the first two years after transplantation.

End-stage chronic liver disease

Recent Italian studies suggest that 68% of patients with cryptogenic cirrhosis have antibodies to HCV, as compared to 5–11% in HBV infected patients and 1% in the normal population [22, 23]. Recurrence of a NANB viral infection after transplantation has been described in this context, in association with a mild hepatitis at 3–4 months and with typical histological features on liver biopsy [24]. Although the reported instances to date have been relatively mild, it is likely that in some cases the infection will persist and histological abnormalities progress even to cirrhosis.

In contrast, HBV recurrence after transplantation is a significant problem increasing morbidity, mortality and the need for retransplantation. In our experience, 82% of patients are chronic HBsAg carriers after liver transplantation [25]. The small number who cleared HBsAg from serum after transplantation had no evidence of active HBV replication (i.e. HBeAg and HBV DNA seronegative) at the time of surgery. The initiation of immunosuppression after transplantation resulted in a massive increase in HBV replication as measured by HBV DNA levels, including those who had no evidence of HBV replication at the time of transplantation. The impact of HBV reinfection was very variable; 45% of the patients were free of HBV related graft disease between 1.7–15 years after transplantation, but 41% had suffered loss of the primary graft as a result of HBV recurrence within 13 months of the time of surgery. A number of these patients had a clinical and histological syndrome which has recently been described as cholestatic fibrosing hepatitis (CFH) [26]. The clinical features of CFH are progressive jaundice associated with a rapidly rising prothrombin time and loss of the

314   R. Williams et al.

graft to death or retransplantation within 4–6 weeks of clinical presentation. In most cases the marked increase in serum transaminase levels seen in conventional hepatitic illnesses is not observed. The histological features are extensive serpiginous periportal fibrosis, canalicular and cellular cholestasis and prominent cytoplasmic HBsAg and HBeAg expression. The only effective treatment for this expression of HBV infection is retransplantation. Delta virus (HDV) infection may offer some medium-term protection from this complication [25].

Immunoprophylaxis with hepatitis B immunoglobulin (HBIg) has been used in an attempt to reduce the incidence and implications of HBV recurrence. Pichlmayr's group combined passive (500–128,000 units HBIg perioperatively and maintaining anti-HBs levels in serum > 100 IU/L) and active immunopropylaxis in 6 patients with active HBV replication at the time of transplantation [27]. One patient remained HBsAg seronegative for 4.8 years but did not develop anti-HBs; the remaining 5 became HBsAg seropositive 6–18 weeks post-transplant. It may be significant that none of the long-term survivors in the German series developed serious graft dysfunction as a consequence of HBV recurrence.

Better results with long-term immuno-prophylaxis were reported by Bismuth's group in a study of 16 patients with HDV infection and without evidence of HBV replication at the time of transplantation [28]. The regimen used by them was 30,000 IU HBIg during the anhepatic phase and 3,000 IU HBIg monthly, combined with active immunisation. Three of the patients became HBsAg seropositive, but 10 of the remaining 13 had HBsAg in liver tissue despite being persistently seronegative. Only two patients, both HBsAg seropositive, demonstrated histological abnormalities on long-term follow-up (chronic active hepatitis at 6 and 18 months respectively). Six of nine HDV patients (66.7%) in an Italian study who were given much lower doses of HBIg (1600–2100 IU in the anhepatic phase), together with active immunization, remained HBsAg seronegative 6–32 months post-transplantation [29].

The role for other antiviral agents has not yet been defined post-transplantation. The theoretical risk that interferon might provoke graft rejection has not been substantiated. A case report from the Mayo Clinic described a patient with active HBV replication who was given 1.5 million units of recombinant interferon for 8 days prior to transplantation, and 3 million units daily for 3 months post surgery, but continued to replicate [30]. A liver biopsy at 6 months showed mild lobular hepatitis and the serum transaminases were moderately elevated.

References

1. Iwatsuki S, Starzl TE, Gordon TE et al (1988) Experience of 1,000 liver transplants under cyclosporine-steroid therapy: a survival report. Trans Proc 20: 498–504

2. Busuttil RW, Colonna JO, Hiatt JR et al (1987) The first 100 liver transplants at UCLA. Ann Surg 206: 387–402

3. Krom RAF, Wiesner RH, Rettke SR et al (1989) The first 100 liver transplantations at the Mayo Clinic. Mayo Clin Proc 94: 84–94

4. Peleman RR, Gavaler JS, Van Thiel DH et al (1987) Orthotopic liver transplantation for acute and subacute hepatic failure in adults. Hepatology 7: 484–489

5. Bismuth H, Samuel D, Gugenheim J et al (1987) Emergency liver transplantation for fulminant hepatitis. Ann Intern Med 107: 337–341

6. Vickers C, Neuberger J, Buckels J, McMaster P, Elias E (1988) Transplantation of the liver in adults and children with fulminant hepatic failure. J Hepatol 7: 143–150

7. O'Grady JG, Alexander GJM, Thick M, Potter D, Calne RY, Williams R (1988) Outcome of orthotopic liver transplantation in the aetiological and clinical variants of acute liver failure. Q J Med 69: 817–824

8. Emond JC, Aran PP, Whitington PF, Broelsch CE, Baker AL (1989) Liver transplantation in the management of fulminant hepatic failure. Gastroenterology 96: 1583–1588

9. O'Grady JG, Gimson AES, O'Brien CJ, Pucknell A, Hughes R, Williams R (1988) Controlled trials of charcoal hemoperfusion and prognostic indicators in fulminant hepatic failure. Gastroenterology 94: 1186–1192

10. O'Grady JG, Alexander GJM, Hallyar K, Williams R (1989) Early indicators of prognosis in fulminant hepatic failure. Gastroenterology 97: 439–435

11. Bernuau J, Goudeau A, Poynard T et al (1986) Multivariate analysis of prognostic factors in fulminant hepatitis B. Hepatology 6: 648–651

12. Fagan E, Brahm J, Mann G, et al (1990) Persistence of hepatitis A virus before and after liver transplantation for fulminant hepatitis. J Med Virol 30: 131–136

13. Gimson AES, Tedder RS, White YS, Eddelston ALWF, Williams R (1983) Serological markers in fulminant hepatitis B. Gut 24: 615–617

14. Brechot C, Bernuau J, Thiers V et al (1984) Multiplication of hepatitis B virus in fulminant hepatitis B. Br Med J 288: 270–271

15. Vogel W, Dietze O, Judmaier G, Then P, Schmid Th, Margreiter R (1988) Delayed clearance of HBsAg after transplantation for fulminant delta-hepatitis (letter). Lancet i: 52

16. Kuo G, Choo Q-L, Alter HJ et al (1989) An assay for circulating antibodies to a major etiologic virus of human non-A, non-B hepatitis. Science 244: 362–364

17. O'Grady J, Smith HM, Sutherland S, Sheron N, Williams R (1990) Low detection of hepatitis C antibodies in serum after liver transplantation. J Hepatol 11: S47 (Abstract)

18. Trainor T, Farr G, Cooper S, et al (1990) Antibody to hepatitis C (Anti-HCV) in orthotopic liver transplant recipients. Hepatology 12: 886 (Abstract)

19. Choo Q-L, Kuo G, Weiner AJ, Overby LR, Bradley DW, Houghton M (1989) Isolation of a cDNA clone derived from a blood-borne non-A, non-B viral hepatitis genome. Science 244: 359–362

20. Fagan EA, Ellis DS, Tovey et al (1989) Toga-like virus as a cause of fulminant hepatitis attributed to sporadic non-A, non-B. J Med Virol 28: 150–155

21. Fagan EA, Ellis DS, Portmann B, Tovey GM, Williams R, Zuckerman AJ (1987) Microbial structures in a patient with sporadic non-A, non-B fulminant hepatitis treated by liver transplantation. J Med Virol 22: 189–198

22. Colombo M, Kuo G, Choo L et al (1989) High prevalence of antibody to hepatitis C virus in patients with hepatocellular carcinoma (abstract). J Hepatol 9

23. Bonino F, Kuo G, Brunetto MR et al (1989) Antibody to hepatitis C virus in the serum of patients with chronic hepatitis (abstract). J Hepatol 9

24. Donovan JP, Markin RS, Zetterman RK, et al (1989) Non-A, non-B hepatitis after orthotopic liver transplantation. Hepatology 10: 569 (Abstract)

25. O'Grady JG, Smith HM, Davies SE, et al (submitted for publication) Hepatitis B virus reinfection after orthotopic liver transplantation: serological and clinical implications
26. Davies S, Portmann B, O'Grady JG, et al (1991) Hepatic histological findings after liver transplantation for chronic hepatitis B virus infection including an unique pattern of fibrosing cholestatic hepatitis. Hepatology 13: 150–157
27. Lauchart W, Müller R, Pichlmayr R (1987) Immunoprophylaxis of hepatitis B virus reinfection in recipients of human liver allografts. Trans Proc 19: 2387–2389
28. Reynes M, Zignego L, Samuel D et al (1989) Graft hepatitis delta virus reinfection after orthotopic liver transplantation in HDV cirrhosis. Trans Proc 21: 2424–2425
29. Colledan M, Grendele M, Gridelli B et al (1989) Long-term results after liver transplantation in B and delta hepatitis. Trans Proc 21: 2421–2423
30. Rakela J, Wooten RS, Batts KP, Perkins JD, Taswell HF, Krom RAF (1989) Failure of interferon to prevent recurrent hepatitis B infection in hepatic allograft. Mayo Clin Proc 64: 429–432

Authors' address: Dr. R. Williams, Institute of Liver Studies, King's College School of Medicine and Dentistry, Denmark Hill, London SE5 9RX, U.K.

XIII Epidemiology of HCV and other blood-transmissible viruses: Selected abstracts

Arch Virol (1992) [Suppl] 4: 319–320
© Springer-Verlag 1992

Prevalence of antibody to hepatitis C virus in acute non-A, non-B hepatitis in patients from different epidemiological categories

M. Rodríguez, S. Riestra, F. San Román, C. A. Navascues, A. Suárez, R. Pérez, J. L. Lombraña, and **L. Rodrigo**

Hospital Covadonga and Transfusion's Communitary Center. Oviedo, Spain

Summary. The prevalence of antibodies to hepatitis C virus (HCV) was determined in 65 patients with acute non-A, non-B hepatitis (NANBH). The results suggest that HCV is the most common causative agent in post-transfusion NANBH and in drug-related hepatitis. Detection of HCV antibodies does not appear to be a particularly useful diagnostic criterion due to the kinetics of the immune response in the course of the disease.

*

A specific enzyme immunoassay (EIA) for non-A, non-B hepatitis virus infection has been developed by recombinant DNA technology [2]. This test detects antibodies to a virus termed hepatitis C virus (HCV) [1].

To evaluate the prevalence of antibodies to hepatitis C virus (anti-HCV) in acute non-A, non-B hepatitis (NANBH), this marker was tested in 65 patients with acute NANBH: 17 post-transfusion, 29 intravenous drug addicts (IVDA) and 19 sporadic cases. Three serum samples from each patient, obtained within the 1st, 3rd and 6th months after the onset of the symptoms, were tested. Anti-HCV was analyzed by ELISA (Ortho). The diagnosis of acute NANBH was based on conventional criteria. The Chi-square test and the Fisher's exact test were used for statistical comparisons.

The prevalence of anti-HCV in the 1st, 3rd and 6th months in each epidemiological group is shown in table 1. From a total of 65 patients, 22 (33.8%) were anti-HCV positive in the first sample. In 10 of these, a progressive increase of the ELISA ratio in the three samples was observed; in the other 12 patients the ELISA ratio was >6 in the first sample and remained stable. Chronicity developed in 9/16 (52.9%) post-transfusion NANBH, in 24/29 (82.8%) IVDA and in 4/19 (21.1%) sporadic cases;

Table 1. Anti-HCV antibody in acute non-A, non-B hepatitis

Months after the onset	Acute non-A, non-B hepatitis			
	Post-transf.	IVDA	Sporadic	Total
First	4/16 (25%)	11/25 (44%)	4/16 (25%)	19/57 (33.3%)
Third	12/16 (75%)[a]	22/25 (88%)[b]	5/16 (31.2%)[c]	39/57 (68.4%)
Sixth	14/16 (87.5%)[d]	24/25 (96%)[e]	5/16 (31.2%)[f]	43/57 (75.4%)

[a] vs. [c] $p < 0.05$ [d] vs. [f] $p < 0.01$
[b] vs. [c] $p < 0.001$ [e] vs. [f] $p < 0.001$

(between 2nd and 3rd groups $p < 0.0001$). In the 6th month the prevalence of anti-HCV in patients who progressed towards chronicity was significantly higher than in those with acute resolving NANBH (94% vs. 53%, $p < 0.001$).

These results suggest that HCV is the main etiological agent in the post-transfusion acute NANBH and in those occurring in drug addicts. The lowest prevalence of anti-HCV detected in sporadic NANBH is probably due to the higher rate of recovery observed in this group. Finally, anti-HCV does not seem to be a good marker for the diagnosis of acute HCV infection, since: 1. – The prolonged window period noted, and 2. – Inability to distinguish between acute and chronic HCV infections in those patients that are anti-HCV positive at the onset of the acute hepatitis.

References

1. Choo Q-L, Kuo G, Weiner AJ, Overby IR, Bradley DW, Houghton M (1989) Isolation of a cDNA clone derived from a blood-borne Non-A, Non-B viral hepatitis genome. Science 244: 359–361
2. Kuo G, Choo Q-L, Alter HJ, Gitnick GL, Redeker AG, Purcell RH, Miyamura T, Dienstag JL, Alter MJ, Stevens CE, Tegtmeier GE, Bonino F, Colombo M, Lee W-S, Kuo C, Berger K, Shuster JR, Overby LR, Bradley DW, Houghton M (1989) An assay for circulating antibodies to a major etiologic virus of human Non-A, Non-B hepatitis. Science 244: 362–364

Authors' address: Dr. M. Rodríguez, Gastroenterology Unit, Hospital Covadonga, Celestino Villamil s/n, E-33006-Oviedo, Spain.

Arch Virol (1992) [Suppl] 4: 321–322
© Springer-Verlag 1992

Prevalence of anti-HCV antibodies in patients with acute nonA-nonB viral hepatitis

T. Giuberti, S. Marchelli, A. Degli Antoni, G. Magnani, A. Cavalli, P. Pizzaferri, C. Schianchi, C. Ferrari, and F. Fiaccadori

Cattedra di Malattie Infettive, Università di Parma, Italy

Summary. In a group of 55 patients with NANBH, 81% were found to be reactive for HCV antibodies. In addition, many patients who had not been subject to parenteral risk of infection were also found to be reactive. Statistically, HCV positive patients have an increased tendency to develop chronic hepatitis.

*

Fifty-five patients with nonA-nonB acute hepatitis were tested for the presence of antibodies to hepatitis C virus (anti-HCV; Ortho Diagnostic System). Samples were taken at the onset of the disease and then serially for a period of time ranging from 12 and 36 months. Twenty-seven patients had previously received blood transfusions (post-transfusional hepatitis; PTH), whereas 28 were negative for previous exposure to infected blood (non-post-transfusional hepatitis; non-PTH).

In the group of PTH patients, anti-HCV was already detectable during the first month of disease in 9 cases; in 4 patients seroconversion occurred within 2 months from the onset; in 9 patients between the third and the sixth month; 5 subjects remained negative during the time of follow-up. No relation was observed between incubation time and detection of anti-HCV. Eleven PTH patients showed normalization of the transaminase values; 9 of them were anti-HCV positive (33%) and 2 negative (7%). In 16 patients serum transaminases were persistently elevated, 13 were anti-HCV positive (49%) and 3 negative (11%).

Among the 28 patients with non-PTH, 16 became anti-HCV positive during the first month; 5 between the second and sixth month; 7 patients remained persistently negative. Of the 10 patients who recovered with normalization of serum transaminases, 5 were anti-HCV positive (18%), 5 anti-HCV negative (18%). Eighteen patients showed evolution toward chronicity; 16 of these were anti-HCV positive (57%) and 2 negative (7%).

In conclusion, 1) HCV seems to be responsible for a high number of acute post-transfusional nonA-nonB hepatitis (81%), but it is also implicated in a large proportion of cases of acute hepatitis without previous exposure to infected blood and without other anamnestic criteria of parenteral transmission (75%); 2) patient to patient variations in the time of anti-HCV seroconversion were observed, even though most of the patients sero-converted within 3 months from the onset (86%); 3) when the entire patient population is considered, evolution toward chronicity is observed more frequently in anti-HCV positive patients.

Authors' address: Dr. Tiziana Giuberti, Cattedra di Malattie Infettive, Università di Parma, I-43100 Parma, Italy.

Arch Virol (1992) [Suppl] 4: 323–324
© Springer-Verlag 1992

Antibody to hepatitis C virus in acute, self-limited, type B hepatitis

M. Rodríguez, A. Suárez, R. Cimadevilla, S. Riestra, C. A. Navascues, P. Sala,
and **L. Rodrigo**

Hospital Covadonda, Oviedo, Spain

Summary. Sera from 104 patients with self-limited, acute type B hepatitis were tested for the presence of anti-HCV antibodies. The results show that especially drug users with acute type B (and occasionally coinfected with type D) hepatitis commonly are infected with HCV. Furthermore, HCV infection may have preceded infection with the other agents and may be responsible for high ALT levels.

*

The cloning and identification of the RNA genome of hepatitis C virus (HCV) [1], was a breakthrough in research on hepatitis viruses. In order to determine the prevalence and the significance of antibody to hepatitis C virus (anti-HCV) in acute B hepatitis, we tested for anti-HCV sera from 104 patients with self-limited disease, in whom HBsAg disappeared within 6 months after the onset of the symptoms.

The anti-HCV was tested by ELISA (Ortho) [2] in a serum sample obtained in the 6th month after the beginning of the acute disease. If the antibody was positive, we then looked for anti-HCV in a sample obtained at the onset of the symptoms.

Out of 104 patients, 50.9% were intravenous drug abusers (IVDA). Hepatitis delta virus (HDV) coinfection was observed in 27/53 (50.9%) IVDA with acute hepatitis B. Persistence of high alanine aminotransferase (ALT) values for more than 6 months after onset of the acute hepatitis occurred in 28/104 (26.9%). Diagnosis of acute hepatitis B and of HDV infection was based on conventional criteria. The anti-HCV was positive in the 6th month in 46 of 104 (44.2%) patients. Of 53 IVDA, 42 (79.2%) were anti-HCV positive and 4/51 (7.8%) non-IVDA were so (p<0.001). Among

IVDA, the prevalence of anti-HCV was higher in those with HDV coinfection (27/30; 90%), than in those without HDV coinfection (15/23; 65.2%) (p < 0.05). The anti-HCV was positive in 25/28 (89.3%) patients whose ALT values were persistently raised for more than 6 months and in 21/76 (27.6%) with normal ALT values (p < 0.001). In 38/46 (82.6%) anti-HCV positive patients, the antibody was already detected at the onset of the symptoms; in the other 8 (17.4%) patients seroconversion was observed and in these patients simultaneous infection by more than one virus probably occurred.

These findings suggest that infection by HCV is present in a high proportion of IVDA with acute hepatitis B or B and D, and that the infection by HCV generally precedes infection by the other viruses. The HCV is probably responsible for the persistently high ALT values in patients with acute B hepatitis and with clearance of HBsAg.

References

1. Choo Q-L, Kuo G, Weiner AJ, Overby LR, Bradley DW, Houghton M (1989) Isolation of a cDNA clone derived from a blood-borne Non-A, Non-B viral hepatitis genome. Science 244: 359–361
2. Kuo G, Choo Q-L, Alter HJ, Gitnick GL, Redeker AG, Purcell RH, Miyamura T, Dienstag JL, Alter MJ, Stevens CE, Tegtmeier GE, Bonino F, Colombo M, Lee W-S, Kuo C, Berger K, Shuster JR, Overby LR, Bradley DW, Houghton M (1989) An assay for circulating antibodies to a major etiologic virus of human Non-A, Non-B hepatitis. Science 244: 362–364

Authors' address: Dr. Manuel Rodríguez, Gastroenterology Unit, Hospital Covadonga, Celestino Villamil s/n, E-33006-Oviedo, Spain.

Arch Virol (1992) [Suppl] 4: 325–326
© Springer-Verlag 1992

HCV infection in HBsAg positive chronic liver disease

F. Fatuzzo, M. T. Mughini, B. Cacopardo, R. La Rosa, L. Nigro, G. Lupo, S. Bruno,
M. Zuccarello, E. Caltabiano, F. Zipper, S. Cosentino, R. Russo, and A. Nunnari

Institute of Infectious Disease University of Catania, Catania, Italy

Summary. The prevalence of anti-HCV antibodies was determined for a group of 68 patients with various forms of chronic liver disease. All patients that were anti-HCV positive but did not show signs of HBV replication had severe liver disease. We therefore suggest that HCV may be responsible for liver damage in HBsAg positive subjects when there are no evident signs of HBV replication.

*

The frequency of anti-HCV antibodies in patients with post-transfusional or cryptogenic chronic liver disease is high, whereas in patients with HBsAg positive chronic liver disease it appears noticeably lower, particularly in those with ongoing HBV replication [1–4]. Further studies showed that in patients with HBsAg positive chronic liver disease without evidence of HBV replication, anti-HCV prevalence rises to 30–50% [2–4]. These data and a recent report on HBsAg clearance during hepatitis C virus infection [3] suggest interference of HCV on HBV replication, which is similar to the negative influence of HBV replicative mechanism exerted by HDV super-infection.

We detected retrospectively, anti-HCV antibodies in sera of 68 patients (50 M, 18 F, mean age 53 years) who were HBsAg positive with a biopsy proved chronic liver disease (14 chronic persistent hepatitis, 5 chronic lobular hepatitis, 30 chronic active hepatitis, 19 liver cirrhosis), and without evidence of HDV infection, autoimmune hepatitis, previous exposure to known hepatotoxic drugs or excessive alcohol consumption.

In all, 18 (26.5%) patients were anti-HCV positive: 2 out of 29 (6.8%) HBeAg positive, 16 out of 32 (50%) antiHBe positive/HBV-DNA negative and 0 out of 7 antiHBe positive/HBV-DNA positive.

Our data confirm the low anti-HCV prevalence in patients with HBsAg positive chronic liver disease, with a significant relationship ($p < 0.01$) between anti-HCV positivity and absence of HBV replication.

In 100% of anti-HCV positive patients without markers of HBV replication, we found a histological pattern of severe liver disease: chronic active hepatitis or cirrhosis. We therefore hypothesize that HCV may play a role in the evolution of liver damage in HBsAg positive patients with no evidence of HBV replication, while its role in liver disease progression among subjects with ongoing HBV replication is difficult to define.

References

1. A. Craxi, V. Di Marco, P. Almasio et al (1989) Hepatitis C virus infection in patients with compensated chronic liver disease from Southern Italy. First International Symposium on Hepatitis C virus. Rome 14–15 Settembre 1989
2. G. Fattovich, A. Tagger, L. Brollo et al (1989) Liver disease in anti HBe positive HBsAg carriers and hepatitis C virus. Lancet ii: 797–798
3. G. Fattovich, G. Giustina, L. Brollo et al (1990) Hepatitis B surface antigen clearance during hepatitis C virus infection. It J Gastroenterol 22: 246
4. JM Sanchez-Tapias, JM Barrera, J Costa et al (1989) Hepatitis C virus infection in non-alcoholic chronic liver disease. First International Symposium on Hepatitis C virus, Rome 14–15 Settembre 1989

Authors' address: Dr. Fatuzzo F., Institute of Infectious Diseases, Via Passo Gravina, 187, I-95125 Catania, Italy.

Arch Virol (1992) [Suppl] 4: 327–328
© Springer-Verlag 1992

Prevalence of antibody to hepatitis C virus in chronic HBsAg carriers

M. Rodríguez, C. A. Navascues, A. Martínez, A. Suárez, S. Riestra, P. Sala,
M. González, and L. Rodrigo

Hospital Covadonda, Oviedo, Spain

Summary. The presence of the anti-HCV antibody was investigated in sera from 102 chronic HBsAg carriers. The subjects varied as to the characteristics of the clinical states. It was found that HCV coinfection was more common in HBsAg positive intravenous drug addicts than in other parenteral risk groups. It also appears that HCV may be the causative agent of chronic liver disease in HBsAg carriers with undetectable HBV (and possibly HDV) replication.

*

Recently, the genome of hepatitis C virus (HCV) was cloned [1] and a specific assay was developed to capture circulating HCV antibodies (anti-HCV) [2].

With the aim of studying, retrospectively, the prevalence of the antibody to hepatitis C virus in chronic HBsAg carriers, serum from 102 patients were tested. Thirty were healthy, chronic HBsAg carriers, 29 were chronic HBsAg carriers with hepatitis B virus (HBV) replication (HBeAg +ve/HBV-DNA +ve), without hepatitis delta virus (HDV) superinfection (antiHD −ve), and with chronic liver disease. Thirteen were chronic HBsAg carriers without HBV replication (antiHBe +ve/HBV-DNA −ve), without HDV superinfection (anti-HD −ve) and with chronic liver disease. Thirty were chronic HBsAg carriers with HDV superinfection (anti-HD +ve) and with chronic liver disease. The histological diagnosis in the 72 patients with chronic liver disease was: chronic persistent hepatitis (CPH) in 7, chronic active hepatitis (CAH) in 49 and active hepatic cirrhosis (AHC) in 16. The presumed source of exposure was: blood transfusion in 12, intravenous drug use in 32, homosexual activity in 5 and unknown in 53.

Anti-HCV was tested by ELISA (Ortho), according to the manufacturer's instructions. Serum markers of HBV and HDV were tested by commercial

Table 1. Prevalence of anti-HCV in chronic HBsAg carriers

Healthy chronic HBsAg carries	Chronic HBsAg carriers with chronic liver disease		
	HBeAg +ve HBV-DNA +ve anti-HD −ve	anti-HBe +ve HBV-DNA −ve anti-HD −ve	anti-HD +ve
1/30 (3.3%)[a]	1/29 (3.4%)[b]	5/13 (38.4%)[c]	21/30 (70%)[d]

[a] vs. [c] $p < 0.01$; [a] vs. [d] $p < 0.001$; [b] vs. [c] $p < 0.01$; [b] vs. [d] $p < 0.001$

assays. HBcAg and HDAg in liver biopsy were detected by immunoperoxidase technique.

The prevalence of anti-HCV in each group is showed in the Table 1.

Of 22 chronic HBsAg carriers anti-HD positive, HBcAg and/or HDAg positive in liver biopsy, 14 (63.6%) were anti-HCV positive, while 7/8 (87.5%) chronic HBsAg carriers anti-HD positive and HBcAg and HDAg negative in liver biopsy were anti-HCV positive.

The prevalence of anti-HCV was significantly higher in intravenous drug addicts (68.7%) than in post-transfusion (16.6%), homosexuals (0%) or in those with unknown source.

Our results show a high spread of HCV infection in intravenous drug addicts that are chronic HBsAg carriers. Infection by HCV may be responsible for the chronic liver disease in some patients with HBsAg positive chronic hepatitis and undetectable HBV replication. Likewise it could take place in some patients with chronic hepatitis B and HDV superinfection in whom replication of HBV and HDV is lacking.

References

1. Choo Q-L, Kuo G, Weiner AJ, Overby LR, Bradley DW, Houghton M (1989) Isolation of a cDNA clone derived from a blood-borne non-A, non-B viral hepatitis genome. Science 244: 359–361
2. Kuo G, Choo Q-L, Alter HJ, Gitnick GL, Redeker AG, Purcell RH, Miyamura T, Dienstag JL, Alter MJ, Stevens CE, Tegtmeier GE, Bonino F, Colombo M, Lee W-S, Kuo C, Berger K, Shuster JR, Overby LR, Bradley DW, Houghton M (1989) An assay for circulating antibodies to a major etiologic virus of a human non-A, non-B hepatitis. Science 244: 362–364

Authors' address: Dr. M. Rodríguez, Gastroenterology Unit, Hospital Covadonga, Celestino Villamil s/n, E-33006-Oviedo, Spain.

Arch Virol (1992) [Suppl] 4: 329–332
© Springer-Verlag 1992

HBV and HCV infection in i.v. drug addicts; coinfection with HIV

P. Botti, A. Pistelli, F. Gambassi, A. M. Zorn, L. Caramelli, S. Peruzzi, C. Smorlesi, E. Masini, and **P. F. Mannaioni**

Toxicological Unit, Department of Preclinical and Clinical Pharmacology, Florence University, Florence, Italy

Summary. A group of 122 drug addict patients were studied to evaluate the incidence of HIV, HBV, HCV infections and of laboratory findings of hepatic damage. Our data show that hepatic damage is more frequent in patients affected by HBV-HCV coinfection than those with HBV or HCV infection alone and that HIV positivity supports HBV-HCV coinfection.

Introduction

Intravenous (i.v.) drug addicts represent a high-risk group for parenteral infections, particularly those due to hepatitis viruses and HIV, of which this group represents a huge reservoir and an unpredictable source of diffusion [1, 2]. Preliminary data on the incidence of infection and coinfection by HBV, HCV and HIV in this group of patients are reported here.

Materials and methods

One hundred twenty two patients (96 males and 26 females, mean age 27.2 years, range 17–42 years, with a mean period of i.v. drug addiction equal to 6.8 years) were studied. They were admitted to the Toxicological Unit of Florence from January 1st 1990 to June 30th 1990 for medical check-up, detoxication or emergency such as overdose or withdrawal syndrome. All were screened with standard blood tests, for HBV antigen and antibody markers (RIA), HCV antibodies (ELISA), HIV antibodies (ELISA), besides undergoing at least 3 samplings for ALT determination.

Results

None of the patients presented clinical and/or humoral signs of acute hepatitis during the observation period, although 62 (50.8%) showed abnormally elevated transaminase levels, suggesting hepatic involvement.

Table 1. Epidemiology of HBV and HCV
infection in i.v. drug addicts .

	N° of patients	%
HBV markers positivity	21	17.2
HCV antibodies positivity	16	13.1
HBV markers + HCV antibodies positivity	64	52.5
HBV markers + HCV antibodies negativity	21	17.2
Total	122	100

Toxicological Unit, Florence University

The following groups were formed (Table 1):

A) Twenty-one patients (17.2%) with exclusive HBV-marker positivity, 11 of which revealing ALT altered values;
B) Sixteen patients (13.1%) with exclusive HCV-antibody positivity, 9 with ALT altered values;
C) Sixty-four patients (52.5%) with both HBV and HCV-marker positivity. Forty-two cases (65.6%) had altered ALT levels;
D) Twenty-one patients without any positive viral markers. The only case with elevated ALT values had developed hepatotoxicity following therapy with naltrexone. The whole group was free of any virus-induced liver damage.

Mean ALT values were not significantly different than in the other groups (A, B and C).

On the whole, 80 patients (65.5%) showed evidence of HCV antibodies and 85 (68.8%) the presence of HBV markers.

Nineteen (15.6%) HIV-positive patients showed evidence of HBV and HCV coinfection in 78.9% of cases vs. 47.6% observed in the HIV- negative group (Fig. 2).

The degree of hepatic damage and/or immunodeficiency (evaluated by the CD4/CD8 ratio) was found to be unaffected by the presence of HBV-HCV coinfection among HIV-positive patients.

Discussion

Our data concerning the wide diffusion of both HBV and HCV observed in i.v. drug addicts are in full agreement with the reported literature [3]. The coinfection with HBV and HCV seems to enhance the rate of occurrence of hepatic damage (Fig. 1), though further data are needed to better evaluate its level, evolution and prognosis.

The lack of hepatic impairment observed in both HBV- and HCV-negative patients may suggest that these are, at present, the only involved pathogenic viruses in such population, even if HCV itself does not represent the whole NANB virus group.

The present data on HIV positivity rate are in agreement with our previous study carried out on 1076 i.v. drug addicts, observed in the period 1985–1989, which resulted to be HIV-positive in 17.1% of cases (185 cases).

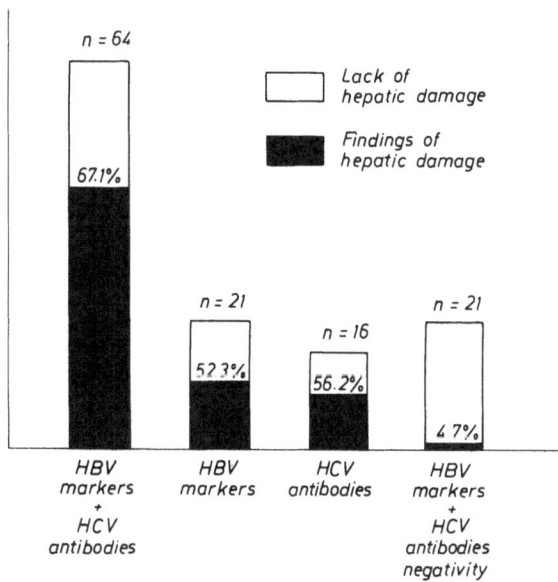

Fig. 1. Laboratory findings of hepatic damage in i.v. drug addicts with HBV and/or HCV infection

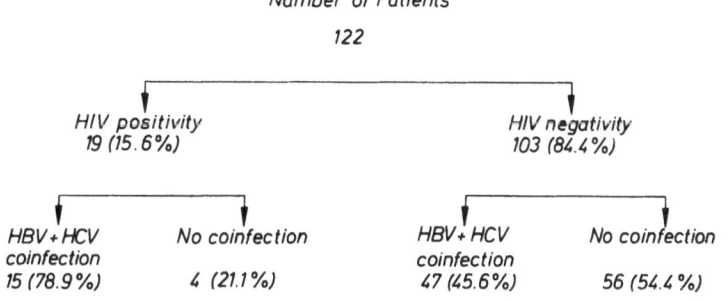

Fig. 2. HBV, HCV coinfection in i.v. drug addicts coinfection with HIV

The patients of our present study with positive HIV antibodies seem to be more susceptible to HBV, HCV coinfection. However, evidence for a negative prognostic role played by any viral infection on the others is still missing [4].

References

1. Kuo G, Choo QL, Alter HJ et al (1989) An assay for circulating antibodies to a major etiologic virus of human NANB hepatitis. Science 244: 362–364
2. Mattson L, Weiland O, Glaumann H (1989) Chronic non-A, non-B hepatitis developed after transfusion, illicit self injection or sporadically. Outcoming during long term follow-up. Liver 9: 120–127
3. Esteban JI, Esteban R, Viladomiu L et al (1989) Hepatitis C virus antibodies among risk groups in Spain. Lancet ii, 8658: 294–297
4. Martin P, Di Bisceglie AM, Kassianides C et al (1989) Rapidly progressive non-A, non-B hepatitis in patients with HIV infection. Gastroenterology 97: 1559–1561

Authors' address: Dr. P. Botti, Department of Preclinical and Clinical Pharmacology, Florence University, USL 10 D, Florence, Italy.

Arch Virol (1992) [Suppl] 4: 333–334
© Springer-Verlag 1992

HCV and HIV infection among intravenous drug abusers in Eastern Sicily

B. Cacopardo, F. Fatuzzo, S. Cosentino, B. M. Celesia, M. T. Mughini, R. La Rosa,
S. Bruno, G. Lupo, F. Zipper, G. La Medica, R. Russo, A. Nunnari

Institute of Infectious Disease, University of Catania, Italy

Summary. In a study of 175 intravenous drug addicts from Eastern Sicily, 58.3% were found to be anti-HCV positive. In this population, the presence of anti-HCV was independent of HIV infection, age, duration of drug use and the practice of needle sharing. This may indicate that HCV is more readily transmitted (or spread earlier in this population) among drug addicts than is HIV.

*

It has been widely demonstrated that HCV infection is frequently associated with high-risk behavior, such as intravenous drug abuse: anti HCV prevalence among intravenous drug addicts (IVDA) is in fact 50–70% [1, 2, 3].

We studied anti-HCV antibody prevalence among 175 (151 M, 24 F) drug abusers from Eastern Sicily; 65 were anti HIV + ve and 110 anti HIV − ve. In all, 102 (58.3%) subjects were found to be anti-HCV positive: 40/65 anti HIV + ve (61.5%) and 62/110 anti HIV − ve (56.5%).

Our results confirmed a high anti-HCV prevalence among IVDAs in our area; in our data anti-HCV prevalence was independent of anti-HIV seropositivity, in accordance with Cadeo et al. [1], but in disagreement with the observations of Huemer et al. [3].

Furthermore, anti-HCV prevalence was unrelated to the IVDA's age, duration of drug abuse and habit of needle sharing, while it is well recognized that anti-HIV seropositivity is strongly related to these cofactors. A possible reason for the above mentioned results could be the earlier and higher spread of HCV in our area, or the fact that HCV is presumably more readily transmitted than HIV among intravenous drug abusers.

Finally, in accord with Esteban et al. [2], we found no significant difference in CD4+ values between anti-HCV positive and anti-HCV negative subjects, meaning that anti-HCV seropositivity was independent of immunological status considered as CD4+ value.

References

1. Cadeo GP, EL Hamond I, Rodelle A, et al (1989) Preliminary report on hepatitis C antibody prevalence among high risk groups and blood donors in the area of Brescia. First International Symposium on Hepatitis C Virus. Rome 14–15 Settembre 1989
2. Esteban JI, Viladomiou L, Esteban R, et al (1988) Hepatitis C virus antibodies among risk groups in Spain. Lancet ii: 249–296
3. Huemer HP, Prodinger WM, Lorcher C, et al (1990) Correlation of hepatitis C virus antibodies with HIV-1 seropositivity in intravenous drug addicts. Infection 18: 122

Address of authors: Dr. B. Cacopardo, Institute of Infectious Diseases, Via Passo Gravina 187, I-95125, Catania, Italy.

Arch Virol (1992) [Suppl] 4: 335–336
© Springer-Verlag 1992

Anti-hepatitis C antibody prevalence among intravenous drug addicts in the Catanzaro area

V. Guadagnino[1], G. Zimatore[2], A. Rocca[3], F. Montesano[2], R. Masciari[3], B. Caroleo, A. Izzi, D. Morabito, E. Naso[2], and R. Scicchitano[2]

[1] Department of Experimental and Clinical Medicine, Catanzaro Medical School, University of Reggio Calabria
[2] Center for Drug addicts, 1st Division of General Medicine, and
[3] Service of Virology, Local Health District 18 of Calabria Region, Catanzaro, Italy

Summary. A higher seroprevalence of anti-HCV antibodies (63.4%) was found in 41 intravenous drug addicts (IVDA) when compared to 220 controls (1.8%). Life style is an important risk factor for HCV transmission among IVDA.

*

Intravenous Drug Addicts (IVDA) are at risk of Hepatitis B Virus (HBV) infection because of multiple factors. Sharing of needles and syringes, the high number of sexual partners, tattooing, promiscuity, poor hygienic habits, etc. favor the virus to penetrate into the organism by the parenteral and the inapparent-parenteral route [1–3]. Due to epidemiological similarities between HBV and Hepatitis C virus (HCV) [3], a study was conducted to investigate the prevalence of anti-HCV antibodies in IVDA in the Catanzaro area.

Forty-one IVDA (34 men and 7 women, mean age: 26.1 years), attending from January to June 1990 for the first time a Center for Drug Addicts in Catanzaro were investigated. Drug addiction lasted at least 2 years (mean 7.7 years). Two-hundred and twenty healthy subjects, from the same geographic area, comparable for age, served as controls. Anti-HCV antibodies were detected in serum samples by ORTHO HCV ELISA TEST.

A statistically significant difference (P < 0.001) was found in prevalence of anti-HCV antibodies between IVDA and control groups. In fact, 26 (63.4%)

out of 41 IVDA studied were positive for anti-HCV antibodies with respect to 4 (1.8%) out of 220 controls.

Serum ALT levels $>2\times$ the upper limit of normal values were found in only 9 out of 26 (34.6%) anti-HCV positive subjects, and thus no correlation was found between anti-HCV positivity and abnormal ALT levels.

Our data indicate that IVDA are also at high risk of HCV infection in our geographic area. A "life style" which increases the risk of HBV infection among IVDA, clearly plays a similar role in the transmission of HCV among this group of subjects.

References

1. Piazza M, Di Stasio G, Maio G, et al (1973) Hepatitis B antigen inhibitor in human faeces and intestinal mucosa. Br Med J 334–337
2. Guadagnino V, Ayala F, Chirianni A, et al (1982) Risk of hepatitis B virus infection in patients with eczema or psoriasis of the hand. Br Med J 254: 84–85
3. Piazza M (1990) Epatite virale acuta e cronica, 5th edn. Ghedini editore, Milano

Authors' address: Dr. V. Guadagnino, Department of Experimental and Clinical Medicine, Catanzaro Medical School, University of Reggio Calabria, Italy.

Arch Virol (1992) [Suppl] 4: 337–338
© Springer-Verlag 1992

Hepatitis C virus (HCV) infection in the Piacenza dialysis center

P. Pizzaferri[2], D. Padrini[1], P. Viale[1], F. Fontana[3], G. P. Poisetti[3], and F. Alberici[1]

[1] Div Mal Infettive, Ospedale Civile di Piacenza
[2] Cattedra di Malattie Infettive, Università di Parma
[3] II°Div. Medicina-Sez Emodialisi, Ospedale Civile di Piacenza, Italy

Summary. In a group of 110 dialysis patients, 21% had anti-HCV antibodies. Statistical differences were noted according to method and duration of dialysis as well as the presence of further risk factors.

*

The present study was undertaken on the patients of the Piacenza dialysis Center to define the prevalence of hepatitis C virus infection and the relative risk factors.

Materials and methods

From November 1989 to May 1990, 110 patients were examined: 76 men and 34 women, average age 57 years (range 21–82). They underwent haemodialysis treatment for 3 to 266 months (average 74), whereby 19% of the patients had been polytransfused, 81% simply dialysed, 1 had received a renal transplant and none were drug addicts.

As far as the dialytic treatment is concerned, 68.2% were treated by haemodialysis, 28% by continued peritoneal dialysis, 3.6% by intermittent peritoneal dialysis. Statistical analysis was done with the Chi Square test. The test for anti-HCV was made by a commercially available immunoenzymatic assay (Ortho Diagnostic System).

Results

Of the dialysed patients, 21% were anti-HCV positive (9 women and 14 men, without a statistically significant difference between the two groups). No acute hepatitis cases were observed and anti-HCV negative patients who were tested more than once, remained persistently negative.

Thirty percent of the haemotransfused, the one transplanted patient, and 18% without any risk factors besides dialysis were anti-HCV positive, without any statistically significant difference between the studied groups.

As far as the dialytic methods are concerned 24% of the patients undergoing haemodialysis, 19% of the patients undertaking continuous peritoneal dialysis and none of the patients undertaking intermittent peritoneal dialysis were anti-HCV positive.

Time of dialysis was significantly higher in the group of anti-HCV positive in comparison with anti-HCV negative patients (mean, 106 vs 65 month, $p < 0.00001$).

References

1. A. M. Couroŭcé et al (1990) Antibodies to hepatitis C virus in haemodialysis patients. Abst. 395, International Symposium on viral hepatitis and liver disease. Houston, Texas, April 4–8, 1990
2. U. Schlipköter et al (1990) Prevalence of anti-HCV in Haemodialysis patients in southern Germany. Abst. 397, International Symposium on viral hepatitis and liver disease. Houston, Texas, April 4–8, 1990

Authors' address: Dr. P. Pizzaferri, Cattedra di Malattie Infettive, Università di Parma, Via Gramsci 14, I-43100 Parma, Italy.

Arch Virol (1992) [Suppl] 4: 339–342
© Springer-Verlag 1992

Prevalence of hepatitis C virus (HCV) antibodies in haemodialysis patients

L. Vandelli[1], **G. Medici**[1], **A. M. Savazzi**[1], **M. De Palma**[2], and **E. Lusvarghi**[1]

[1] Nephrology and Dialysis Department and [2] Blood Transfusion Service University of Modena, Italy

Summary. The prevalence of antibodies to HCV and the course of hepatitis have been determined in 357 haemodialysed patients treated at a single institution.

The prevalence of HCV infection increases with the duration of haemodialysis and with the use of blood transfusions, yet there is high frequency of HCV seropositivity even without blood transfusions. Evolution of HCV hepatitis to chronicity is frequent and biological signs of chronic hepatopathy can coexist with absence of alanine aminotransferase (ALT) abnormalities.

*

Non A, non B hepatitis (NANBH) in dialysis units continues to be an important disease both for its incidence and for some aspects of its natural history [1]. Hepatitis C virus (HCV) is the agent of 80% of post-transfusional NANBH and 60% of sporadic NANBH [2].

The recent introduction of an assay for hepatitis C antibodies (anti-HCV) allows for NANB diagnosis in addition to the criteria of exclusion, but since this test represents only an indirect marker for infection, the correlation between anti-HCV positivity and blood infectivity is not absolute. HCV infection is becoming an emerging problem in haemodialysis units, not only because of its prevalence, but because of the doubts about the route of transmission of the virus in patients with known immunodeficiency.

Several reports recently published are based on only a few cases or refer to multicentric studies that do not always guarantee a methodological uniformity. While these studies are useful in quantifying the phenomenon, they are less useful in identifying routes of transmitting the infection, other than blood transfusion.

However, there is no argument as to the high frequencies of HCV infection observed in dialysis units in the patients but not in the staff members; this suggests that HCV is not as readily transmitted as HBV [3]. This observation, plus the appreciable amount of HCV seropositivities not transfusion-related, suggest a cross infection in the dialysis setting by "inapparent" parenteral transmission of the virus.

In this report we describe the anti-HCV seroprevalence and the course of hepatitis in patients treated in a large dialysis unit, and discuss the importance of preventive measures.

HCV seroprevalence (Ortho Diagnostic System) was assessed in 357 patients treated at a single institution: 38% on centre haemodialysis and 62% on self-care haemodialysis. The mean age of the patient, was 62 ± 14 years and the mean duration of maintenance dialysis was 63 ± 50 months. Dialysis records of all cases were reviewed in a retrospective analysis. The statistical analysis was performed using chi-square and Student t-test.

Anti-HCV was found in 90 patients, which represents a prevalence of 25.2%, very close to the national data: 24% is the mean prevalence of anti-HCV found in 1188 patients from 13 Italian haemodialysis units (data from Abstracts of XXXI National Congress of the Italian Society of Nephrology—Siena 1990).

31 patients (22.9%) were dialysed at the centre and 59 (26.8%) at self-care units, 1 at home. No difference was observed regarding sex. In only 2 patients was HCV associated with HBsAg.

Anti-HCV positivity was correlated with duration of dialysis (89.81 vs 53.62 months, $p < 0.0001$) and with the use of blood transfusions:

Non-transfused	antiHCV+	20 (14.3%)	Transfused	antiHCV+	70 (32.3%)	$p < 0.002$
	antiHCV−	120 (85.7%)		antiHCV−	147 (67.7%)	

However, 22.2% of the anti-HCV patients had never received blood transfusion. Anti-HCV positivity in non-transfused patients was observed in all classes of age and without significant differences (p: ns):

Months	0–12	13–26	37–60	61–84	85–108	108
Transfused	3 (60%)	8 (72.7%)	14 (93.3%)	7 (63.6%)	12 (80%)	26 (78.8%)
Non-transfused	2 (40%)	3 (27.3%)	1 (6.7%)	4 (36.4%)	3 (20%)	7 (21.2%)

Anti-HCV positivity was more commonly found in patients with a history of elevated ALT levels than in those without [63/85 (74%) vs 27/272 (9.9%)] and reached 82% (27/33) in those patients with persistently elevated ALT levels.

For 34 anti-HCV positive patients, NANBH had been diagnosed in the past (69.3% of all the NANBH observed at the follow up). The course of hepatitis and of ALT activity was very capricious. Basically, however, we observed 3 patterns, so we divided patients into groups: 1) patients (N = 27) without history of elevated ALT; 2) those (N = 36) with ALT exceeding 2 times the normal value and then normalized for at least 1 year; 3) those (N = 27) with ALT exceeding 2 times the normal value and fluctuating for at least 6 months. Biological signs of chronic liver disease (gammaglobulins > 20%, Platelets < 150000 mm^3, Cholesterol < 150 mg/dl) were estimated and found in all 3 groups with different, though not always significant, incidence. The smallest incidence in group 2 is, however, suggestive.

	Group 1	Group 2	Group 3	
Gammaglobulins	16 (59.3%)	14 (38.9%)	16 (59.3%)	
Platelets	6 (22.2%)	5 (13.9%)	10 (37%)	
Cholesterol	14 (51.9%)	10 (27.8%)	10 (37%)	p: ns

According to the frequency of signs of chronic liver disease, we created a score and formed 2 classes:

A) with at least 2 signs; B) with 1 or no signs.

A	13 (48.1%)	7 (19.4%)	11 (40.7%)	
B	14 (51.9%)	29 (80.6%)	16 (59.3%)	$p < 0.05$

A significant difference exists between the groups; the highest incidence remains in groups 1 and 3, while the lowest frequency occurs in patients of group 2 who could be considered as cured. We conclude that the prevalence of HCV infection increases with the duration of maintenance dialysis and with the use of blood transfusions, yet an appreciable frequency of HCV positivity is present even in patients without blood transfusions. We think HCV infection can be acquired within dialysis units and the contamination of environmental surfaces seems to be responsible for patient-to-patient transmission of the virus.

Evolution of HCV hepatitis to chronicity is frequent and biological signs of chronic hepatopathy can coexist with absence of ALT abnormalities. Given the high incidence of chronicity, a preventive programme for C hepatitis should be set up, particularly to protect patients awaiting renal transplant. At present, we think that patients with acute suspected NANBH and HCV positive patients with previous hepatitis and fluctuating ALT should be dialysed in separate sections.

References

1. Marchesi D, Arici C, Poletti E, Mingardi G, Minola E, Mecca G (1988) Outbreak of non-A, non-B hepatitis in center hemodialysis patients: A retrospective analysis. Nephrol Dial Transplant 3: 795–799
2. Hopf U, Möller B, Küther D, Stemerowicz R, Lobeck H, Lüdtke-Handjery A, Walter E, Blum HE, Roggendorf M, Deinhardt F (1990) Long-term follow-up of posttransfusion and sporadic chronic hepatitis non-A, non-B and frequency of circulating antibodies to hepatitis C virus (HCV). Hepatology 10: 69–76

3. Jeffers LJ, Perez GO, De Medina MD, Ortiz-Interian CJ, Schiff ER, Rajender Reddy K, Jimenes M, Bourgoignie JJ, Vaamonde CA, Duncan R, Houghton M, Choo G-L, Kuo G (1990) Hepatitis C infection in two urban hemodialysis units. Kidney Int 38: 320–322

Authors' address: Dr. L. Vandelli, Nephrology and Dialysis Department, University of Modena, Modena, Italy.

Arch Virol (1992) [Suppl] 4: 343–344
© Springer-Verlag 1992

Preliminary investigation on intrafamilial spread of hepatitis C virus (HCV)

M. T. Mughini, B. Cacopardo, F. Fatuzzo, B. M. Celesia, R. La Rosa, S. Bruno, E. Oddo, S. Tosto, S. Cosentino, L. Nigro, R. Russo, and **A. Nunnari**

Institute of Infectious Disease, University of Catania, Italy

Summary. To determine the risk of cohabitant HCV infection, we investigated the sera of 101 family members of 53 anti-HCV antibody positive chronic liver disease patients. Altogether 14.8% of the cohabitants were also anti-HCV antibody positive, compared to a prevalence of 1.4% in the general population. These results suggest that hepatitis-C-virus may spread by person-to-person infection.

*

The high rate of anti-HCV positive cryptogenic chronic hepatitis, in contrast with anti-HCV prevalence among blood donors (1%), clearly indicates the existence of additional modalities of HCV transmission apart from those well recognized (blood or blood-derivative transfusion, drug abuse). Some authors have demonstrated the possibility of intrafamiliar circulation [2] or sexual transmission [3] of hepatitis C virus.

We evaluated the prevalence of anti-HCV in 101 cohabitant contacts of 53 patients with anti-HCV positive chronic liver disease (CLD).

Out of 101 family members, 15 (14.8%) were anti-HCV positive: 8/39 (20.5%) spouses, 5/20 (25%) brothers or sisters, 2/39 (5.1%) offspring, 0 of 3 parents; the overall anti-HCV prevalence did not differ widely from that described by other authors [1, 4]. Furthermore, we observed that no concomitant risk factors influenced these results.

Certainly, such a prevalence is higher than that reported among blood donors in our area (1.4%) [4], inducing us to conclude that family environment plays a role in the transmission of HCV infection. As in Caporaso et al. [1], we found that anti-HCV prevalence was higher in spouses and siblings; no significant difference was found when comparing anti-HCV prevalence in spouses or in other family members.

The high anti-HCV prevalence among brothers and sisters bears out the role of person to person contact in the spread of hepatitis C virus further; in fact sharing of toothbrush, nail scissors, combs, etc. is often described among young cohabitants.

We could not demonstrate a significant role of family socio-educational level on the spread of HCV in a household: in any case, the problem deserves further study either with a higher number of families or with more parameters of evaluation of the socio-educational status.

Duration of elevated ALT in anti-HCV positive patients with CLD was not significantly related to anti-HCV reactivity in family members. We think that the availability of other more specific markers of HCV infection will help us to better outline intrafamiliar HCV spread, as the antibody we are looking for in samples, is of little epidemiological utility.

References

1. Caporaso N, Morisco F, Romano M et al (1990) Intrafamiliar transmission of hepatitis C virus infection and of HCV related chronic liver disease. It J Gastroenterol 22: 239
2. Hess G, Massing A, Rossol S et al (1989) HCV and sexual transmission. Lancet ii: 987
3. Kamitsukasa H, Harada H, Yakura M et al (1989) Intrafamiliar transmission of hepatitis C virus. Lancet ii: 987
4. Sirchia G, Bellobuono A, Giovannezzi A et al (1989) Antibodies to hepatitis C virus in italian blood donors. Lancet ii: 797

Address of authors: Dr. M. T. Mughini, Institute of Infectious Diseases, Via Passo Gravina 187, I-95125 Catania, Italy.

Arch Virol (1992) [Suppl] 4: 345–346
© Springer-Verlag 1992

Involved factors in the intrafamilial spread of hepatitis C virus

S. Riestra, M. Rodríguez, S. Suárez, C. A. Navascués, F. Tevar[1]
J. L. S. Lombraña, R. Pérez, and **L. Rodrigo**

Gastroenterology Section, Covadonga Hospital, [1]Transfusion Communitary Center,
Oviedo, Spain

Summary. To investigate the risk of non-parenteral HCV infection, sera from 302 relatives of 120 anti-HCV positive subjects were tested for the presence of anti-HCV antibodies. For the sake of comparison, sera from 17,000 blood donors were also assayed. The prevalence of HCV positivity was 4.3% in household contacts, compared to 0.78% in the donor population, indicating a significantly higher risk of infection for family members. Close personal contact may not be as critical a factor for infection as is duration of the disease.

*

The hepatitis C virus (HCV) is efficiently transmitted by a parenteral mechanism [1]. However, there is little information on the mode of transmission in cases without history of parenteral inoculation. Epidemiological similitude with HBV suggested the possibility of intrafamilial transmission of hepatitis C virus.

With the aim of determining the prevalence of antibody to HCV (anti-HCV) among household contacts of anti-HCV positive subjects (index-cases: IC), we tested for anti-HCV by ELISA (Ortho) serum samples of 302 relatives of 120 (93 with anti-HCV-positive chronic liver disease and 27 anti-HCV-positive "healthy" blood donors). Sera obtained from 17000 consecutive blood donors were assayed for comparison.

Thirteen (4.3%) of 302 household contacts had anti-HCV, a prevalence significantly higher than that found among blood donors (132 of 17000; 0.78%), (p < 0.001). The positive relatives included 1 mother, 7 sexual partners and 5 sons; their mean age was higher than that of the negative-relatives (45.7 ± 18.5 vs. 32.2 ± 19.3 years; p < 0.05); the mean contact time with the index case was greater for the anti-HCV-positive contacts (29.2 ± 7.6 vs 19.5 ± 11.5 years; $p < 0.01$). The prevalence of anti-HCV was higher among relatives of patients with cirrhosis than that found among relatives of patients with other chronic non-cirrhotic liver disease (10.3 vs 2%; $p < 0.05$). The anti-HCV ELISA ratio (OD/cutoff) was higher in the IC of anti-HCV-positive relatives than that found in IC of negative relatives (5.6 vs 4.7; $p = 0.08$).

In conclusion, the prevalence of anti-HCV among household contacts of subjects with anti-HCV is higher than that found among blood donors [2]. The anti-HCV-positive relatives are older and the time of contact with the IC is longer than that observed in the negative relatives. The presence of anti-HCV among relatives is associated mainly with the existence of cirrhosis in the IC, which suggests a long-term infection. Probably, the risk for infection through close contact is not high [3], and a longer duration of contact increases the probability of transmission.

References

1. Esteban JI, Esteban R, Viladomiu L, López-Talavera JC, González A, Hernández JM, Roget M, Vargas V, Genescá J, Buti M, Guardia J, Houghton M, Choo Q-L, Kuo G (1989) Hepatitis C virus antibodies among risk groups in Spain. Lancet ii: 294–297
2. Idéo G, Bellati G, Pedraglio E, Bottelli R, Donzelli T, Putignano G (1990) Intrafamilial transmission of hepatitis C virus. Lancet 335: 353
3. Everhart JE, Di Bisceglie AM, Murray LM, Alter HO, Melpolder JJ, Kuo G, Hoofnagle JH (1990) Risk for Non-A, Non-B (type C) hepatitis through sexual or household contact with chronic carriers. Ann Intern Med 112: 544–545

Authors' address: Dr. S. Riestra, Gastroenterology Section, Covadonga Hospital, Oviedo, Spain.

Arch Virol (1992) [Suppl] 4: 347–348
© Springer-Verlag 1992

Prevalence of anti-HCV in two Tanzanian villages

I. Ilardi[1], G. Errera[1], G. M. De Sanctis[1], I. G. Barbacini[1], A. Madera[1], F. Leone[1], A. Nderingo[2], A. Antognoli[1], and L. V. Chircu[1]

[1] Istituto Malattie Infettive. Università di Roma "La Sapienza", Roma, Italy
[2] St. Joseph Medical Service, Magu, Muanza (Tanzania)

Summary. The presence of anti-HCV antibodies was investigated in sera from a total of 123 inhabitants of two Tanzanian villages. In one of the villages, 72.2% of the sera and in the other village, 82.6% of the sera were found to be anti-HCV positive. These values are dramatically higher than other reported prevalences, whereby cross-reactivity between HCV and Flaviviruses as well as possible transmission by arthropod vectors cannot be ruled out.

Introduction and objective

Recent demonstration of the worldwide distribution of hepatitis C virus and its very different prevalence in various regions [1], prompted us to compare sera from two Tanzanian villages and from adult subjects with chronic liver disease who had spent some time in tropical regions.

Materials and methods

The sera were collected at random in the villages of Nyomikoma (4700 inhabitants) and Ihale (6000 inhabitants) located near Lake Victoria in the Magu district (region of Mwanza). Agriculture is the main activity. Nutrition and sanitation are very poor. Fifty-four sera (9 from children, 34 adult females and 11 males) were examined in the first village and 69 (all children 6–12 years old) in the second. In addition, we studied 87 adults, living in Italy, suffering from chronic liver disease (47 Italians who spent 6 months to 20 years in Africa, Asia or South America and 40 natives of these areas). The detection of antibody was performed employing a qualitative ELISA based on the use of a recombinant polypeptide (ORTHO).

Results

— Nyomikoma: 39 out of 54 sera (72.2%) anti-HCV positive.
— Ihale: 57 out of 69 sera (82.6%) anti-HCV positive.
— patients with chronic liver disease: 16 out of 47 Italians (34%) and 14 out of 40 natives of tropical regions (35%) anti-HCV positive.

Conclusions

The prevalence of HCV positivity in the Tanzanian villages appears strikingly high if compared to other data from the literature. Transmission by mosquito bites and cross-reaction of HCV to Flavivirus antigens are not excluded by Wong et al. owing to some common physical properties [2]. The positivity of the other patients agrees with what is reported by other authors [3].

References

1. Esteban R (1990) Epidemiology of hepatitis C virus infection. International symposium on viral hepatitis and liver disease. Houston, Texas, April 4–8th 1990
2. Wong DC et al (1990) The non specificity of Anti-HCV in a South Pacific Melanesian population. Houston, Texas, April 4–8th 1990
3. Coursaget P et al (1990) HCV Infection in liver diseases in tropical africa (Senegal) Houston, Texas, April 4–8th 1990

ISSN 0304-8608
Title No. 705

Archives of Virology

Official Journal of the Virology Division
of the International Union of Microbiological Societies

Editorial Board

J. W. Almond, Reading
D. R. Lowy, Bethesda, Md.
F. A. Murphy, Davis, Calif. (Editor-in-Chief)
Y. Nagai, Nagoya
C. Scholtissek, Giessen
J. H. Strauss, Pasadena, Calif.
A. Vaheri, Helsinki
M. H. V. Van Regenmortel, Strasbourg
D. O. White, Melbourne

Virology Division

M. C. Horzinek, Utrecht

Special Issues

C. H. Calisher, Fort Collins, Colo.
H.-D. Klenk, Marburg

Archives of Virology publishes original contributions from all branches of research on viruses, viruslike agents, and virus infections of humans, animals, plants, insects, and bacteria. Coverage includes the broadest spectrum of topics, from initial descriptions of newly discovered viruses, to studies of virus structure, composition, and genetics, to studies of virus interactions with host cells, host organisms, and host populations. Multidisciplinary studies are particularly welcome, as are studies employing molecular biologic, molecular genetic, and modern immunologic and epidemiologic approaches. For example, studies on the molecular pathogenesis, pathophysiology, and genetics of virus infections in individual hosts, and studies on the molecular epidemiology of virus infections in populations, are encouraged. Studies involving applied research, such as diagnostic technology, development, monoclonal antibody panel development, vaccine development, and antiviral drug development, are also encouraged. However, such studies are often better presented in the context of a specific application or as they bear upon general principles of interest to many virologists. In all cases, it is the quality of the research work, its significance, and its originality which will decide acceptability.

As a new opportunity for publication of proceedings of meetings, treatises, and large reviews the series of Special Issues of *Archives of Virology* was initiated in 1990. Individuals who are organizing a meeting, symposium, conference, or congress, and individuals who would like to organize a treatise or large review are invited to communicate directly with one of the Special Issues Editors for further information: Dr. C. H. Calisher, Fort Collins, Colorado, or Dr. H.-D. Klenk, Marburg, FRG.

Subscription Information:
1992. Vols. 122–127 (4 issues each): DM 1.908,–, öS 13.356,–, plus carriage charges

E. Kurstak

Viral Hepatitis

Current Status and Issues

1992. Approx. 30 figures. Approx. 240 pages.
Soft cover approx. DM 120,-, öS 840,-
ISBN 3-211-82387-5

Prices are subject to change withouth notice

In the 1990's significant advances in the understanding of viral hepatitis have been observed. In particular, our knowledge of the nature and diversity of viruses causing hepatitis in humans have substantially increased.

"Viral Hepatitis: Current Status and Issues" comprehensively and uniquely presents these valuable information all in a single volume for the utmost benefit of medical practitioners, microbiologists as well as those actively involved in health administration world-wide.

The virological, clinical epidemological, diagnostic, therapeutic, and preventive aspects pertaining to all the types of hepatitis known to date including hepatitis C and E are thoroughly discussed.

From the contents:

Hepatitis A Virus Properties and Replication. - Hepatitis B Virus and Disease. - Hepatitis D Virus and Disease. - Hepatitis C Virus, Hepatitis E Virus and Disease. - Different Forms of Viral Hepatitis

(To be published in Fall 1992)

Springer-Verlag Wien New York